Difference Equations and Discrete Dynamical Systems

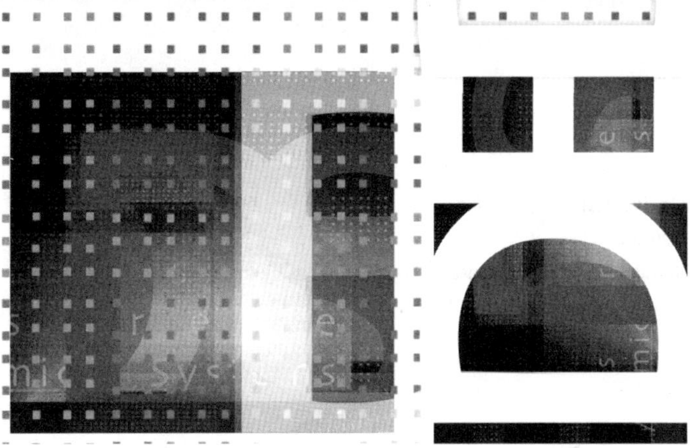

Difference Equations and Discrete Dynamical Systems

Proceedings of the 9th International Conference

University of Southern California, Los Angeles, California, USA 2 – 7 August 2004

Editors

Linda J S Allen
Texas Tech University, USA

Bernd Aulbach
University of Augsburg, Germany

Saber Elaydi
Trinity University, USA

Robert Sacker
University of Southern California, Los Angeles, USA

International Society of Difference Equations

ISDE

World Scientific

NEW JERSEY • LONDON • SINGAPORE • BEIJING • SHANGHAI • HONG KONG • TAIPEI • CHENNAI

Published by

World Scientific Publishing Co. Pte. Ltd.

5 Toh Tuck Link, Singapore 596224

USA office: 27 Warren Street, Suite 401-402, Hackensack, NJ 07601

UK office: 57 Shelton Street, Covent Garden, London WC2H 9HE

British Library Cataloguing-in-Publication Data
A catalogue record for this book is available from the British Library.

DIFFERENCE EQUATIONS AND DISCRETE DYNAMICAL SYSTEMS
Proceedings of the 9th International Conference

ISBN 981-256-520-5

Printed in Singapore by World Scientific Printers (S) Pte Ltd

Preface

Difference Equations or Discrete Dynamical Systems is a diverse field which impacts almost every branch of pure and applied mathematics. Not surprisingly, the techniques that are developed vary just as broadly. The annual International Conference on Difference Equations and Applications is organized under the auspices of the International Society of Difference Equations. The Conferences have world wide attendance and cover a wide range of topics.

It is through this Proceedings that the mathematical community can (1) obtain an overview of all the presentations at the 9th International Conference on Difference Equations and Applications, (2) see a variety of problems and applications having one ingredient in common, the Discrete Dynamical System and (3) keep abreast of the many new techniques and developments in the area.

The emphasis of the meeting was on Mathematical Biology and accordingly about half of the presentations were in that area, Mathematical Ecology and Mathematical Medicine.

Contents

A discrete-time Beverton-Holt competition model

Azmy S. Ackleh, Youssef M. Dib, Sophia R.-J. Jang

Department of Mathematics, University of Louisiana at Lafayette, Lafayette, LA 70504-1010

Abstract. A model of competition between multiple populations in discrete time is proposed and studied. It is assumed that individuals within a single population are identical and therefore there is no structuring variable within each population. Under the assumption that population growth is modeled by Beverton-Holt functionals, it is shown that the population with maximal fitness will out compete the other population. While global stability results are provided for the case of two populations, only local stability results are obtained for the model with more than two populations.

1 Introduction

Interspecific competition occurs when individuals of one population suffer a reduction in fecundity, growth or survivorship as a result of resource exploitation or interference by individuals of another population. This kind of interaction between populations has been frequently observed in nature and it can affect population dynamics and abundance in many different ways. Among these is the well established competitive exclusion principle, which states that at most one population can survive when several populations compete for the same limiting resource [3, 4].

Many experiments and mathematical models have supported the well known ecological theory of competitive exclusion. For example, the classical continuous-time Lotka-Volterra competition model demonstrates that there is no coexistence between populations under reasonable assumptions on the model parameters [3]. Chemostat models have also validated the principle very successfully [8]. See also [1] for a model with competitive exclusion.

More recently, Cushing et al. [5] studied a discrete-time competition model of two populations and showed similar results as those of continuous-time Lotka-Volterra system. We refer the reader to [6] for general theory of difference equations. Our aim of this manuscript is to propose a simple discrete-time competition model of multiple populations and to study its competitive outcome. Specifically, Beverton-Holt functional forms are used to model population growth in the absence of other populations.

The remainder of this manuscript is organized as follows. The model and mathematical analysis are presented in the next section. The final section provides discussion and concluding remarks.

2 A Beverton-Holt competition model

Let $n \geq 2$ be a positive integer denote the number of populations, and $x_i(t)$ be the population size at time t, $t = 0, 1, 2, \cdots$ and $i = 1, 2, \cdots, n$. Each isolated population without interaction is modeled by the well-known Beverton-Holt equation [3, 4]. The equation shares the same simple asymptotic dynamics as the continuous-time logistic equation, and it has been used to study many populations such as fish and insects. The Beverton-Holt equation is also often referred to as the discrete logistic equation [3]. The interaction between these n populations is given by the following system of first order nonlinear difference equations

$$\begin{cases} x_i(t+1) = \dfrac{a_i x_i(t)}{1 + b_i \sum_{j=1}^{n} x_j(t)} \\ x_i(0) \geq 0 \\ i = 1, 2, \cdots, n, . \end{cases} \tag{2.1}$$

where parameters $a_i > 0$ and $b_i > 0$ are assumed to satisfy the following conditions

$$a_i > 1 \text{ for } i = 1, 2, \cdots, n \tag{2.2}$$

and

$$\frac{a_1 - 1}{b_1} > \frac{a_i - 1}{b_i} \text{ for } i = 2, \cdots, n. \tag{2.3}$$

Under the assumptions of (2.2) and (2.3), we see that each individual population can persist in the absence of other populations and population $i = 1$ has the largest fitness. In the following, we shall study system (2.1).

It is easy to see for the following first order scalar difference equation

$$y(t+1) = \frac{a y(t)}{1 + b y(t)}, \tag{2.4}$$

all solutions with $y(0) > 0$ converge to the steady state $\dfrac{a-1}{b}$ if $a > 1$ and $b > 0$ [3]. Using the asymptotic dynamics of equation (2.4), it can be easily shown that $\limsup_{t \to \infty} x_i(t) \leq \dfrac{a_i - 1}{b_i}$ for $i = 1, 2, \cdots, n$ for any solution of (2.1). On the other hand, letting $a_0 = \min\{a_1, a_2, \cdots, a_n\}$, $b_0 = \max\{b_1, b_2, \cdots, b_n\}$ and $P(t) = \sum_{i=1}^n x_i(t)$, we have for $t \geq 0$

$$P(t+1) \geq \frac{a_0 P(t)}{1 + b_0 P(t)}. \tag{2.5}$$

Consequently, solutions $(x_1(t), x_2(t), \cdots, x_n(t))$ of (2.1) with $P(0) > 0$ satisfy $\liminf_{t\to\infty} P(t) \geq \dfrac{a_0 - 1}{b_0}$. We summarize our discussion in the following lemma.

Lemma 2.1 *Solutions of system (2.1) exist and remain nonnegative for $t > 0$. Moreover, let $m = \dfrac{a_0 - 1}{b_0} > 0$ and $M = \dfrac{\hat{a} - 1}{\hat{b}} > 0$ where $\hat{a} =$ max $\{a_1, a_2, \cdots, a_n\}$ and $\hat{b} =$ min $\{b_1, b_2, \cdots, b_n\}$, then*

$$m \leq \liminf_{t\to\infty} P(t) \leq \limsup_{t\to\infty} P(t) \leq M$$

if $P(0) > 0$, where $P(t) = \sum_{i=1}^n x_i(t)$ and $(x_1(t), x_2(t), \cdots, x_n(t))$ is the solution of (2.1).

As solutions of (2.1) are bounded, we proceed to discuss its equilibria. Clearly (2.1) always has a trivial steady state $E_0 = (0, 0, \cdots, 0)$ which exits for all parameter values. The nontrivial equilibrium equations satisfy

$$\sum_{j=1}^n x_j = \frac{a_i - 1}{b_i} \text{ for } i = 1, 2, \cdots, n.$$

As a result, there are no interior equilibria, and in addition to E_0, there are n boundary equilibria: $E_1 = (\dfrac{a_1 - 1}{b_1}, 0, \cdots, 0)$, $E_2 = (0, \dfrac{a_2 - 1}{b_2}, 0, \cdots, 0)$, $\cdots$, and $E_n = (0, \cdots, 0, \dfrac{a_n - 1}{b_n})$.

The local stability of each of these steady states can be determined straightforwardly. Indeed, the Jacobian matrix at E_0 is a diagonal matrix with diagonal entries a_i, $i = 1, \cdots, n$, which implies that E_0 is a repeller by our assumption (2.2). The Jacobian matrix of (2.1) at E_1 has an upper triangular form with diagonal entries $\dfrac{1}{a_1}, \dfrac{a_2}{1 + b_2 \frac{a_1 - 1}{b_1}}, \cdots, \dfrac{a_n}{1 + b_n \frac{a_1 - 1}{b_1}}$. Since (2.3) holds, we have

$$\frac{a_i}{1 + b_i \dfrac{a_1 - 1}{b_1}} = \frac{a_i b_1}{b_1 + b_i a_1 - b_i} < 1$$

for $i = 2, \cdots, n$. This shows that E_1 is locally asymptotically stable. Furthermore, the Jacobian matrix at E_2 has the following form

$$J_2 = \begin{pmatrix} \frac{a_1}{1+b_1\frac{a_2-1}{b_2}} & 0 & 0 & 0 & \cdots & 0 \\ \frac{-(a_2-1)}{a_2} & \frac{1}{a_2} & \frac{-(a_2-1)}{a_2} & \frac{-(a_2-1)}{a_2} & \cdots & \frac{-(a_2-1)}{a_2} \\ 0 & 0 & \frac{a_3}{1+b_3\frac{a_2-1}{b_2}} & 0 & \cdots & 0 \\ \vdots & \vdots & \vdots & \vdots & \vdots & 0 \\ 0 & 0 & 0 & \cdots & 0 & \frac{a_n}{1+b_n\frac{a_2-1}{b_2}} \end{pmatrix}.$$

By using a first row expansion, we see that eigenvalues of J_2 are: $\dfrac{a_i}{1+b_i\frac{a_2-1}{b_2}}$ for $i \neq 2$ and $\dfrac{1}{a_2}$. Therefore condition (2.3) imposed on the parameters implies that E_2 is unstable. Similar arguments can be shown that E_i is unstable for $i = 2, \cdots, n$.

In the following we prove that when $n = 2$ solutions of (2.1) with $x_1(0) > 0$ converge to $E_1 = (\frac{a_1-1}{b_1}, 0)$. Notice that system (2.1) is competitive and it is known that bounded solutions of planar competitive systems either converge to a steady state or have a 2-cycle as its ω-limit set [9, Theorem 4.2]. An explicit calculation was carried out in [7] to eliminate the existence of an interior 2-cycle and one can thus conclude that the interior steady state is globally asymptotically stable. Our analysis performed here is different from that given in [7].

System (2.1) with $n = 2$ can be written explicitly as

$$\begin{cases} x_1(t+1) = \dfrac{a_1x_1(t)}{1+b_1(x_1(t)+x_2(t))} \\ x_2(t+1) = \dfrac{a_2x_2(t)}{1+b_2(x_1(t)+x_2(t))} \\ x_1(0), x_2(0) \geq 0, \end{cases} \tag{2.6}$$

where $a_1, a_2 > 1$, $b_1, b_2 > 0$ and $\dfrac{a_1-1}{b_1} > \dfrac{a_2-1}{b_2}$. Let $f : \mathbb{R}^2 \to \mathbb{R}^2$ denote the map induced by system (2.6), i.e., $f(x_1, x_2) = (f_1(x_1, x_2), f_2(x_1, x_2))$, where $f_1(x_1, x_2) = \dfrac{a_1x_1}{1+b_1(x_1+x_2)}$ and $f_2(x_1, x_2) = \dfrac{a_2x_2}{1+b_2(x_1+x_2)}$. The

Jacobian matrix associated with the system is given below.

$$J = \begin{pmatrix} \dfrac{a_1 + a_1 b_1 x_2}{[1 + b_1(x_1 + x_2)]^2} & \dfrac{-a_1 b_1 x_1}{[1 + b_1(x_1 + x_2)]^2} \\ \dfrac{-a_2 b_2 x_2}{[1 + b_2(x_1 + x_2)]^2} & \dfrac{a_2 + a_2 b_2 x_1}{[1 + b_2(x_1 + x_2)]^2} \end{pmatrix}.$$

Define a partial ordering $\leq_K$ on $\mathbb{R}^2_+$ by $(x_1, x_2) \leq_K (z_1, z_2)$ if $x_1 \leq z_1$ and $x_2 \geq z_2$. Then the map J is K-positive as its diagonal entries are nonnegative and its off diagonal entries are non-positive on $\mathbb{R}^2_+$, and the determinant of J is also positive on $\mathbb{R}^2_+$. Moreover, $\mathbb{R}^2_+$ contains order intervals and is $\leq_K$-convex. It remains to verify that the map f is one-to-one by [9, Corollary 4.4].

Let $(y_1, y_2) \in \mathbb{R}^2_+$ be given arbitrarily. If $y_1 = y_2 = 0$, then it has a unique preimage $(0, 0)$. If $y_1 = 0$, then $(0, y_2)$ has at most one preimage as $g(x) = \dfrac{a_2 x_2}{1 + b_2 x_2}$ is strictly increasing. The same is true if $y_2 = 0$. Suppose now $y_1, y_2 > 0$. If there are $(\bar{x}_1, \bar{x}_2) \neq (\hat{x}_1, \hat{x}_2)$ such that $(y_1, y_2) = f(\bar{x}_1, \bar{x}_2) = f(\hat{x}_1, \hat{x}_2)$, we may assume $\bar{x}_2 < \hat{x}_2$ as the other cases can be proved similarly. Let x_2^* with $\bar{x}_2 \leq x_2^* \leq \hat{x}_2$ be given. Notice $y_1 = \dfrac{a_1 x_1}{a + b_1(x_1 + x_2^*)}$ has a unique solution $x_1^* > 0$, where $\dfrac{a_1 \bar{x}_1}{1 + b_1(\bar{x}_1 + x_2^*)} \leq \dfrac{a_1 \bar{x}_1}{1 + b_1(\bar{x}_1 + \bar{x}_2)} = y_1$ and $\dfrac{a_1 \hat{x}_1}{1 + b_1(\hat{x}_1 + x_2^*)} \geq \dfrac{a_1 \hat{x}_1}{1 + b_1(\hat{x}_1 + \hat{x}_2)} = y_2$. Hence $\bar{x}_1 \leq x_1^* \leq \hat{x}_1$. Denote this x_1^* by $X_1^*(x_2^*), \bar{x}_2 \leq x_2^* \leq \hat{x}_2$. Since $\dfrac{a_1 X_1^*(x_2^*)}{1 + b_1(X_1^*(x_2^*) + x_2^*)} = y_1$, we have $\dfrac{dX_1^*(x_2^*)}{dx_2^*} = \dfrac{b_1 X_1^*(x_2^*)}{1 + b_1 x_2^*} > 0$. Define $Y_2(x_2^*) = \dfrac{a_2 x_2^*}{1 + b_2(X_1^*(x_2^*) + x_2^*)}$ for $\bar{x}_2 \leq x_2^* \leq \hat{x}_2$. Using $\dfrac{dX_1^*}{dx_2*}$ given above, a direct computation yields $\dfrac{dY_2}{dx_2^*} = \dfrac{a_2 + a_2 b_1 x_2^* + a_2 b_2 X_1^*(x_2^*)}{1 + b_2 x_2^*} > 0$. Since $\bar{x}_2 \leq x_2^* \leq \hat{x}_2$ and $\bar{x}_2 < \hat{x}_2$, we must have $y_2 = Y_2(\bar{x}_2) < Y_2(\hat{x}_2) = y_2$ and obtain a contradiction. Therefore, f is one-to-one and solutions of (2.6) with $x_1(0) > 0$ converge to the steady state E_1 as solutions are bounded and E_1 is locally asymptotically stable [9, Corollary 4.4].

The asymptotic dynamics of (2.6) is given below.

Theorem 2.2 *Solutions $(x_1(t), x_2(t))$ of system (2.6) with $x_1(0) > 0$ converge to the steady state $E_1 = (\dfrac{a_1 - 1}{b_1}, 0)$.*

We conjecture for $n > 2$ that the equilibrium E_1 is globally asymptotically stable provided that (2.3) is satisfied. We shall use a numerical example to demonstrate this conjecture. Specifically, $n = 3$, $a_1 = 1.5, a_2 =$

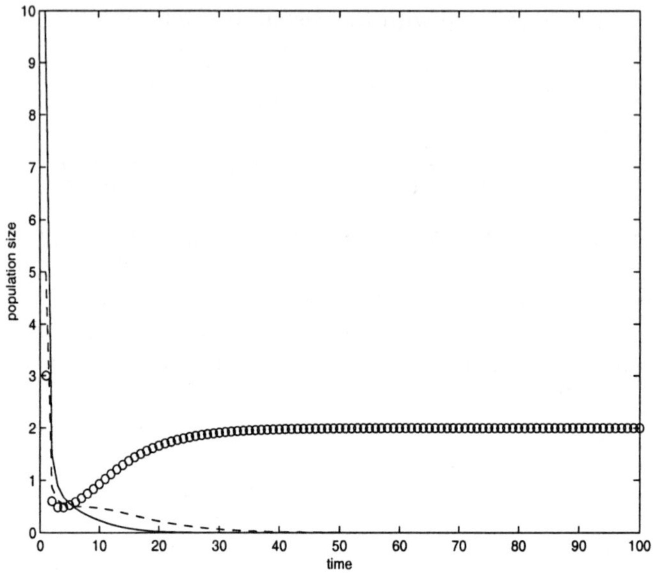

Figure 1: This is a competition model between three populations. We can see from the graph that population 1 drives the other populations to extinction.

$2, a_3 = 1.75$, $b_1 = 0.25$, $b_2 = 0.6$ and $b_3 = 0.8$. The system with these parameters is given below.

$$\begin{cases} x_1(t+1) = \dfrac{1.5x_1(t)}{1+0.25(x_1(t)+x_2(t)+x_3(t))} \\ x_2(t+1) = \dfrac{2x_2(t)}{1+0.6(x_1(t)+x_2(t)+x_3(t))} \\ x_3(t+1) = \dfrac{1.75x_2(t)}{1+0.8(x_1(t)+x_2(t)+x_3(t))} \\ x_1(0), x_2(0), x_3(0) \geq 0. \end{cases} \tag{2.7}$$

Therefore population one has the largest fitness than other two populations. Numerical simulation presented in Figure 1 with initial population size $x_1(0) = 3, x_2(0) = 10$ and $x_3(0) = 5$ shows that only population one survive while others become extinct. Simulations with the same parameter values and different initial conditions also confirm our observation.

In fact, one can indeed see that this is the case under the conditions $a_i < a_1$ and $b_i \geq b_1$, $i = 1, 2, \ldots, n$ (which are stronger than (2.3)).

To this end, using Lemma 2.1 we see that for $i = 2, \cdots, n$

$$\frac{x_i(t+1)}{x_1(t+1)} = \frac{x_i(t)}{x_1(t)} \frac{a_i(1 + b_1 \sum_{j=1}^n x_j(t))}{a_1(1 + b_i \sum_{j=1}^n x_j(t))} \leq \frac{a_i}{a_1} \frac{x_i(t)}{x_1(t)}.$$

From this we deduce that $\frac{x_i(t)}{x_1(t)} \to 0$ as $t \to \infty$. Since solutions are bounded this implies that $x_i(t) \to 0$ as $t \to \infty$ for $i = 2, \cdots, n$, and therefore we see that $x_1(t) \to \frac{a_1 - 1}{b_1}$ as $t \to \infty$. The above discussion can be summarized below.

Lemma 2.3 *In addition to (2.2) and (2.3) we assume that $a_i < a_1$ and $b_i \geq b_1$ for $i = 2, \cdots, n$, then solutions of (2.1) with $x_1(0) > 0$ converge to the steady state $E_1 = (\frac{a_1 - 1}{b_1}, 0, \cdots, 0)$.*

3 Concluding remarks

In this manuscript a simple competition model between populations is proposed and analyzed. It is demonstrated that competitive exclusion principle holds for our model. Specifically, the population with the maximal fitness will out compete other populations. For two populations, it is shown that the other population is driven to extinction.

We point out that the model (2.1) considered here is a discrete version of the distributed rate model developed in [2] which is given by the following integro-differential equation:

$$\begin{cases} \dfrac{\partial x(t,q)}{\partial t} = x(t,q)\left[q_1 - q_2 X(t)\right], \\ x(0,q) = x_0(q). \end{cases} \tag{3.1}$$

Here the parameter $q = (q_1, q_2)$ lies in the set $Q = [\alpha_1, \beta_1] \times [\alpha_2, \beta_2]$, where $\alpha_1, \alpha_2, \beta_1, \beta_2 \in \mathbb{R}_+$. The function $x(t,q)$ is the density of the subpopulation having parameter q. Also, $X(t) = \int_Q x(t,q)dq$ is the total population at time $t \geq 0$. The initial data $x_0(q)$ is a nonnegative continuous function satisfying $x_0(q^*) > 0$ with $q^* = (\beta_1, \alpha_2)$.

To show that (2.1) can be obtained from discretizing (3.1), we begin by choosing a partition V of the parameter space Q. Let $V = V_1 \times V_2$, where

$$\begin{aligned} V_1 &: \alpha_1 = p_1^0 < p_1^1 < \cdots < p_1^{n_1} = \beta_1, \\ V_2 &: \alpha_2 = p_2^0 < p_2^1 < \cdots < p_2^{n_2} = \beta_2. \end{aligned}$$

Let $n = n_1 \cdot n_2$. Define $\{Q^i\}, i = 1, ..., n$, to be the family of subrectangles of Q resulting from the above defined partition, and set $q^i = (q_1^i, q_2^i)$ to be the midpoint of the rectangle $Q^i, i = 1, ..., n$.

With these specifications we approximate the continuous model (3.1) by the following system of ordinary differential equations:

$$\begin{cases} \dfrac{dz^i(t)}{dt} = z^i(t)\left(q_1^i - q_2^i \displaystyle\sum_{j=1}^{n} \mu(Q^j) z^j(t)\right), \\ z^i(0) = \dfrac{1}{\mu(Q^i)} \displaystyle\int_{Q^i} x_0(q) dq, \end{cases} \tag{3.2}$$

where $\mu(Q^i)$ denotes the measure of the rectangle Q^i. Making the substitution $y^i = \mu(Q^i) z^i$, (3.2) becomes

$$\begin{cases} \dfrac{dy^i(t)}{dt} = y^i(t)\left(q_1^i - q_2^i \displaystyle\sum_{j=1}^{n} y^j(t)\right), \\ y^i(0) = \displaystyle\int_{Q^i} x_0(q) dq. \end{cases} \tag{3.3}$$

Next we approximate $\dfrac{dy^i(t)}{dt}$ by $\dfrac{y^i(t+h) - y^i(t)}{h}$ and use a mixed implicit explicit approximation for the quadratic term in (3.3) to obtain the following fully discretized version of (3.1):

$$\begin{cases} \dfrac{y^i(t+h) - y^i(t)}{h} = q_1^i y^i(t) - q_2^i y^i(t+h) \displaystyle\sum_{j=1}^{n} y^j(t), \\ y^i(0) = \displaystyle\int_{Q^i} x_0(q) dq. \end{cases} \tag{3.4}$$

Rearranging terms we obtain:

$$\begin{cases} y^i(t+h) = \dfrac{(1 + hq_1^i) y^i(t)}{1 + hq_2^i \sum_{j=1}^{n} y^j(t)}, \\ y^i(0) = \displaystyle\int_{Q^i} x_0(q) dq. \end{cases} \tag{3.5}$$

Letting $h = 1$, $a_i = 1 + q_1^i > 1$ and $b_i = q_2^i$ results in (2.1).

Using the theory of weak convergence of probability measures, the authors in [2] showed that essentially the solution to the model (3.1) $x(t, q) \to c\delta_{q^*}(q)$ as $t \to \infty$, where δ_{q^*} denotes the Dirac delta function concentrated at q^*. The constant c turns out to be the ratio $\frac{\beta_1}{\alpha_2}$. This proves that all subpopulations with parameters $q \neq q^*$ go to extinction as $t \to \infty$ while the subpopulation with parameter q^* persists. Hence, the results obtained here for the model (2.1) indicate (at least locally) that the dynamics of the discrete model preserve those of the continuous one.

References

[1] Ackleh, A.S., Marshall, D., Heatherly, H., Extinction in a generalized Lotka-Volterra predator-prey model, *J. App. Math. Stoch. Anal.*, **13**:287-297, 2000.

[2] Ackleh, A.S., Marshall, D., Heatherly, H., Fitzpatrick, B.G., Survival of the fittest in a generalized logistic model, *Mathematical Models and Methods in Applied Sciences*, **9**:1379-1391, 1999.

[3] Allen, L., *An Introduction to Deterministic Models in Biology*, Prentice-Hall, 2004.

[4] Begon, M., Harper, J., Jownsend, C., *Ecology: Individuals, Populations and Communities*, Blackwell Science Ltd, NY, 1996.

[5] Cushing, J.M., Levarge, S., Chitnis, N., Henson, S., Some discrete competition models and the competitive exclusion principle, *J. Diff. Equ. Appl.*, to appear.

[6] Elaydi, S., *An Introduction to Difference Equations*, Second Edition, Springer, New York, 1999.

[7] Jang, S. R.-J., Ackleh, A. S., Discrete-time, discrete stage-structured predator-prey models, *J. Diff. Equ. Appl.*, to appear.

[8] Smith, H.L., Waltman, P., *The Theory of the Chemostat*, Cambridge University Press, 1995.

[9] Smith, H.L., Planar competitive and cooperative difference equations, *J. Diff. Eqns. Appl.* **3**:335-357, 1998.

A Dynamic Analysis of the Bush Fiscal Policy*

Richard H. Day and Chengyu Yang
University of Southern California

Abstract

An empirical, dynamic aggregate demand and supply (DADS) model is used to explain the policy paradox associated with the Kennedy, Reagan and Bush II administrations: that is, the possibility that *tax rate decreases* could in principle—and might in practice—*increase tax revenue*. A nonlinear elaboration is introduced to explain some aspects of monetary policy. Several factors explaining why employment lags behind the recovery of output after a recession are identified.

Contents

*Plenary paper presented at the Ninth International Conference on Difference Equations and Applications.

1 Introduction

If you include verbal description and numerical examples, discrete dynamics has been used in economics more or less explicitly since at least the Classical Era. By the 1930s the Swedish school (e.g., Olin, Lundberg) was basing its analyses on "period" or "sequence" analysis, although without the help of much explicit mathematics. Perhaps the first formal use of difference equations occurred in the 1930s with Leontief's nonlinear analysis of the "cobweb cycle," Samuelson's linear analysis of Alvin Hansen's accelerator and Metzler's study of inventory cycles. By mid century numerous examples emerged such as Hicks' analysis of Harrod's model, Smithies' attempt to explain growth and cycles, and Samuelson's very general treatment in his *Foundations of Economic Analysis.* By now discrete dynamics has become a standard working tool in macroeconomic theory, especially following the application of chaos and ergodic theory by Benhabib, Day, Grandmont, Shafer, and others and the reformulation of optimal growth theory in terms of Bellman's discrete dynamic programming by Lucas, Stokey, and the members of the Real Business Cycle (RBC) school of Cooley, Prescott, Sargent, and others.

The present discussion uses a discrete dynamic, aggregate demand and supply (DADS) growth model to examine what is widely presumed to be a paradoxical or self–contradictory fiscal policy proposition: namely, a *reduction in tax rates* can both stimulate economic growth and *increase tax revenues.* This proposition was argued explicitly in the early 1960s by President Kennedy, subsequently by Ronald Reagan, and most recently by George W. Bush. In each case the proposal, which we refer to as the KRB tax policy paradox, has been widely viewed by politicians and media pundits with skepticism. Many of our colleagues in economics react in a similar manner. Although the model presented cannot accommodate all the complications of economic growth and fluctuations, it clearly provides a useful approximation and a clear cut explanation of the KRB paradox.[1]

Some of you may be surprised by the results.

2 The Basic DADS Growth Model

2.1 Aggregate Demand

The basic linear dynamic aggregate demand and supply (DADS) model incorporates aggregate consumption and investment functions, which are,

$$c_{t+1} = \alpha_0 + \alpha(1 - \tau_{t+1})y_{t+1}, \tag{1}$$

where c_{t+1} and y_{t+1} are per capita private consumption and income in period $t+1$, respectively, and

$$j_{t+1} = \phi(y_t - y'), \tag{2}$$

where j_{t+1} is per capita private investment. The coefficients α_0, α, ϕ, and y' are parameters and the exogenous variable τ_{t+1} is the average per capita net tax rate. The parameter α is the *marginal propensity to consume* (MPC), $\sigma = (1-\alpha)$ is the *marginal propensity to save* (MPS), and ϕ is the *marginal propensity to invest* (MPI). We assume throughout that $0 < \alpha < 1$ and $\phi > 0$.

Fitting these functions to the data for 1929–2002 (excluding the war 1942–1947) yields the highly significant parameter estimates

$$\alpha_0 = 0, \ \alpha = .69, \ \phi = .28, \ y' = 3673.$$

The explained variation in the data for the consumption function is 98.5% and 95.6% for the investment function.

Per capita aggregate demand is given by

$$z_{t+1} = c_{t+1} + j_{t+1} + g_{t+1} + x_{t+1} \tag{3}$$

where g_{t+1} equals government spending (public investment + government services), and x_{t+1} is net exports (exports – imports). It is assumed that aggregate supply (production), aggregate demand, and aggregate income generated by production satisfy the balance equation,

$$y_{t+1} = q_{t+1} = z_{t+1}. \tag{4}$$

As a result,

$$q_{t+1} = \alpha_0 + \alpha(1-\tau_{t+1})q_{t+1} + \phi(q_t - y') + g_{t+1} + x_{t+1} \tag{5}$$

or, solving for q_{t+1}, we get the non–autonomous first difference equation,

$$q_{t+1} = \frac{1}{\sigma + \alpha\tau_{t+1}}\Big(\gamma + g_{t+1} + x_{t+1} + \phi q_t\Big) \tag{6}$$

where $\gamma = \alpha_0 - \phi y'$. Henceforth, the argument will be conducted in terms of output.

2.2 Production and Productivity

The supply side of the macro economy is represented by the aggregate production function,

$$Q_t = F(K_t^u, B_t E_t) := K_t^{u\beta}(B_t E_t)^{1-\beta}, \tag{7}$$

where Q_t is GDP, K_t^u is utilized aggregate social (private plus public) capital, B_t is a productivity term, E_t is employment. It is here assumed for simplicity that private and public capital are perfect substitutes. The productivity term is defined by an autonomous difference equation,

$$B_{t+1} = (1+b)B_t, \tag{8}$$

where b is a productivity coefficient, a form that says that productivity feeds on itself. The productivity term cannot be observed directly but must be inferred from the other data. This can be done by using the implied trend

$$B_t = (1+b)^t B_0. \tag{9}$$

Given this, production per employed worker can be expressed by

$$q_t^u = f(k_t^u, t) := (1+b)^{(1-\beta)t} A(k_t^u)^\beta, \tag{10}$$

where q_t^u = GDP per employed worker, k_t^u is utilized social capital per employed worker or capital/labor ratio, and $A = B_0^{(1-\beta)}$.

The parameter estimates using data for the entire period are

$$b = .0186 \qquad \beta = .358$$

Both are highly significant and the percentage of data variability explained is .9989.

2.3 Capital Accumulation

Social capital consists of private and government components, say K^p and K^g or in per capita terms, k^p and k^g. The equation of government capital accumulation is

$$k_{t+1}^g = \frac{1}{1+n_t}\left[(1-\delta^g)k_t^g + j_t^g\right] \tag{11}$$

where j^g is the public component of investment and δ^g the depreciation rate on government capital. Public investment is exogenous. Private investment is estimated by the private investment function, so the equation of private capital accumulation equation is

$$k_{t+1}^p = \frac{1}{1+n_t}\left[(1-\delta^p)k_t^p + \phi[q_t - y']\right] \tag{12}$$

where δ^p is the depreciation rate for private capital. Social capital per capita is

$$k_{t+1} = k_{t+1}^p + k_{t+1}^g. \tag{13}$$

2.4 Employment

By definition, $Q_t = N_t q_t = q_t^e E_t$, where N_t is the total population in period t, which implies that the demand for employment is

$$E_{t+1} = N_{t+1} q_{t+1} / q_{t+1}^u. \tag{14}$$

3 Fiscal Policy: Short Run Effects

Let us suppose government spending or tax changes could be implemented at the beginning of a new year. From (6)

$$\begin{aligned} \frac{\partial q_{t+1}}{\partial g_{t+1}} &= \frac{1}{\sigma + \alpha \tau_{t+1}} > 1 \qquad \text{and} \\ \frac{\partial q_{t+1}}{\partial \tau_{t+1}} &= -\frac{\alpha}{\sigma + \tau_{t+1}\alpha} q_t < 0. \end{aligned}$$

Tax revenue in a given year will be

$$T_{t+1} = \tau_{t+1} Q_{t+1} = \tau_{t+1} N_{t+1} q_{t+1}.$$

The Net government budget (revenue − expenditures) is

$$T_{t+1} - G_{t+1} = N_{t+1}\Big(\tau_{t+1} q_{t+1} - g_{t+1}\Big)$$

where G_{t+1} is $N_{t+1} g_{t+1}$ is aggregate government spending. Changes in government expenditure and the tax rate have, respectively, the effects

$$\begin{aligned} \frac{\partial (T_{t+1} - G_{t+1})}{\partial g_{t+1}} &= \frac{\sigma(\tau - 1)}{\sigma + \tau_{t+1}\alpha} N_{t+1} q_{t+1} < 0 \quad \text{and} \\ \frac{\partial (T_{t+1} - G_{t+1})}{\partial \tau_{t+1}} &= \frac{\sigma}{\sigma + \tau_{t+1}\alpha} N_{t+1} q_{t+1} > 0. \end{aligned}$$

From these inequalities we find the short run effects of the fiscal instruments.

A *(i) A government spending increase (decrease) increases (decreases) income but reduces the net budget balance.*

(ii) An increase (decrease) in the average tax rate decreases (increases) income but increases (decreases) tax revenue.

4 Steady State Growth

The average tax rate, government spending, net exports, and wages, capital value, interest, and depreciation change somewhat from period to period, but the model does not explain their movement. It merely explains how GDP per capita changes given whatever values those exogenous variables assume. It is of considerable interest, however, to investigate a situation in which the average tax rate is constant and exogenous demand grows at the same rate as productivity. Given these assumptions, aggregate demand (6) becomes

$$q_{t+1} = \frac{1}{\sigma + \alpha\tau}\Big(\gamma + (1+b)^{t+1}(g+x) + \phi q_t\Big). \tag{15}$$

Divide both sides through by B_{t+1} and define per capita efficiency income in productivity units $q_t^e := q_t/B_t$. Then

$$q_{t+1}^e = \frac{1}{\sigma + \alpha\tau}\left(\frac{\gamma}{(1+b)^{t+1}B_0} + (g+x) + \frac{\phi q_t^e}{1+b}\right) \tag{16}$$

or

$$q_{t+1}^e = \frac{A}{(1+b)^{t+1}B_0} + B + Cq_t^e \tag{17}$$

where $A = \gamma/(\sigma+\alpha\tau)$, $B = (g+x)/(\sigma+\alpha\tau)$ and $C = \phi/\big[(1+b)(\sigma+\alpha\tau)\big]$. Generating the left side recursively, the first term is a transient that vanishes as time passes so the behavior of efficiency income should be approximated with increasing accuracy by the difference equation

$$x_{t+1} = B + Cx_t. \tag{18}$$

Recursively generate its solution to get the series

$$x_t = B(1 + C + C^1 + \cdots + C^{t-1}) + C^t x_0.$$

If $C > 1$, this series is unbounded. If $0 < C < 1$, it converges. If $x_0 = q_0^e$, then

$$\lim_{t\to\infty} x_t^e = \tilde{q}^e = B/(1-C) = \frac{g+x}{(\sigma+\alpha\tau) - \phi/(1+b)}. \tag{19}$$

Assume $0 < C < 1$.

Assume γ is negative, so A is also. Let $\{x_s\}_0^t$ be a series generated recursively by equation (18), and $\{q_s^e\}_0^t$ a series generated by equation (17). Assume $x_0 = q_0^e < \tilde{q}^e$. The former is a positive, eventually monotonically increasing series converging to $\tilde{q}^e$. The latter is also a positive, eventually monotonically increasing series, each element of which, $q_t < x_t < \tilde{q}, t > 1$, so it converges also. It is always eventually increasing, however, so it has the same upperbound $\tilde{q}^e$ given by (19). If $x_0 = q_0^e > \tilde{q}^e$, the argument is identical except the series are eventually decreasing and the inequalities reversed.[2] An analogous argument is followed for γ positive. Consequently, we have proposition:

B *Suppose exogenous demand* $(g_{t+1}+x_{t+1})=(1+b)^{t+1}(g+x)$ *where* g, x *and* b *are positive constants for an arbitrary base year. Assume the average tax rate is a constant,* τ. *If*

$$(a) \quad \alpha_0 - \phi y' + g + x > 0$$

and

$$(b) \quad \phi < (1+b)(\sigma+\alpha\tau),$$

then there exists a unique positive, asymptotically stable, efficiency income $\tilde{q}^e$,

$$\tilde{q}^e = \frac{g+x}{\sigma+\alpha\tau-\phi/(1+b)}.$$

The implication is that per capita and aggregate income converge to exponential growth paths given, respectively, by

$$q_t \longrightarrow (1+b)^t B_0 \tilde{q}^e \quad \text{and} \quad Q_t \longrightarrow \Big((1+b)(1+n)\Big)^t B_0 N_0 \tilde{q}^e.$$

5 Fiscal Policy: Long Run Effects

To consider the longer run influence of fiscal policy, we look at the effects of government spending and taxation on steady state growth. In such a situation the influence of a given tax or spending change continues throughout the future. Given Proposition **B**,

$$\begin{aligned}
\frac{\partial Q_t}{\partial g} &= \frac{1}{\sigma+\alpha\tau-\phi/(1+b)}\Big((1+b)(1+n)\Big)^t B_0 N_0 q^e > 0,\\
\frac{\partial Q_t}{\partial \tau} &= \left(\frac{-\alpha}{(\sigma+\alpha\tau)-\phi/(1+b)}\right)\Big((1+b)(1+n)\Big)^t B_0 N_0 q^e < 0,\\
\frac{\partial (T_t-G_t)}{\partial g} &= \frac{\sigma(\tau-1)+\phi/(1+b)}{\sigma+\tau\alpha-\phi/(1+b)}\Big((1+b)(1+n)\Big)^t B_0 N_0 \tilde{q}^e\\
\frac{\partial (T_t-G_t)}{\partial \tau} &= \left(\frac{\sigma-\phi/(1+b)}{\sigma+\alpha\tau-\phi/(1+b)}\right)\Big((1+b)(1+n)\Big)^t B_0 N_0 \tilde{q}^e.
\end{aligned}$$

These inferences imply the following policy repercussions.

C *Assume*

$$\begin{aligned}
&(a) \qquad g_{t+1}+x_{t+1} = (1+b)^{t+1}(g+x)\\
&(b) \qquad \gamma+g+x>0.
\end{aligned}$$

(i) If

$$\text{(c)} \quad \frac{\phi}{1+b} < \sigma + \tau\alpha,$$

a permanent increase (decrease) in government spending increases (decreases) the steady state growth path;

a permanent increase (decrease) in the tax rate decreases (increases) the growth path of income.

(ii) If

$$\sigma < \frac{\phi}{1+b} < \sigma + \tau\alpha,$$

then

a permanent increase (decrease) in government spending increases (decreases) the budget surplus; and

a permanent increase (decrease) in the average tax rate decreases (increases) the budget surplus.

(iii) If

$$\sigma(1-\tau) < \frac{\phi}{1+b} < \sigma,$$

then

a permanent increase (decrease) in government spending increases (decreases) the budget surplus; and

a permanent increase (decrease) in the average tax rate increases (decreases) the budget surplus.

(iv) If

$$\frac{\phi}{1+b} < \sigma - \tau\sigma,$$

then

a permanent increase (decrease) in government spending decreases (increases) the budget surplus; and

a permanent increase (decrease) in the average tax rate increases (decreases) the budget surplus.

The part of these results that pertains to the title of this paper is **C**(ii) which can be expressed as:

The Two–Fold KRB Tax Policy Paradox: *If*

$$(1+b)\ MPS\ <\ MPI\ <(1+b)\ (MPS\ +\tau\ MPC),$$

then a permanent <u>*decrease*</u> *in the tax rate will* <u>*increase*</u> *the entire future balanced growth path of income, tax revenue, and the net budget.*

Thus, the policy position taken in common by Kennedy, Reagan, and Bush rests on a logical theoretical basis.

5.1 Empirical Results for the Consolidated Era: 1929–1941, 1948–2002

The aggregate demand parameter estimates and the average net tax rate obtained for the Consolidated Era together with the terms that enter the stability and KRB tax policy conditions are the following.

Table 1: Parameter Estimates for the Linear Model

ϕ	σ	α	τ	$\tau\alpha$	$\sigma+\tau\alpha$
.28	.31	.69	.16	.11	.42

Using these estimates,

$$\phi = .28 < (1+b)\sigma = .32$$

so the Standard Conditions, **C**(i), (ii), and (iv) apply. The KRB Tax Condition of Proposition **C**(iii) does not appear to hold.

The anti–recession policy of the current administration and that of Kennedy, G.W. Bush, and Reagan, who all took the same position, appears to be supported in that tax decreases should be expected and seem, indeed, to have stimulated output in the short run and growth in the longer run. On the other hand, their argument that tax revenues should increase as a result of tax reductions is not supported by these parameter estimates. But let us not jump to conclusions prematurely. Further empirical evidence on the issue gives a very different inference.

6 A Nonlinear Complication

6.1 Piecewise Linear Consumption Function

A close inspection shows that in spite of the overall correlation, the data exhibit systematic departures from linearity. These suggest structural changes or possibly a smooth nonlinearity in the consumption and savings relationships. For the purpose of testing this possibility we divided the period 1939–2002 into the following growth Eras.

I	1929–1941	The Great Depression Era
II	1942–1947	World War II and Post War Recovery
III	1948–1980	The Cold War Era
IV	1981–2002	The Computer Global Information Era

Excluding Era II, we estimated the piecewise linear function separately for periods I, III, and IV,

$$f(q_t, \tau_t) = \begin{cases} \alpha_0^1 + \alpha^1 y_t^d & , \ y_t^d \in S^I := (0, y^I) \\ \alpha_0^3 + \alpha^e y_t^d & , \ y_t^d \in S^{II} := (y^{II}, y^{III}) \\ \alpha_0^4 + \alpha^4 y_t^d & , \ y_t^d \in S^{III} := (y^{III}, y^{IV}) \end{cases} \tag{20}$$

where $y_t^d := (1 - \tau_t)q_t$, obtaining the following parameter estimates.

Piece I	$\alpha_0^1 = 1827$	$\alpha^1 = .4378$	$y^I = 6777$	$R^2 = .966$
Piece III	$\alpha_0^3 = 492$	$\alpha^3 = .6348$	$y^{III} = 20047$	$R^2 = .998$
Piece IV	$\alpha_0^4 = -3871$	$\alpha^4 = .8674$		$R^2 = .995$

All coefficients are highly significant. The total amount of variation in the data explained is close to unity in Eras III and IV. Note that the marginal propensity to consume increased and the marginal propensity to save decreased successively through periods I, III, and IV. These changes are substantial. Indeed, the increase in the MPC and decline in the MPS has been drastic—the former almost double, the latter less than a fourth—in Era IV compared to Era I.

At this point one may ask, "Why did these parametric changes occur?" A possible answer comes from the microeconomic theory of consumption and savings. Using a utility representation of people's preferences, one can derive nonlinear consumption and savings fractions that can be approximated by the piecewise linear functions used here. Of course, the national economy is made up of people of great diversity in age, race, and various social characteristics. The data suggest that their aggregate behavior per capita behaves like a representative "average" individual who at low income levels spend a high fraction of income increases on housing, durable goods, and savings but who spend an increasing fraction of income on luxury consumables, such as eating out, vacations, fashionable clothing, etc., as income increases, a phenomenon found in the data by Hildenbrand (1995).

6.2 A Nonlinear DADS Model

Except for two exceptional years at the bottom of the Great Depression, we have not found evidence of a similar nonlinearity in the investment income relationship, so we retain for the time being equation (2) and its parameter estimates. But aggregate demand using the nonlinear consumption function is also nonlinear and also has three pieces,

$$q_{t+1} = \begin{cases} \alpha_0^I + \alpha^I(1-\tau_{t+1})q_{t+1} & + \quad \phi(q_t - y') + \alpha_0 + g_{t+1} + x_{t+1} \\ \alpha_0^{III} + \alpha^{III}(1-\tau_{t+1})q_{t+1} & + \quad \phi(q_t - y') + \alpha_0 + g_{t+1} + x_{t+1} \\ \alpha_0^{IV} + \alpha^{IV}(1-\tau_{t+1})q_{t+1} & + \quad \phi(q_t - y') + \alpha_0 + g_{t+1} + x_{t+1} \end{cases} \tag{21}$$

which, solving for q_{t+1}, gives the reduced form non–autonomous difference equation,

$$q_{t+1} := \frac{1}{\sigma^i + \alpha^i \tau_{t+1}} \left[\alpha_0^i - \phi y' + g_{t+1} + x_{t+1} + \phi q_t \right], \; q_t \in S_t^i, i = I, III, IV \tag{22}$$

where

$$\begin{aligned} S_t^I &:= (1+\tau_{t+1})S^I \\ S_t^{III} &:= (1+\tau_{t+1})S^{III} \\ S_t^{IV} &:= (1+\tau_{t+1})S^{IV}. \end{aligned}$$

As before, we consider only incomes $y \geq y'$.

6.3 Stability and Tax Implications

The qualitative stability and KRB conditions (7(a)) and (7(b)) must now be applied separately to each Era. Using the new parameter estimates for the piecewise segments, the picture changes drastically as shown in Table 2.

Comparing column **(a)** with **(e)** and **(f)** to determine which of Propositions **B** or **C** apply, we get the results summarized in Table 3.

Table 2: Key Coefficients

Periods	(a) ϕ	(b) α	(c) τ	(d) $(1+b)$	(e) $(1+b)\cdot\sigma$	(f) $(1+b)\cdot$ $(\sigma+\alpha\tau)$
(0) 1929–1941, 1948–2002	.26	.69	.16	1.0186	.32	.43
(I) 1929–1941	.44	.44	.12	1.0321	.58	.64
(III) 1948–1980	.28	.63	.18	1.0241	.38	.50
(IV) 1981–2002	.25	.87	.16	1.0176	.13	.31

Table 3: The Stability and KRB Tax Policy Conditions

Era	**Stability**	**KRB Condition**
(0) 1929–1941, 1948–2002	Satisfied	Not Satisfied
(I) 1929–1941	Satisfied	Not Satisfied
(III) 1948–1980	Satisfied	Not Satisfied
(IV) 1981–2002	Satisfied	Satisfied

All three Eras were stable and the KRB condition was not satisfied in Eras I and III. Having incorporated the evident nonlinearity in the consumption and savings relationships, however, the KRB condition *is* satisfied during Era IV. Indeed, if the drastic change in consumer behavior is incorporated, it is clear Bush's arguments rest on a sound theoretical and empirical basis. But Kennedy's hopes were apparently not justified. He died too soon to see the vindication that emerged during the Reagan years due to the substantial increase in the MPC and corollary decline in the MPS.

6.4 Productivity and Employment

We now consider the implications of the DADS model for the employment paradox: a rising rate of GDP with a lagging increase in employment. For this purpose we draw on the following definitions.

N	population
E	employment
L^s	labor supply (employed and unemployed workers)
L	work force (working age — institutional population)

$$\begin{aligned} \varepsilon_{t+1} &:= (E/L^s)_{t+1} &=& \text{ the employment rate} \\ \lambda_{t+1} &:= (L^s/L)_{t+1} &=& \text{ the labor participation rate} \\ \omega_{t+1} &:= (L/N)_{t+1} &=& \text{ the work force fraction of population.} \end{aligned}$$

(Bear in mind that the unemployment rate is $u = 1 - \varepsilon$.) For GDP we have the identity

$$Q_{t+1} = E_{t+1} q^e_{t+1} = N_{t+1} q_{t+1},$$

and since by assumption, $q_t = z_t$, we have for the ratio of employment to population,

$$E_{t+1}/N_{t+1} = z_{t+1}/q^e_{t+1}, \tag{23}$$

which depends on per capita demand divided by productivity per employed worker. Evidently, if the latter were to grow at a rate faster than that of demand, the employment/population ratio would decline. That ratio, however, is not the crucial one for macroeconomic policy. Rather, it is the ratio of employment to the labor supply, E_t/L^s_t, and how it is influenced by the labor participation rate, L^s_t/L_t, and the work force/population ratio, L_t/N_t.

Multiply the left side of (23) by $\frac{L^s_{t+1}}{L^s_{t+1}} \cdot \frac{L_{t+1}}{L_{t+1}}$ and do some re–arranging to get the employment rate,

$$\varepsilon_{t+1} = \frac{1}{q^e_t} \cdot \frac{1}{\lambda_{t+1} \cdot \omega_{t+1}} \cdot \frac{(\gamma + g_{t+1} + x_{t+1} + \phi q_t)}{\sigma^i + \alpha \tau_{t+1}}. \tag{24}$$

From this equation we infer the following model estimates.

D *The employment rate increases (decreases) if*

(i) aggregate demand increases (decreases);

(ii) labor productivity decreases (increases);

(iii) the labor participation rate decreases (increases).

In terms of the instrumental variables of fiscal policy we get

E *The employment rate increases (decreases) if*

(i) government spending increases (decreases);

(ii) the average tax rate decreases (increases).

Fiscal policy is often discussed vis á vis a target unemployment rate, say $\bar{u}$. The employment rate equation can be used to estimate the government spending and taxation policy that would be required to achieve such a rate. Set the left side at $\bar{\varepsilon}_{t+1} = 1 - \bar{u}$ and adjust τ and g to find a desired combination. Fixing τ, for example, and solving for g, we get

$$g_{t+1} = \bar{\varepsilon} \cdot \lambda_{t+1} \cdot \omega_{t+1} \cdot q^e_{t+1} \cdot (\sigma + \alpha\tau) - (\gamma + x_{t+1} + \phi q_t). \qquad (25)$$

F *The level of government spending required to achieve a given target employment rate increases if*

(i) the labor participation rate increases;

(ii) the work force/population ratio increases;

(iii) labor productivity increases.

7 Discussion

The model presented here cannot be taken as a perfect tool for policy repercussion analysis. Important variables have not all been taken into account. However, as far as empirical macroeconomics is concerned, it is evidently a good first approximation. The results would seem to compel serious attention. Of special interest is the KRB tax policy paradox. Evidently, tax reductions should be expected to stimulate the economy and—given the drastic reduction in the economy's per capita marginal propensity to consume—a rise in tax revenues should accompany the economic growth that follows—with approximately a year's delay.

The work force and employment data also support the inferences obtained in the preceding section. Increases in the labor participation rate were substantial in much of the post war period, just as with the work force as a percentage of the population. Together with a continued upward trend in productivity, these facts explain why unemployment lags behind economic recovery after a recession and constitute a continuing problem. The lesson of equation (25)—given fixed labor participation rates and the work force/population ratio—is basically that government must expect to grow along with population if a reasonable level of employment is to be sustained. But options also exist: labor participation can be discouraged by delaying entry into the labor force (military or other service, extended education) by reducing the full employment fraction of the year (shorter hours, longer vacations). A reversion to the traditional family structure of one bread winner, one family manager would have a similar effect. Eventually, in the U.S. as well as other developed countries, the need for one or more of these possibilities is being reduced by, and maybe eliminated by, the graying of the population which will lower the work force/population ratio.

Notes

1. An early theoretical analysis of the Kennedy tax policy will be found in Day, Richard H., 1970, "An Elementary Analysis of the Kennedy Tax Program," Chapter 29 in W.L. Johnson and D.S. Kammerschen, *Macroeconomics: Selected Readings*, Boston: Houghton Mifflin. The present discussion introduces critical nonlinearities, distinguishes between short and long run growth effects, and presents empirical evidence relevant to the period 1929–2002.

2. An alternative method of proof is by direct iteration.
Since $q^e_{t+1} = \frac{A}{(1+b)^{t+1}} + B + Cq^e_t$ by definition, we have by iteration that

$$\begin{aligned} q^e_1 &= \tfrac{A}{1+b} + B + Cq^e_0 \\ q^e_2 &= \tfrac{A}{(1+b)^2}\Big[1 + C(1+b)\Big] + B(1+C) + C^2 q^e_0 \end{aligned}$$

and

$$q^e_3 = \frac{A}{(1+b)^3}\Big[1 + C(1+b) + C^2(1+b)^2\Big] + B(1+C+C^2) + C^3 q^e_0.$$

Suppose

$$\begin{aligned} q^e_{t-1} = {} & \frac{A}{(1+b)^{t-1}}\Big[1 + C(1+b) + \cdots + C^{t-2}(1+b)^{t-2}\Big] \\ & + B(1 + C + \cdots + C^{t-2}) + C^{t-1} q^e_0. \end{aligned}$$

As

$$q^e_t = \frac{A}{(1+b)^t} + B + Cq^e_{t-1},$$

it follows that

$$q^e_t = \frac{A}{(1+b)^t}\Big[1{+}C(1{+}b){+}\cdots{+}C^{t-1}(1{+}b)^{t-1}\Big]{+}B(1{+}C{+}\cdots{+}C^{t-1}){+}C^t q^e_0.$$

Therefore,

$$q^e_t = \frac{A}{(1+b)^t}\sum_{j=0}^{t-1} C^j(1+b)^j + B\sum_{j=0}^{t-1} C^j + C^t q^e_0,\ t = 1, 2, 3, \ldots.$$

If $0 < C < 1$ and $b > 0$, we have $\lim_{t\to\infty} q^e_t = \dfrac{B}{1-C} < \infty$. Consequently, the series $\{q^e_t\}$ converges to $\frac{B}{1-C}$ as t goes to infinity. □

A HYBRID APPROXIMATION TO CERTAIN DELAY DIFFERENTIAL EQUATION WITH A CONSTANT DELAY

GEORGE SEIFERT

IOWA STATE UNIVERSITY, AMES, IOWA 50011

ABSTRACT. We study a delay differential equation with piecewise constant delays which could serve as an approximation to a corresponding delay differential equation with a finite constant delay. We consider a number of special cases.

1. INTRODUCTION

We consider a differential equation of the form

$$x'(t) = Ax(t) + g(x(-1 + k/N) \tag{1.1}$$

on the interval $k/N \leq t < (k+1)/N$, N a fixed positive integer, and $k = 0, 1, 2, \ldots$; here $x(t) : R \to R^n$, A is a constant $n \times n$ nonsingular matrix, $g(x) : R^n \to R^n$, and n is a positive integer. The function $x(t) : R \to R^n$ is said to be a solution of (1.1) on $[0, \infty)$ if it satisfies (1.1) on each interval $k/N < t < (k+1)/N, k = 0, 1, 2, \ldots,$ is continuous on $[0, \infty)$, and satisfies an initial condition of the form $x(-1 + k/N) = \phi_k$ where $\phi_k \in R^n$, $k = 0, 1, \ldots, N-1$. it is easy to see that in a sense such a solution approximates a solution on $[0, \infty)$ of the equations

$$x'(t) = Ax(t) + g(x(t-1)), x(t) \in R^n \tag{1.2}$$

with initial condition $x(t) = \phi(t), -1 \leq t \leq 0$, and where $\phi(-1 + k/N) = \phi_k, k = 0, 1, 2, \ldots, N-1$, and where the approximation become better as $N \to \infty$.

We assume that $g(0) = 0 \in R^n$ which is clearly no loss of generality. We also assume that the $n \times n$ Jacobian matrix $\partial g/\partial x$ is nonsingular at $x = 0 \in R^n$. The special case of (1.1) where

$$A = -\begin{pmatrix} \alpha & 0 \\ 0 & \alpha \end{pmatrix}, \quad g(x) = b(h(x_2), h(x_1))^T, \quad \alpha > 0, b > 0, x \text{ real}$$

and $h(x) = (1 - e^{-x})(1 + e^{-x})^{-1}$ can model the interactions of two identical neurons; [2], [3]. For the case $N = 1$, (1.1) can be put in for form $x'(t) = Ax(t) + g(x([t]))$, where $[\cdot]$ is the greatest integer function, and a special case of this for $n = 2$ is discussed in [4].

2. The General Case

It is easy to see that for $k/N \le t < (k+1)/N, \quad k = 0, 1, 2, \ldots$, we have from (1.1)

$$x(t) = \left(\int_{k/N}^{t} e^{A(t-s)} ds \right) (g(x(-1 + k/N))) + e^{A(t-k/N)} x(k/N), \tag{2.1}$$

and by continuity of $x(t)$ at $t = (k+1)/N$, we must have

$$x((k+1)/N) = Bg(x(-1 + k/N)) + Cx(k/N) \tag{2.2}$$

where $B = A^{-1}(e^{A/N} - E), \quad C = e^{A/N}, E$ the $n \times n$ identity matrix; here $x(-1 + k/N) = \phi_k \in R^n$ are given $k = 0, 1, 2, \ldots, N-1$, a so called initial condition.

We can express (2.2) as a system in R^{nN} as follows:

$$\begin{aligned} x_1((k+1)/N) &= Bg(x_N(k/N) + Cx_1(k/N)) \\ x_2((k+1)/N) &= x_1(k/N) \\ x_3((k+1)/N) &= x_2(k/N) \\ &\vdots \\ x_N((k+1)/N) &= x_{N-1}(k/N), \end{aligned} \tag{2.3}$$

$k = 0, 1, 2, \ldots$; here $x_1(t) = x(t), x_2(t) = x_1(t - 1/N) \ldots, \ x_N(t) = x_{N-1}(t - 1/N), x(t) \in R^n$.

Define the function $F(x) : R^{nN} \to R^{nN}$ by

$$F(x) = ((Cx_1 + Bg(x_N), x_1, x_2, \ldots, x_{N-1})^T \tag{2.4}$$

where B and C are as in (2.2), and $x_j \in R^n, j = 1, 2, \ldots, N$, $x = (x_1, x_2, \ldots, x_N)^T$. It then follows from (2.3) that solutions of (2.2), and hence (1.1), can be obtained by using iterations of the function F; i.e., $F^j(F(x)) = F^{j+1}(x), j = 0, 1, \ldots,$ where $F^\circ(x) = x$.

Clearly, much of the qualitative behavior of the solution of (1.1) can be determined from the corresponding solutions of (2.3). For example, if for some $x_0 \in R^{nN}$, $F^k(x_0) \to \bar{0}$, the zero in R^{nN}, as $k \to \infty$, and $x(t)$ is the corresponding solution of (1.1) then $x(t) \to 0 \in R^n$ as $t \to \infty$. Also if for some positive integer p, $F^p(x_0) = x_0 \in R^{nN}$, the corresponding solution $x(t)$ of (1.1) will satisfy $x(t + p/N) = x(t)$ for all $t \geq 0$.

We now state without proof three other obvious results. Here and henceforth we define $||x|| = \max\{|x_j| j = 1, 2, \ldots, m\}$ for any $x = (x_1, \ldots, x_m)^T$, m a positive integer.

Lemma 2.1. If $x = (x_1, \ldots x_N)^T$, $x_j \in R^n, j = 1, \ldots, N$, is a fixed point of F, then $Cx_1 + Bg(x_1) = x_1$.

Lemma 2.2. If all the eigenvalues λ of the Jacobian matrix $\partial F/\partial x$ of F at $x = 0 \in R^{nN}$ satisfy $|\lambda| < 1$, the solutions $x(t)$ of (1.1) are locally attracted to the constant solution (or critical point) $x = 0 \in R^n$ of (1.1). If $|\lambda| > 1$ for all such λ, $x = 0$ is a repeller, or ejective point; cf[1].

Lemma 2.3. If we define $R_+^k = \{x = (x_1, \ldots, x_k)^T, x_j \geq 0, j = 1, \ldots, k\}$ k a positive integer and suppose A diagonal, and $g(x) \in R_+^n$ for $x \in R_+^n$, then $F(x) \in R_+^{nN}$ where $x = (x_1, x_2, \ldots, x_N)^T$, provided $x_j \in R_+^n, j = 1, \ldots, N - 1$.

We also have

Lemma 2.4. Let there exist constants $b_0 > 0$ and $c_0, 0 < c_0 < 1$, such that $||Cx|| < c_0||x||$ and $||Bg(x)|| \leq b_0$ for all $x \in R^n$, there we have $||F(\bar{x})|| \leq b_0/(1 - c_0)$ whenever $||\bar{x}|| \leq b_0/(1 - c_0), \bar{x} \in R^{nN}$.

Proof. If $\bar{x} = (x_1, x_2, \dots, x_N)^T, \quad x_j \in R^n, j = 1, 2, \dots, N,$

$$\begin{aligned} ||F(\bar{x})|| &= \max\{||Bg(x_N) + Cx_1||, ||x_1||, ||x_2||, \dots, ||x_{N-1}||\} \\ &\le \max\{b_0 + c_0 b_0/(1-c_0), b_0/(1-c_0), \dots, b_0/(1-c_0)\} \\ &= b_0/(1-c_0). \quad \square \end{aligned}$$

Theorem 1. *Let the conditions in Lemma 2.3 and Lemma 2.4 hold, and let all the eigenvalues λ of the Jacobian matrix $\partial F/\partial x$ at $x = \bar{0} \in R^{nN}$ satisfy $|\lambda| > 1$. Then there exists a $x_c \in R_+^{nN}, ||x_c|| > 0$ such that $F(x_c) = x_c$. Thus if $x_c = (x_{c_1}, x_{c_2}, \dots, x_{c_N})^T, x_{c_j} \in R_+^n$ for $j = 1, \dots, N, x(t) = x_{c_1}$ is a constant solution of (1.1) for $t > 0$.* $\square$

Proof. This can be proved using an ejective fixed point theorem such as Theorem 2.1 on p. 249 in [1]. Clearly $\bar{0} \in R^{nN}$ is an ejective fixed point of F, and by Lemma 4, F maps the closed bounded convex set $K = \{x \in R_+^{nN} : ||x|| \le b_0/(1-c_0)\}$ into itself with $\bar{0}$ an extreme point of K. We omit the further details.

3. Some Special Cases

We first consider the case where (1.1) is a scalar equation; i.e., where $n = 1, A = -\alpha, \alpha > 0$, and the solution $x(t) \in R, t \ge 0$, is determined by the known initial values $x(-1 + k/N), k = 0, 1, \dots, N-1$. If F is as in (2.4), $F : R^N \to R^N$. Thus the Jacobian $\partial F/\partial x$ at $\bar{0} \in R^N$ is a $N \times N$ matrix of scalars given by

$$\begin{pmatrix} c_0 & 0 & 0 & \dots & 0 & b_0 \\ 1 & 0 & 0 & \dots & 0 & 0 \\ 0 & 1 & 0 & \dots & 0 & 0 \\ \vdots & & & & & \\ 0 & 0 & 0 & \dots & 1 & 0 \end{pmatrix} \tag{3.1}$$

where $c_0 = \exp(-\alpha)$, and $b_0 = g'(0)(1 - \exp(-\alpha))/\alpha$. As before, we assume $g'(0) > 0$.

It is easy to see that the eigenvalues of this matrix are roots of

$$\lambda^{N-1}(\lambda - c_0) - b_0 = 0, \tag{3.1}$$

and we have the following result:

Lemma 3.1. Suppose $c_0 + b_0 < 1$. Then all the roots of (3.1) satisfy $|\lambda| < 1$, and so any solution $x(t)$ of (1.1) with $|x(t)|$ sufficiently small at $t = -1 + \frac{1}{N}, -1 + \frac{2}{N} \ldots, 0$ satisfies $x(t) \to 0$ as $t \to \infty$.

Proof. Suppose some root λ_0 of (3.1) satisfies $|\lambda_0| \geq 1$. Then $b_0 \geq |\lambda_0 - c_0| \geq |\lambda_0| - c_0$, and so $|\lambda_0| \leq c_0 + b_0 < 1$. This contradiction proves Lemma 3.1.

We consider next the special case of (1.1) where $n = 2$, and $g = (g_1, g_2)^T$ where $g_1 = G(x_2), g_2 = G(x_1), x = (x_1, x_2)^T$ and

$$A = -\begin{pmatrix} \alpha & 0 \\ 0 & \alpha \end{pmatrix}; \quad \alpha > 0.$$

This can model the interaction of two identical neurons, and for the case $N = 1$ was studied extensively in [4]. We assume that G is continuously differentiable on R, satisfies $xG(x) > 0$ for $x \neq 0$, and $G'(0) = b_0 > 0$. The Jacobian matrix for the corresponding $F : R^{2N} \to R^{2N}$ of (2.4) is the form of (3.1) where $c_0 = A$, and b_0 is replaced by the 2×2 matrix

$$\begin{pmatrix} 0 & b_0 \\ b_0 & 0 \end{pmatrix}.$$

It follows easily that the eigenvalues λ of this $2N \times 2N$ Jacobian matrix satisfy

$$|\lambda - \alpha|^2 |\lambda|^{2N+2} = b_0^2 \ . \tag{3.2}$$

Lemma 3.2. If $b_0 > \alpha > 0$ where α and b_0 are as above, then the zero solution of (1.1) is a local attractor when all roots of λ of (3.2) satisfy $|\lambda| < 1$, and an ejective point if they all satisfy $|\lambda| > 1$ and in that case if $G(x)$ is bounded on $x \geq 0$, and $b_0 - \alpha > 1$, there exists a $x_c = (x_{c_1}, x_{c_2})^T, x_{c_1} > 0$ for $j = 1, 2$ such that $x_c = Ax_c + g(x_c)$; i.e., $x(t) = x_c, t \geq 0$, is a constant solution of (1.1).

Proof. This follows immediately from Theorem 1.

We next consider a case of (1.1) where $n = 1, N = 4$, and give conditions under which such an equation has a nonconstant periodic solution.

Our $F : R^4 \to R^4$ is now given by

$$(ax_1 + g(x_4), x_1, x_2, x_3)^T$$

where $0 < a < 1$. By direct calculation we have F^2 given by

$$(a^2x_1 + ag(x_4) + g(x_3), ax_1 + g(x_4), x_1, x_2),$$

and so the condition $F^2(x) = x, x \in R^4$, is just

$$\begin{aligned} a^2x_1 + ag(x_4) + g(x_3) &= x_1 \\ ax_1 + g(x_4) &= x_2 \\ x_1 &= x_3 \\ x_2 &= x_4\,. \end{aligned} \tag{3.3}$$

Hence we seek solutions of the system

$$\begin{aligned} a^2x_1 + ag(x_2) + g(x_1) &= x_1 \\ ax_1 + g(x_2) &= x_2 \end{aligned}$$

and for such a solution (x_1, x_2), we get $x_3 = x_1, x_4 = x_2$. Multiplying the second equation by a, and subtracting it from the first, we get $g(x_1) = x_1 - ax_2$ which together with the second leads to the system

$$\begin{aligned} ax_2 &= x_1 - g(x_1) \\ ax_1 &= x_2 - g(x_2)\,. \end{aligned} \tag{3.4}$$

Lemma 3.3. Suppose $g(x) = -g(-x), g'(x) > 0$ for all x, $g'(0) > 1+a$, and $g(x)$ is bounded for all $x \geq 0$. Then there exists $(\bar{x}_1, \bar{x}_2), \bar{x}_1 > 0 > \bar{x}_2$, such that

$$\begin{aligned} F^2(\bar{x}_1, \bar{x}_2, \bar{x}_1, \bar{x}_2) &= (\bar{x}_1, \bar{x}_2, \bar{x}_1, \bar{x}_2)^T \quad \text{and} \\ F(\bar{x}_1, \bar{x}_2, \bar{x}_1, \bar{x}_2) &= (\bar{x}_2, \bar{x}_1, \bar{x}_2, \bar{x}_1)^T. \end{aligned} \tag{3.5}$$

Proof. From the conditions on g, we see that the graph Γ_1 of the first equation in (3.4) crosses the x_1-axis at some point $(r_1, 0), r_1 > 0$ and is below the x_1-axis for $0 < x_1 < r_1$.

The graph Γ_2 of the second equation is the reflection of Γ_1 at the line $x_1 = x_2$. Since its slope at the origin is the reciprocal of the slope of Γ_1 there; i.e., $a/(1 - g'(0))$, which satisfies $a/(1 - g'(0) > (1 - g'(0)/a$, Γ_2 is above Γ_1 for $0 < x < \delta, \delta$ sufficiently small, is below the x_1-axis for and to the right of the x_2 axis for $0 < x_2 < r_1$ it intersects Γ_1 at, say, $(\bar{x}_1, \bar{x}_2), \bar{x}_1 > 0 > \bar{x}_2$. This essentially proves Lemma 3.3. The value of $F(x)$ for $x = (\bar{x}_1, \bar{x}_2, \bar{x}_3, \bar{x}_4)^T$ as in (3.5) follows easily. We note that $\bar{x}_1 - \bar{x}_2 \to 0$ as $g'(0) \downarrow 1 + a$. □

Weaker conditions on g for Lemma (3.3) to hold can be given at at the expense of more complicated arguments. We note that the $g(x) = b(1 - e^{-x}(1 + e^{-x})^{-1}$ mentioned in the introduction satisfies the conditions of Lemma 3.3 if $b > 2a + 2$. The simpler special case where $g(x) = b \tan^{-1} x$ where $b > a + 1$ also satisfies the conditions of Lemma 3.3.

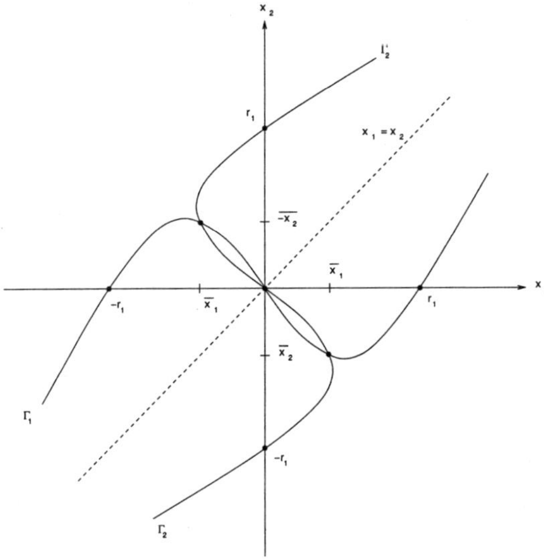

Figure 1

The following are direct consequences of Lemma 3.3 and Lemma 3.2.

Theorem 2. *Let* $n = 1, N = 4$ *and* $g(x)$ *be as in Lemma 3.3. Then there exists a solution* $x(t)$ *of (1.1) such that* $x(t + 1/2) = x(t)$ *for* $t \geq 0$ *where* $\bar{x}(0) = x_1, x(\frac{1}{2}) = \bar{x}_2$, *and where* $(\bar{x}_1, \bar{x}_2)$ *solves (3.4) with* $\bar{x}_1 > 0 > \bar{x}_2$.

Theorem 3. *Let $n = 2$ and N be any fixed positive integer. Let* $A = -\begin{pmatrix} \alpha & 0 \\ 0 & \alpha \end{pmatrix}$, $\alpha > 0$, *and* $g(x) = (G(x_2), G(x_1))^T$ *for* $x = (x_1, x_2)^T$, *and* $xG(x) > 0$, *where* $G(x)$ *is continuously differentiable and bounded on* R. *Let* $b_0 = G'(0)(1 - \exp(-\alpha))/\alpha > 0$. *Then if all roots* λ *of*

$$|\lambda - \alpha||\lambda|^{N+1} = b_0$$

satisfy $|\lambda| > 1$, *there exists a constant solution* $x(t) = (x_1, x_2), t \geq 0$, *of (1.1) where* $(x_1, x_2) \neq (0, 0), (x_1, x_2) \in R^2_+$.

Equation (1.1) for $t > 0$ can in fact be more compactly expressed as

$$x'(t) = Ax(t) + g(x([N(t-1)]/N)), \quad t > 0,$$

with initial conditions $x(-1 + k/N) = \phi_k$, $k = 0, 1, \ldots, N - 1$.

References

1. J.K. Hale, *Theory of Functional Differential Equations*, Springer-Verlag, New York, 1977.
2. C.M. Markus and R. M. Westervelt, *Stability of analog neural networks with delays*, Phys. Rev. **39** (59), 342-354.
3. K. Pakdaman, C. Gota-Ragazzo, C.P. Malta, O. Arino, J.-F. Vert, *Effect of delay on the boundary of the basin of attraction in a system of two neurons*, Neural Networker **11** (1998), 509-519.
4. G. Seifert, *Qualitative behavior and stability of solutions of a hybrid system*, Analysis and Applications (to appear).

Compulsory asymptotic behavior of solutions of two-dimensional systems of difference equations*

Josef Diblík

Department of Mathematics and Descriptive Geometry, Faculty of Civil Engineering, Brno University of Technology, Žižkova 17, 662 37 Brno, Czech Republic, E-mail address: diblik.j@fce.vutbr.cz

Irena Růžičková

Department of Mathematics, Faculty of Electrical Engineering and Communication, Brno University of Technology, Technická 8, 616 00 Brno, Czech Republic, E-mail address: ruzickov@feec.vutbr.cz

Abstract

This contribution is devoted to a discussion of the asymptotic behavior of solutions of systems of two difference equations. General principles giving a guarantee that the graph of at least one solution stays in a prescribed domain were given in previous papers of the first author. Two techniques – the so called retract type technique, and Liapunov type approach – were used separately. In the present contribution we try to connect both principles to investigate the asymptotic behavior of solutions of systems consisting of two equations. We show that under appropriate conditions containing conditions typical for the retract technique approach, and conditions typical for the Liapunov type approach, there exists at least one solution of the system considered the graph of which stays in a prescribed domain. Illustrative examples are given too. Moreover, as an application, a difference equation of Emden-Fowler type is considered.

1 Introduction

In this contribution we investigate the asymptotic behavior for $k \to \infty$ of the solutions of two difference equations

$$\begin{aligned} \Delta u_1(k) &= f_1(k, u_1(k), u_2(k)), \\ \Delta u_2(k) &= f_2(k, u_1(k), u_2(k)) \end{aligned} \tag{1}$$

with $f_1, f_2 : N(a) \times \mathbb{R} \times \mathbb{R} \to \mathbb{R}$, $\Delta u_i(k) = u_i(k+1) - u_i(k)$, $i = 1, 2$, where $k \in N(a) = \{a, a+1, \dots\}$, $a \in \mathbb{N}$ is fixed and $\mathbb{N} = \{0, 1, \dots\}$.

*1991 Mathematics Subject Classifications: 39A10, 39A11.

Key words and phrases: Discrete equation, asymptotical behaviour of solution, bounded solution, retract approach, Liapunov type technique, Emden-Fowler equation.

1.1 Description of the Problem Considered

Let us define a set $\Omega \subset N(a) \times \mathbb{R} \times \mathbb{R}$ as

$$\Omega := \bigcup_{k \in N(a)} \Omega(k)$$

where $\Omega(k)$ is an *open bounded* and *simply connected* set in the space

$$S(k) := \{(k, u_1, u_2) : u_i \in \mathbb{R}, i = 1, 2\}$$

for every $k \in N(a)$. Supposing, for every $k \in N(a)$, the set $\Omega(k)$ well-defined, we pose the following problem.

Problem 1 Find sufficient conditions with respect to the right-hand side of system (1) in order to *guarantee the existence of at least one solution* $u(k) = (u_1^*(k), u_2^*(k))$, $k \in N(a)$ satisfying

$$(k, u_1^*(k), u_2^*(k)) \in \Omega(k), \tag{2}$$

for every $k \in N(a)$. □

Since the asymptotic behavior of a solution satisfying (2) is in some sense conditioned by the sets $\Omega(k)$, we call the asymptotic behavior of such solutions *compulsory* asymptotic behavior. Investigation of asymptotic behavior of solutions of discrete equations is in the centre of interest of many investigators. Let us mention e.g. the book [1] and papers [3], [5]–[9].

1.2 Preliminary

Let us recall some known facts. The *existence and uniqueness* of the solution of initial problem (1), (3) with prescribed value

$$(u_1(a+s), u_2(a+s)) = (u_1^s, u_2^s) \in \mathbb{R} \times \mathbb{R}, \ \ s \in \mathbb{N}, \ \ (s \text{ is fixed}) \tag{3}$$

on $N(a+s)$ is obvious. The solution $u = u(k)$, $k = a+s, a+s+1, \dots$ of initial problem (1), (3) is defined as an infinite sequence of number vectors

$$\{u(a+s) = u^s, u(a+s+1), u(a+s+2), \dots, u(a+s+n), \dots\}$$

with $u(\cdot) = (u_1(\cdot), u_2(\cdot))$ such that for any $k \in N(a+s)$, equalities (1) hold. The sequence $\{(k, u_1(k), u_2(k))\}$, $k \in N(a+s)$, is called the *graph* of the solution $(u_1, u_2) = (u_1(k), u_2(k))$ for $k \in N(a+s)$ of initial problem (1), (3).
If for every fixed $k \in N(a)$ the right hand sides $f_i(k, u_1, u_2)$, $i = 1, 2$, are continuous on $(u_1, u_2) \in \mathbb{R} \times \mathbb{R}$, then the solution of initial problem (1), (3)

depends continuously on the initial data.

A point $M_1^c(k, u_1, u_2)$ is called the *first consequent point* of a point $M = (k^*, u_1^*, u_2^*) \in N(a) \times \mathbb{R} \times \mathbb{R}$ (and we write $M_1^c = \mathcal{C}[M]$) if its first coordinate is shifted to $k^* + 1$ and

$$u_1 = u_1^* + f_1(k^*, u_1^*, u_2^*),$$
$$u_2 = u_2^* + f_2(k^*, u_1^*, u_2^*).$$

1.2.1 Polyfacial set

Let auxiliary functions $b_1(k), b_2(k), c_1(k), c_2(k) : N(a) \to \mathbb{R}$ with $b_1(k) < c_1(k)$ and $b_2(k) < c_2(k)$ be given. Define functions

$$B_i := -u_i + b_i(k), \quad C_i := u_i - c_i(k), \quad i = 1, 2$$

and sets

$$\Omega_B^1 := \{(k, u_1, u_2) : k \in N(a), B_1 = 0, B_2 \le 0, C_1 \le 0, C_2 \le 0\},$$
$$\Omega_B^2 := \{(k, u_1, u_2) : k \in N(a), B_1 \le 0, B_2 = 0, C_1 \le 0, C_2 \le 0\},$$
$$\Omega_C^1 := \{(k, u_1, u_2) : k \in N(a), B_1 \le 0, B_2 \le 0, C_1 = 0, C_2 \le 0\},$$
$$\Omega_C^2 := \{(k, u_1, u_2) : k \in N(a), B_1 \le 0, B_2 \le 0, C_1 \le 0, C_2 = 0\}.$$

In the following we use sets $\Omega(k)$, $k \in N(a)$, having the form

$$\Omega(k) := \{(k, u_1, u_2) : B_1 < 0, B_2 < 0, C_1 < 0, C_2 < 0\}.$$

Then the corresponding set Ω can be written as

$$\Omega = \{(k, u_1, u_2) : k \in N(a),\ B_1 < 0, B_2 < 0, C_1 < 0, C_2 < 0\} \tag{4}$$

and

$$\partial\Omega = \Omega_B^1 \cup \Omega_B^2 \cup \Omega_C^1 \cup \Omega_C^2.$$

Definition 1 (Polyfacial set) The set Ω, written in the form (4), is called a *polyfacial set.*

1.2.2 Points of strict egress and their geometrical sense

In the paper [5] Problem 1 is solved with the aid of the retract type technique (in the theory of ordinary differential equations it goes back to Ważewski [4]). An important role is played by the so called strict egress points. In accordance with [5, Lemmas 1,2], a point $(k, u_1, u_2) \in \partial\Omega$ is a *point of the type of strict egress* for the polyfacial set Ω with respect to discrete system (1) if and only if

$$(k, u_1, u_2) = (k, b_1(k), u_2) \quad \text{and} \quad f_1(k, b_1(k), u_2) < b_1(k+1) - b_1(k), \tag{5}$$

or

$$(k, u_1, u_2) = (k, u_1, b_2(k)) \quad \text{and} \quad f_2(k, u_1, b_2(k)) < b_2(k+1) - b_2(k), \tag{6}$$

or

$$(k, u_1, u_2) = (k, c_1(k), u_2) \quad \text{and} \quad f_1(k, c_1(k), u_2) > c_1(k+1) - c_1(k), \tag{7}$$

or, finally,

$$(k, u_1, u_2) = (k, u_1, c_2(k)) \quad \text{and} \quad f_2(k, u_1, c_2(k)) > c_2(k+1) - c_2(k). \tag{8}$$

Geometrically inequalities (5)–(8) mean the following:
If a point $M = M(k, u_1, u_2) \in \partial\Omega$ is a point of the type of strict egress, then the first consequent point $\mathcal{C}[M] \notin \overline{\Omega}(k+1)$.

1.2.3 Points of strict ingress and their geometrical sense

In the paper [7] Problem 1 is solved with the aid of the Liapunov type technique. For this purpose the so called strict ingress type points are defined. Corresponding inequalities characterizing such points are not only the opposite inequalities to (5)–(8). Namely, in accordance with [7, Lemmas 1,2], a point $(k, u_1, u_2) \in \partial\Omega$ is a *point of the type of strict ingress* for the polyfacial set Ω with respect to discrete system (1) if and only if

$$(k, u_1, u_2) = (k, b_1(k), u_2) \quad \text{and}$$
$$b_1(k+1) - b_1(k) < f_1(k, b_1(k), u_2) < c_1(k+1) - b_1(k), \tag{9}$$

or

$$(k, u_1, u_2) = (k, u_1, b_2(k)) \quad \text{and}$$
$$b_2(k+1) - b_2(k) < f_2(k, u_1, b_2(k)) < c_2(k+1) - b_2(k), \tag{10}$$

or

$$(k, u_1, u_2) = (k, c_1(k), u_2) \quad \text{and}$$
$$b_1(k+1) - c_1(k) < f_1(k, c_1(k), u_2) < c_1(k+1) - c_1(k), \tag{11}$$

or, finally,

$$(k, u_1, u_2) = (k, u_1, c_2(k)) \quad \text{and}$$
$$b_2(k+1) - c_2(k) < f_2(k, u_1, c_2(k)) < c_2(k+1) - c_2(k). \tag{12}$$

Geometrically inequalities (9)–(12) mean: if $M = M(k, u_1, u_2) \in \partial\Omega$ is a point of the type of strict ingress, then for $M \in \Omega_B^1 \cup \Omega_C^1$ the coordinate u_1 of $\mathcal{C}[M]$ lies between $b_1(k+1)$ and $c_1(k+1)$ and for $M \in \Omega_B^2 \cup \Omega_C^2$ the coordinate u_2 of $\mathcal{C}[M]$ lies between $b_2(k+1)$ and $c_2(k+1)$.

2 Main Result

Suppose that Ω is a polyfacial set given by (4). In this part we will solve Problem 1 supposing that a part of the set $\partial\Omega$ consists of points of the type of strict egress and the remaining part is created by points of the type of strict ingress. Since the set Ω was specified by relation (4), embedding (2) expresses that every solution $u(k) = (u_1^*(k), u_2^*(k))$, $k \in N(a)$ of Problem 1 satisfies the inequalities

$$\begin{cases} b_1(k) < u_1^*(k) < c_1(k), \\ b_2(k) < u_2^*(k) < c_2(k) \end{cases} \tag{13}$$

for every $k \in N(a)$.

Theorem 1 (Main Result) *Let $b_i(k)$, $c_i(k)$, $b_i(k) < c_i(k)$, $i = 1, 2$ be real functions defined on $N(a)$ and let $f_i : N(a) \times \mathbb{R} \times \mathbb{R} \to \mathbb{R}$, $i = 1, 2$ be functions that are continuous with respect to their last two arguments. Suppose that all points of the sets Ω_B^1, Ω_C^1 are points of strict egress, i.e., if $(k, u_1, u_2) \in \Omega_B^1$ then*

$$f_1(k, b_1(k), u_2) < b_1(k+1) - b_1(k), \tag{14}$$

and if $(k, u_1, u_2) \in \Omega_C^1$ then

$$f_1(k, c_1(k), u_2) > c_1(k+1) - c_1(k), \tag{15}$$

and all points of the sets Ω_B^2, Ω_C^2 are points of strict ingress, i.e., if $(k, u_1, u_2) \in \Omega_B^2$ then

$$b_2(k+1) - b_2(k) < f_2(k, u_1, b_2(k)) < c_2(k+1) - b_2(k) \tag{16}$$

and if $(k, u_1, u_2) \in \Omega_C^2$ then

$$b_2(k+1) - c_2(k) < f_2(k, u_1, c_2(k)) < c_2(k+1) - c_2(k). \tag{17}$$

Let, moreover, the function

$$F(w) := w + f_2(k, u_1, w) \tag{18}$$

be monotone for every fixed

$$(k, u_1) \in \Omega_1 := \{(k, u_1) : k \in N(a), b_1(k) \le u_1 \le c_1(k)\}$$

on the interval $b_2(k) \le w \le c_2(k)$.
Then there exists a solution $u = (u_1^*(k), u_2^*(k))$ of system (1) satisfying inequalities (13) for every $k \in N(a)$.

2.1 Proof of the Main Result

In the proof of the theorem the known fact that a closed n-dimensional ball cannot be continuously mapped onto its (entire) boundary is used. In our case we use this fact for $n = 1$, i.e. a segment, say, $[\alpha, \beta]$ with $\alpha < \beta$ cannot be continuously mapped onto (both) two points α, β. The general scheme of the proof is based, in general, on a contrary supposition with respect to the statement of Theorem 1. Namely we suppose that there exists no solution satisfying inequalities (13) for every $k \in N(a)$. In such situation we prove that the above mentioned mapping (the mapping $\mathcal{R}$ below) exists, i.e. we will get a contrary which proves our result. As a tool for the construction of such mapping we use the so called retract technique (see, e.g., [5]). Divide the proof into several steps.

2.1.1 Retract and retraction

Definition 2 If $A \subset B$ are any two sets of a topological space and $\pi : B \to A$ is a continuous mapping from B onto A such that $\pi(p) = p$ for every $p \in A$, then π is said to be a *retraction* of B onto A. If there exists a retraction of B onto A, A is called a *retract* of B.

2.1.2 Specification of the general scheme of the proof

Fix values $k := a$ and $u_2 := u_2^0$ with $u_2^0 \in (b_2(a), c_2(a))$ and define the closed interval

$$I := \{(a, u_1, u_2^0) : u_1 \in [b_1(a), c_1(a)]\}. \tag{19}$$

Hereafter we show that there exists a retraction $\mathcal{R}$ (which will be a composition of two auxiliary mappings) of the set $B := I$ onto the set

$$A := \partial I = \{(a, b_1(a), u_2^0), (a, c_1(a), u_2^0)\}.$$

This fact will prove our result.

2.1.3 A property of the last coordinates of consequent points

Consider points

$$\begin{array}{lcl} M_b(k, u_1, b_2(k)) & \in & \overline{\Omega}, \\ M(k, u_1, u_2) & \in & \overline{\Omega}, \\ M_c(k, u_1, c_2(k)) & \in & \overline{\Omega} \end{array}$$

and their consequent points

$$\begin{aligned}\mathcal{M}_b &= (k+1, u_1 + f_1(k, u_1, b_2(k)), b_2(k) + f_2(k, u_1, b_2(k))),\\ \mathcal{M} &= (k+1, u_1 + f_1(k, u_1, u_2), u_2 + f_2(k, u_1, u_2)),\\ \mathcal{M}_c &= (k+1, u_1 + f_1(k, u_1, c_2(k)), c_2(k) + f_2(k, u_1, c_2(k))).\end{aligned}$$

Obviously $b_2(k) \le u_2 \le c_2(k)$. Moreover, due to inequalities (16), (17) and the monotonicity of function (18) we conclude

$$\begin{aligned}&b_2(k+1) < b_2(k) + f_2(k, u_1, b_2(k))\\ &\qquad \le u_2 + f_2(k, u_1, u_2) \le c_2(k) + f_2(k, u_1, c_2(k)) < c_2(k+1)\end{aligned}$$

if function (18) is increasing, and

$$\begin{aligned}&c_2(k+1) > b_2(k) + f_2(k, u_1, b_2(k))\\ &\qquad \ge u_2 + f_2(k, u_1, u_2) \ge c_2(k) + f_2(k, u_1, c_2(k)) > b_2(k+1)\end{aligned}$$

if function (18) is decreasing.
It means that the last coordinate of $\mathcal{M}$ lies between the last coordinates of points $\mathcal{M}_b$ and $\mathcal{M}_c$ and, moreover

$$b_2(k+1) < u_2(k+1) < c_2(k+1). \tag{20}$$

Due to the generality of our consideration we state that the consequent point $\mathcal{C}[M]$ of every point $M(k, u_1, u_2) \in \overline{\Omega}$ has the last coordinate satisfying inequality (20).

2.1.4 Auxiliary mapping $\mathcal{R}_1$

Let a point $\tilde{M} := (a, \tilde{u}_1, u_2^0) \in I$ be fixed. This point defines the corresponding solution $\tilde{u} = (\tilde{u}_1, \tilde{u}_2)$ which accordingly to our supposition does not satisfy inequalities

$$\begin{cases} b_1(k) < \tilde{u}_1(k) < c_1(k),\\ b_2(k) < \tilde{u}_2(k) < c_2(k)\end{cases} \tag{21}$$

for every $k \in N(a)$. Let for $k = a, a+1, \dots, a+s-1$ with a fixed integer $s \ge 1$ inequalities (21) be valid, but violated for $k = a+s$. Since in view of the property of the last coordinates described in 2.1.3

$$b_2(a+s) < \tilde{u}_2(a+s) < c_2(a+s),$$

we conclude that either $\tilde{u}_1(a+s) \le b_1(a+s)$ or $\tilde{u}_1(a+s) \ge c_1(a+s)$. If an equality occurs in one of the last two relations, then for $k = a+s+1$ only strong inequalities are possible. Let, e.g., $\tilde{u}_1(a+s) = b_1(a+s)$. Then (in view of (14))

$$\begin{aligned}\tilde{u}_1(a+s+1) &= u_1(a+s) + f_1(a+s, b_1(a+s), u_2)\\ &< u_1(a+s) + b_1(a+s+1) - b_1(a+s) = b_1(a+s+1).\end{aligned}$$

The second case can be excluded with the aid of (15). Our discussion leads to the following conclusion:
There exists such k^* that

$$(k, u_1(k), u_2(k)) \in \overline{\Omega}(k) \quad \text{for} \quad k = a, a+1, \dots, k^*$$

and

$$(k^*+1, u_1(k^*+1), u_2(k^*+1)) \notin \overline{\Omega}(k^*+1).$$

Define a mapping $\mathcal{R}_1$ as

$$\mathcal{R}_1 : \tilde{M} \to \tilde{M}_1 := (k^*+1, u_1(k^*+1), u_2(k^*+1)).$$

2.1.5 Auxiliary mapping $\mathcal{R}_2$

For the second coordinate of the point $\tilde{M}_1$ either $u_1(k^*+1) > c_1(k^*+1)$ or $u_1(k^*+1) < b_1(k^*+1)$. On this base we define

$$\mathcal{R}_2 : \tilde{M}_1 \to \begin{cases} (a, b_1(a), u_2^0) & \text{if} \quad u_1(k^*+1) < b_1(k^*+1),\\ (a, c_1(a), u_2^0) & \text{if} \quad u_1(k^*+1) > c_1(k^*+1).\end{cases}$$

2.1.6 Resulting mapping $\mathcal{R}$ and its properties

Define $\mathcal{R} := \mathcal{R}_2 \circ \mathcal{R}_1$. Due to construction we have

$$\begin{aligned}\mathcal{R}(a, b_1(a), u_2^0) &= (a, b_1(a), u_2^0),\\ \mathcal{R}(a, c_1(a), u_2^0) &= (a, c_1(a), u_2^0)\end{aligned}$$

and $\mathcal{R}(I) = \partial I$. The mapping $\mathcal{R}$ is also continuous. If not, then in every neighbourhood $\mathcal{O} \subset I$ of a point $\tilde{M} := (a, \tilde{u}_1, u_2^0) \in I$ with e.g. $\mathcal{R}(\tilde{M}) = (a, c_1(a), u_2^0)$ there exists a point $\tilde{M}_{\mathcal{O}}$ such that $\mathcal{R}(\tilde{M}_{\mathcal{O}}) = (a, b_1(a), u_2^0)$. Let $\mathcal{R}_1(\tilde{M}) = (k^*+1, \tilde{u}_1(k^*+1), u_2^0(k^*+1))$. Then consider two possibilities: either $(k^*, \tilde{u}_1(k^*), u_2^0(k^*)) \in \Omega$ or $(k^*, \tilde{u}_1(k^*), u_2^0(k^*)) \in \partial\Omega$. In the first case due to the property of the continuous dependence of solutions on initial data we have for points $\tilde{M}_o$ from a sufficiently small neighbourhood: $\mathcal{R}_1(\tilde{M}_o) = (k^*+1, \tilde{u}_{1o}(k^*+1), \tilde{u}_{2o}(k^*+1))$ with $\tilde{u}_{1o}(k^*+1) > c_1(k^*+1)$, so $\mathcal{R}(\tilde{M}_o) = (a, c_1(a), u_2^0)$. If the second possibility holds, then for points $\tilde{M}_o$ from a sufficiently small neighbourhood either the same property is valid or $\mathcal{R}_1(\tilde{M}_o) = (k^*, \tilde{u}_{1o}(k^*), \tilde{u}_{2o}(k^*))$ with $\tilde{u}_{1o}(k^*) > c_1(k^*)$. So $\mathcal{R}(\tilde{M}_o) = (a, c_1(a), u_2^0)$ again. Our initial supposition is false. This proves the continuity of $\mathcal{R}$.

2.1.7 Final contradiction

Now we are in a position to state that the set ∂I is a retract of the set I and the corresponding retraction is given by the mapping $\mathcal{R}$. This contradiction ends the proof. □

3 Remarks

Remark 1 The main result was formulated under supposition that

$$f_i : N(a) \times \mathbb{R} \times \mathbb{R} \to \mathbb{R}, \quad i = 1, 2. \tag{22}$$

It is easy to see that all steps of the proof have used only the property

$$f_i : \Omega \to \mathbb{R}, \quad i = 1, 2 \tag{23}$$

with Ω defined by (4). Therefore a generalization of Theorem 1 with condition (23) instead of (22) holds.

Remark 2 The affirmation of the main result can be improved in the following sense. In definition of closed interval I by (19) the value $u_2 := u_2^0$ with $u_2^0 \in (b_2(a), c_2(a))$ was fixed. Obviously, if we fix a different value $u_2 := u_2^{00} \neq u_2^0$ with $u_2^{00} \in (b_2(a), c_2(a))$ the statement of Theorem 1 remains valid, but the corresponding solution $u = (u_1^{**}(k), u_2^{**}(k))$ of system (1) will be different from the previous one $u = (u_1^*(k), u_2^*(k))$. The difference will be (at least) in the first points of their graphs. Therefore, the conclusion of main result can be improved in the following sense: *Then there exists at least one-parametric family $\mathcal{F} \subset S(a)$ of initial data such that each solution $u = (u_1^*(k), u_2^*(k))$ of system* (1) *with $(a, u_1^*(a), u_2^*(a)) \in \mathcal{F}$ satisfies inequalities* (13) *for every $k \in N(a)$.*

4 Examples

Let us illustrate main result of the paper on two examples.

Example 1 Let the system (1) be of the form

$$\begin{cases} \Delta u_1(k) &= \quad u_1(k) - \dfrac{1}{9}u_2(k) - k, \\ \Delta u_2(k) &= -\dfrac{1}{9}u_1(k) - \dfrac{1}{3}u_2(k) + k \end{cases} \tag{24}$$

where $k \in N(4)$. Put $b_1(k) = b_2(k) = k$, $c_1(k) = 2k$ and $c_2(k) = 3k$. Then all conditions of Theorem 1 are valid and there exists a solution $u = (u_1^*(k), u_2^*(k))$ of the system (24) satisfying inequalities

$$\begin{cases} k < u_1^*(k) < 2k, \\ k < u_2^*(k) < 3k \end{cases}$$

for every $k \in N(4)$.

Example 2 Let the system (1) be of the form

$$\begin{cases} \Delta u_1(k) &= u_1^2(k) - u_2^2(k) - 1, \\ \Delta u_2(k) &= \dfrac{1}{2k^3} \cdot (u_1(k)u_2(k) + 1) \end{cases} \tag{25}$$

where $k \in N(2)$. Put $b_1(k) = b_2(k) = 0$, $c_1(k) = k^2$ and $c_2(k) = k$. Then all conditions of Theorem 1 are valid and there exists a solution $u = (u_1^*(k), u_2^*(k))$ of the system (25) satisfying inequalities

$$\begin{cases} 0 < u_1^*(k) < k^2, \\ 0 < u_2^*(k) < k \end{cases}$$

for every $k \in N(2)$.

5 One Difference Equation of Emden-Fowler Type

We will investigate the asymptotic behavior of solutions of second-order equation

$$\Delta^2 v(k) + \frac{(\alpha - 1)(\Delta v(k))^2}{v(k)} - \Delta v(k) + 1 = 0, \tag{26}$$

where $\alpha \in \mathbb{R}$, $\alpha < 0$, $k \in N(a)$ and a is sufficiently large. This equation is an equation of Emden-Fowler type. The properties of its solutions are shown in the following section.

5.1 Emden-Fowler Differential Equation and its Difference Analogy

The well-known Emden-Fowler equation is investigated e.g. in the book [2, Chapter VII]. Consider equation

$$u''(t) - t^{-2}u^n(t) = 0 \tag{27}$$

with $n > 1$ which is its special case. Substitute $t = \exp(s)$ in (27). Then

$$u'(t) = u'(s)\exp(-s), \quad u''(t) = u''(s)\exp(-2s) - u'(s)\exp(-2s)$$

and

$$u''(s) - u'(s) - u^n(s) = 0. \tag{28}$$

The substitution

$$u(s) = Cz^{\alpha}(s) \quad \text{with} \quad \alpha = -1/(n-1) \quad \text{and} \quad C^{n-1} = -\alpha$$

in (28) leads (under the supposition $z \neq 0$) to the equation

$$z'' + \frac{(\alpha-1)(z')^2}{z} - z' + 1 = 0. \tag{29}$$

Obviously, our equation (26) is a discrete analogy of equation (29).

5.2 Investigation of Difference Equation (26)

Let us put

$$v(k) = k(1 + u_1^*(k)), \quad \Delta v(k) = 1 + u_2^*(k) \tag{30}$$

in (26). We get a system

$$\begin{aligned} \Delta u_1^*(k) &= \frac{1}{k+1}[-u_1^*(k) + u_2^*(k)], \\ \Delta u_2^*(k) &= u_2^*(k) - \frac{\alpha - 1}{k(1+u_1^*(k))} \cdot [1 + u_2^*(k)]^2. \end{aligned} \tag{31}$$

In the following we put

$$u_1^* := u_2, \quad u_2^* := u_1 \tag{32}$$

in (31) and system of the form (1) with

$$\begin{aligned} f_1(k, u_1(k), u_2(k)) &:= u_1(k) - \frac{\alpha-1}{k(1+u_2(k))} \cdot [1 + u_1(k)]^2, \\ f_2(k, u_1(k), u_2(k)) &:= \frac{1}{k+1}[-u_2(k) + u_1(k)] \end{aligned}$$

i.e. the system

$$\begin{aligned} \Delta u_1(k) &= u_1(k) - \frac{\alpha-1}{k(1+u_2(k))} \cdot [1 + u_1(k)]^2, \\ \Delta u_2(k) &= \frac{1}{k+1}[-u_2(k) + u_1(k)] \end{aligned} \tag{33}$$

will be investigated instead of equation (26).

Theorem 2 *Let numbers* ν_1, ν_2, $0 < \nu_2 < \nu_1 < 1$ *be given. Then the system of difference equations* (33) *has for sufficiently large* $a \in \mathbb{N}$ *a solution* $u(k) = (u_1(k), u_2(k))$ *such that*

$$-\left(\frac{1}{k}\right)^{\nu_i} < u_i(k) < \left(\frac{1}{k}\right)^{\nu_i} \tag{34}$$

for $i = 1, 2$ *and* $k \in N(a)$.

PROOF. Define

$$b_i(k) := -\left(\frac{1}{k}\right)^{\nu_i}, \quad c_i(k) := \left(\frac{1}{k}\right)^{\nu_i}, \quad i = 1, 2$$

and verify the conditions of Theorem 1. With respect to condition (14) we have to verify that

$$f_1(k, b_1(k), u_2) = b_1(k) - \frac{\alpha - 1}{k(1 + u_2(k))} \cdot [1 + b_1(k)]^2 < b_1(k+1) - b_1(k), \tag{35}$$

if $(k, u_1, u_2) \in \Omega_B^1$.
To prove this inequality, first of all notice that its left-hand side has to be negative for sufficiently large k:

$$b_1(k) - \frac{\alpha - 1}{k(1 + u_2(k))} \cdot [1 + b_1(k)]^2 \leq -\left(\frac{1}{k}\right)^{\nu_1} + c \cdot \left(\frac{1}{k}\right) < 0,$$

(c is a positive constant), meanwhile its right-hand side is positive:

$$b_1(k+1) - b_1(k) = -\left(\frac{1}{k+1}\right)^{\nu_1} + \left(\frac{1}{k}\right)^{\nu_1} > 0.$$

That proves inequality (35).
Similarly, inequality (15) from Theorem 1:

$$f_1(k, c_1(k), u_2) = c_1(k) - \frac{\alpha - 1}{k(1 + u_2(k))} \cdot [1 + c_1(k)]^2 > c_1(k+1) - c_1(k)$$

holds for $(k, u_1, u_2) \in \Omega_C^1$ since its left-hand side

$$\left(\frac{1}{k}\right)^{\nu_1} + \frac{1 - \alpha}{k(1 + u_2(k))} \cdot \left[1 + \left(\frac{1}{k}\right)^{\nu_1}\right]^2$$

is positive and its right-hand side

$$c_1(k+1) - c_1(k) = \left(\frac{1}{k+1}\right)^{\nu_1} - \left(\frac{1}{k}\right)^{\nu_1}$$

is negative.
Let us verify (if $(k, u_1, u_2) \in \Omega_B^2$) inequalities (16):

$$b_2(k+1) - b_2(k) < \frac{1}{k+1}\left[-b_2(k) + u_1(k)\right] \tag{36}$$

and

$$\frac{1}{k+1}\left[-b_2(k) + u_1(k)\right] < c_2(k+1) - b_2(k). \tag{37}$$

With the help of the Mean Value Theorem we can estimate the left-hand side of (36) as follows:

$$b_2(k+1) - b_2(k) = -\left(\frac{1}{k+1}\right)^{\nu_2} + \left(\frac{1}{k}\right)^{\nu_2} = \nu_2\left(\frac{1}{\xi}\right)^{\nu_2+1},$$

where $k \le \xi \le k+1$, i.e.

$$b_2(k+1) - b_2(k) \le \nu_2\left(\frac{1}{k}\right)^{\nu_2+1}.$$

Further,

$$\frac{1}{k+1}\left[-b_2(k) + u_1(k)\right] > \frac{1}{k+1}\left[\left(\frac{1}{k}\right)^{\nu_2} - \left(\frac{1}{k}\right)^{\nu_1}\right] > \nu_2\left(\frac{1}{k}\right)^{\nu_2+1}$$

for k sufficiently large (because $0 < \nu_2 < \nu_1 < 1$). Really, it is easy to see that

$$\frac{1}{k+1} > \nu_2 \cdot \frac{1}{k}.$$

That proves inequality (36).
For the right-hand side of (37) the inequality

$$c_2(k+1) - b_2(k) = \left(\frac{1}{k+1}\right)^{\nu_2} + \left(\frac{1}{k}\right)^{\nu_2} > 2\left(\frac{1}{k+1}\right)^{\nu_2}$$

holds. For the left-hand side we have the estimates

$$\frac{1}{k+1}\left[-b_2(k) + u_1(k)\right] < \frac{1}{k+1}\left[\left(\frac{1}{k}\right)^{\nu_2} + \left(\frac{1}{k}\right)^{\nu_1}\right] < \frac{2}{k+1}\left(\frac{1}{k}\right)^{\nu_2}.$$

Finally,

$$\frac{2}{k+1}\left(\frac{1}{k}\right)^{\nu_2} < 2\left(\frac{1}{k+1}\right)^{\nu_2}$$

for k sufficiently large, which ends the proof of inequality (37).
Also inequalities (16):

$$b_2(k+1) - c_2(k) < f_2(k, u_1, c_2(k)) = \frac{1}{k+1}[-c_2(k) + u_1(k)]$$

and

$$\frac{1}{k+1}[-c_2(k) + u_1(k)] < c_2(k+1) - c_2(k)$$

if $(k, u_1, u_2) \in \Omega_C^2$ hold for $k \to \infty$. That can be shown similarly as in the case of (36) and (37).
Moreover, the function (18)

$$F(w) := w + f_2(k, u_1, w) = w + \frac{1}{k+1}[-w + u_1(k)]$$

is monotone for every fixed

$$(k, u_1) \in \Omega_1 := \{(k, u_1) : k \in N(a), b_1(k) \leq u_1 \leq c_1(k)\}$$

on the interval $b_2(k) \leq w \leq c_2(k)$ since its derivative

$$F'(w) = 1 - \frac{1}{k+1} = \frac{k}{k+1}$$

remains (for sufficiently large $a \in \mathbb{N}$) always positive. Then, in accordance with the statement of Theorem 1 there exists a solution of system (33) satisfying inequalities (34).
Let us end this section with a corresponding statement concerning equation (26). We omit its proof since it is a straightforward consequence of Theorem 1 and transformations (30), (32).

Theorem 3 *Let numbers ν_1, ν_2, $0 < \nu_2 < \nu_1 < 1$ be given. Then the difference equation* (26) *has for sufficiently large $a \in \mathbb{N}$ a solution $v(k)$ such that*

$$|v(k) - k| < k \cdot \left(\frac{1}{k}\right)^{\nu_2}$$

and

$$|\Delta v(k) - 1| < \left(\frac{1}{k}\right)^{\nu_1}$$

for $k \in N(a)$.

Acknowledgment

The first author was supported by the Council of Czech Government MSM 0021630503 and by the Grant 201/04/0580 of Czech Grant Agency (Prague). The second author was supported by the Grant No 1/0026/03 of the Grant Agency of Slovak Republic (VEGA).

References

[1] AGARWAL R. P., *Differential Equations and Inequalities, Theory, Methods, and Applications,* Marcel Dekker, Inc., 2nd ed., 2000.

[2] BELLMAN R., Stability theory of differential equations (Internat. Ser. in Pure and Appl. Math.), *New York-London: McGraw-Hill Book Company, XIII,* 1953.

[3] BOICHUK A. & RŮŽIČKOVÁ M., Solutions of nonlinear difference equations bounded on the whole line, *Folia Facultatis Scientarium Naturalium Universitatis Masarykiana Brunensis,* **13** (2002), 45 – 60.

[4] WAŻEWSKI T., Sur un principe topologique de l'examen de l'allure asymptotique des intégrales des équations différentielles ordinaires, *Ann. Soc. Polon. Math.* **20** (1947), 279–313.

[5] DIBLÍK J., Anti-Lyapunov method for systems of discrete equations, *Nonlinear Anal., Theory Methods Appl.,* **57** (2004), 1043–1057.

[6] DIBLÍK J., Asymptotic behaviour of solutions of discrete equations, *Funct. Differ. Equ.,* **11** (2004), 37–48.

[7] DIBLÍK J., Asymptotic behaviour of solutions of systems of discrete equations via Liapunov type technique, *Comput. Math. Appl.* **45** (2003), 1041–1057.

[8] GYÖRI I. & PITUK M., Asympotic formulae for the solutions of a linear delay difference equation, *J. Math. Anal. Appl.* **195** (1995), 376–392.

[9] GYÖRI I. & PITUK M., Comparison theorems and asymptotic equilibrium for delay differential and difference equations, *Dyn. Systems and Appl.* **5** (1996), 277–302.

Discrete Models of Differential Equations: The Roles of Dynamic Consistency and Positivity

RONALD E. MICKENS*
Department of Physics, Clark Atlanta University,
Atlanta, GA 30314, USA

ABSTRACT: The numerical integration of differential equations begins with the construction of appropriate discrete models. A major difficulty is that these models may have solutions not corresponding to any of those of the differential equations. Such solutions are called numerical instabilities (NI) and their elimination is of prime importance. Nonstandard finite difference (NSFD) schemes, as developed by Mickens and others, can be used to design schemes for which the elementary NI's do not occur. We show that the principle of dynamic consistency (DC) can be used to restrict the possible forms of the discrete models based on NSFD modeling rules. The existence of such schemes is illustrated using examples from heat transfer and cancer dynamics.

Keywords: Dynamic consistency; Nonstandard finite difference schemes; Positivity; Mathematical modeling; Numerical analysis.

1991 *Mathematics Subject Classifications:* 39A10; 39A12; 65L12; 65M06.

1. INTRODUCTION

The purpose of numerical analysis, with regard to differential equations, is to provide accurate numerical solutions [1, 2]. Since complete analytical solutions are not known *a priori*, generally the value of the calculated numerical solutions is determined by how well the properties of such solutions mimic the properties of the system modeled by the original differential equations. A major difficulty is the occurrence of numerical instabilities; these are solutions of the discrete equations having no mathematical or physical correspondence to the continuous differential

* Tel: +1-404-880-6923. E-mail: rohrs@math.gatech.edu

equations. An elementary example of how this can take place is seen by considering the decay equation

$$\frac{dx}{dt} = -x, \tag{1.1}$$

and the following two discretizations

$$\frac{x_{k+1} - x_k}{h} = -x_k, \tag{1.2a}$$

$$\frac{x_{k+1} - x_{k-1}}{2h} = -x_k, \tag{1.2b}$$

where $t \to t_k = (\Delta t)k = hk$, $x(t) \to x_k$, and $k =$ integer. The first scheme uses a forward-Euler discrete representation for the derivative, while the second applies a central difference approximation. An easy calculation shows that for $h > 1$, the first discrete model gives numerical solutions that oscillate, with the oscillations becoming unbounded for $h > 2$.

The second discrete model gives eventually unbounded solutions for all $h > 0$. Since the decay equation has all solutions monotonically going to zero, it follows that the scheme of Eq. (1.2a) should only be used when $0 < h < 1$, while the scheme given by Eq. (1.2b) should never be used (in spite of the fact that it is of higher order in h in terms of the truncation error [1]). Thus, both discretizations produce numerical instabilities if certain restrictions on the step-size are violated.

The nonstandard finite difference schemes investigated by Mickens [3, 4] were created to eliminate (at least the elementary) numerical instabilities. For example, the nonstandard methods give for the decay differential equation the discrete model

$$\frac{x_{k+1} - x_k}{\phi(h)} = -x_k, \qquad \phi(h) = 1 - e^{-h}. \tag{1.3}$$

This representation not only gives numerical solutions that are qualitatively correct, but, provide an "exact" finite difference scheme [3].

The main purpose of this review is to demonstrate that the application of the principle of dynamic consistency [5, 6] in the process to generate nonstandard discretizations greatly restricts the possible finite difference models that need to be considered for a given set of differential equations.

The next section defines the concept of dynamic consistency and briefly outlines the major features of the nonstandard finite difference methodology for constructing discrete models of differential equations for the purpose of obtaining numerical solutions. The next three sections illustrate the joint use of dynamic consistency and nonstandard methods for the construction of schemes for the diffusion equation (section 3), the damped wave equation (section 4), and cancer dynamics (section 5). The final section gives a summary and presents several issues whose needed resolutions can form the basis for future research efforts.

In the discussion to come, the following notation will be used:

DC – dynamic consistency

DE – differential equation

NI – numerical instability

NSFD – nonstandard finite difference

PDE – partial differential equations.

2. DYNAMIC CONSISTENCY AND NSFD RULES

Systems modeled by DE's have special properties that must be reflected in the mathematical consequences of the corresponding discrete difference equations. For the natural and engineering sciences particular properties of importance include [7, 8, 9]:

- positivity
- boundedness
- monotonicity

- correct number and (linear) stability of fixed-points
- integer valued populations
- the existence of special solutions.

A given system may include some or none of the above stated requirements. However, the central issue is that the discrete model of the DE's will not be a valid representation of the original equations unless all of the essential properties of the original system are incorporated into it. A "physical" interpretation of NI's is to realize that they are a reflection of the violation of some physical principle that is satisfied by the DE's, but not for the discrete model.

Dynamic consistency (DC) is an important concept to help in the construction of valid discrete models of DE's. It is defined with respect to the particular properties of a system.

Definition: Dynamic Consistency

Consider the DE

$$\frac{dx}{dt} = f(x, t, \lambda), \tag{2.1}$$

and a discrete model of it given by

$$x_{k+1} = F(x_k, t_k, h, \lambda), \tag{2.2}$$

where $f(\cdot, \cdot, \cdot)$ is such that an existence-uniqueness theorem holds; λ represents the m-parameters defining the system, i.e., $\lambda = (\lambda_1, \lambda_2, \ldots, \lambda_m)$; the time step is $h = \Delta t$, with $t_k = hk$, $k =$ integer; and x_k is an approximation to $x(t_k)$. Let the DE and/or its solutions have property P. The discrete model, Eq. (2.2), is dynamically consistent with the DE, Eq. (2.1), if it and/or its solutions also have property P.

Note that while the above definition of DC is given in terms of a scalar ordinary DE, this definition can be generalized to systems of both DE's and PDE's.

An interesting and important situation occurs when the discrete model is such that

$$x(t_k) = x_k, \tag{2.3}$$

for initial condition $x(0) = x_0$. The case implies that Eq. (2.2) is an *exact difference model* for Eq. (2.1) [4].

Nonstandard finite difference (NSFD) schemes had their genesis in the work of Mickens during the time period 1982–1992 [3, 4]. These procedures are based on three fundamental rules:

i) the discrete first-derivative has the following representation

$$\frac{dx}{dt} \to \frac{x_{k+1} - \psi(h)x_k}{\phi(h)}, \qquad h = \Delta t, \tag{2.4}$$

where

$$\psi(h) = 1 + O(h^2), \qquad \phi(h) = h + O(h^2); \tag{2.5}$$

ii) functions of x may require a nonlocal representation on the discrete computational lattice;

iii) the discrete model should be constructed such that it is DC with regard to all the important properties of the original DE.

For many DE's the function $\psi(h) \equiv 1$ [4] and $\phi(h)$, called the denominator function, is not equal to h, i.e., $\phi(h) \neq h$. There exists a procedure for calculating $\phi(h)$ for systems of ordinary DE's [4] and a typical functional form is given by the expression

$$\phi(h, \lambda) = \frac{1 - e^{-\lambda h}}{\lambda}, \tag{2.6}$$

where λ is related to the inverse of the smallest time-scale appearing in the equations. Complete details for determining valid denominator functions are given in Mickens [4].

By nonlocal discrete representations of functions of x, we mean expressions such as the following [4]

$$\text{(1st order DE)} \quad x \to 2x_k - x_{k+1}, \tag{2.7a}$$

$$\text{(1st order DE)} \quad x^2 \to x_{k+1}x_k, \tag{2.7b}$$

$$\text{(2nd order DE)} \quad x^2 \to \left(\frac{x_{k+1} + x_k + x_{k-1}}{3}\right) x_k, \tag{2.7c}$$

$$\text{(2nd order DE)} \quad x^3 \to \left(\frac{x_{k+1} + x_{k-1}}{2}\right) x_k^2. \tag{2.7d}$$

Note that NSFD rule (iii) incorporates the principle of DC. We now give two examples of the elementary application of these rules. The first equation is the linear decay equation

$$\frac{dx}{dt} = -\lambda x, \tag{2.8}$$

where the parameter λ is positive. Using the qualitative theory of DE's [10], it can be easily demonstrated, even without knowledge of the actual solution, that all solutions monotonically go to zero, i.e., $x(t) = 0$ is a globally stable fixed-point. Also a given solution maintains the sign of the initial conditions, i.e., $x_0 \neq 0$, implies $x_0 x(t) > 0$ for $0 < t < \infty$. Thus, the solutions of this DE have the properties:

P_1 : $x(t) = 0$ is a fixed-point;

P_2 : given x_0, $x_0 x(t) > 0$ for $0 \leq t < \infty$;

P_3 : $x(t)$ monotonically decreases in magnitude to zero for any x_0.

Two standard, explicit discrete models were given in section 1; see Eqs. (1.2). A NSFD scheme with properties P_1, P_2, and P_3 is [4]

$$\frac{x_{k+1} - x_k}{\left(\frac{1-e^{-\lambda h}}{\lambda}\right)} = -\lambda x_k, \tag{2.9}$$

where the denominator function $\phi(h)$ has been calculated according to the procedure given in Mickens [4]. Note that Eq. (2.9) can be rewritten to the form

$$x_{k+1} = x_k e^{-\lambda h}. \tag{2.10}$$

This expression can be used to directly show that the scheme has properties P_1, P_2, and P_3. Thus, the finite difference scheme given by Eq. (2.9) or Eq. (2.10) is DC with the linear decay equation with respect to these three properties. Since all the essential features of the solutions to the DE are also features of the NSFD scheme, we expect this discrete model to provide an excellent representation of the actual solutions. (Actually, Eq. (2.9) is an exact scheme.)

The second example is the normalized Fisher PDE, i.e.,

$$u_t = u_{xx} + u(1-u). \tag{2.11}$$

The physically relevant solutions are those that satisfy the condition

$$0 \leq u(x,0) \leq 1 \Rightarrow 0 \leq u(x,t) \leq 1, \qquad t > 0. \tag{2.12}$$

Note that this requirement combines the properties of positivity and boundedness. Further, Eq. (2.11) has two fixed-points or constant solutions; they are $u^{(1)}(x,t) = 0$ and $u^{(2)}(x,t) = 1$, where the first is (linearly) unstable and the second is (linearly) stable. Thus, any NSFD scheme that we construct should manifest the following properties:

P_1 : two fixed-points at $u_m^k(1) = 0$ and $u_m^k(2) = 1$;

P_2 : $u_m^k(1)$ and $u_m^k(2)$ should be, respectively, linearly unstable and stable;

P_3 and P_4 : positivity and boundedness, i.e.,

$$\underset{\substack{k-\text{fixed}\\ \text{all } m}}{0 \leq u_m^k \leq 1} \Longrightarrow \underset{\text{all } m}{0 \leq u_m^{k+1} \leq 1}. \tag{2.13}$$

In the above, $t \to t_k = (\Delta t)k$, $x \to x_m = (\Delta x)m$, and u_m^k approximates $u(x,t)$.

A possible NSFD scheme for the Fisher equation is

$$\frac{u_m^{k+1} - u_m^k}{\Delta t} = \frac{u_{m+1}^k - 2u_m^k + u_{m-1}^k}{(\Delta x)^2} + 2\bar{u}_m^k - u_m^{k+1} - \bar{u}_m^k u_m^{k+1}, \tag{2.14}$$

where the denominator functions are taken to be Δt and Δx for the discrete first derivatives; a forward Euler is used for u_t and a central scheme is employed for u_{xx}; the linear $u(x,t)$ term is replaced by the nonlocal expression

$$u = 2u - u \longrightarrow 2\bar{u}_m^k - u_m^{k+1}, \tag{2.15}$$

where

$$\bar{u}_m^k \equiv \frac{u_{m+1}^k + u_m^k + u_{m-1}^k}{3}, \tag{2.16}$$

and the $u(x,t)^2$ term is represented as

$$-u^2 \rightarrow -\bar{u}_m^k u_m^{k+1}. \tag{2.17}$$

Note that the discrete, nonlocal form for u^2 is needed to insure that its contribution to the numerical scheme is positive. Likewise, as we shall see later, the nonlocal form for the linear term u is required to have the correct boundedness property. Now u_m^{k+1} appears linearly in Eq. (2.14); thus, it can be solved for and the following expression is obtained

$$u_m^{k+1} = \frac{(u_{m+1}^k + u_{m-1}^k)/2 + (2\Delta t)\bar{u}_m^k}{1 + \Delta t + (\Delta t)\bar{u}_m^k}, \tag{2.18}$$

where we have replaced $R = \Delta t/(\Delta x)^2$ by $R = 0.5$. This last procedure is one which enforces the positivity condition; other values of R also can do this.

We can now use this expression to prove the boundedness condition given in Eq. (2.13). Assume u_m^k to have the property $0 \le u_m^k \le 1$, for fixed k and all m. It follows that

$$\left(\frac{1}{2}\right)\left(u_{m+1}^k + u_{m-1}^k\right) \le 1, \tag{2.19}$$

and

$$(2\Delta t)\bar{u}_m^k = (\Delta t)\bar{u}_m^k + (\Delta t)\bar{u}_m^k \le \Delta t + (\Delta t)\bar{u}_m^k. \tag{2.20}$$

Adding the last two equations gives

$$\left(\frac{1}{2}\right)(u_{m+1}^k + u_{m-1}^k) + (2\Delta t)\bar{u}_m^k \leq 1 + \Delta t + (\Delta t)\bar{u}_m^k, \tag{2.21}$$

and

$$\frac{(u_{m+1}^k + u_{m-1}^k)/2 + (2\Delta t)\bar{u}_m^k}{1 + \Delta t + (\Delta t)\bar{u}_m^k} \leq 1. \tag{2.22}$$

Since the left-side of Eq. (2.22) is the right-side of Eq. (2.18), it follows that the positivity and boundedness result in Eq. (2.13) hold.

It is easy to check that conditions P_1 and P_2 also hold. Note that the general restriction on R for positivity is

$$R \equiv \frac{\Delta t}{(\Delta x)^2} \leq \frac{1}{2}\,. \tag{2.23}$$

3. DIFFUSION EQUATION

The linear diffusion PDE provides an excellent model for a large class of heat transfer phenomena in the natural and engineering sciences [7, 9]. One consequence of this fact is that a great effort has been made to construct valid numerical integration methods [1, 2, 7, 9, 11] for the equation. Since the temperature can always be measured with respect to the absolute Kelvin scale, it follows that only non-negative solutions to discrete models of the diffusion equation have physical relevance [9]. Perhaps the simplest discretization of the diffusion equation

$$u_t = Du_{xx}, \tag{3.1}$$

where the parameter D is positive, is the scheme

$$\frac{u_m^{k+1} - u_m^k}{\Delta t} = D\left(\frac{u_{m+1}^k - 2u_m^k + u_{m-1}^k}{(\Delta x)^2}\right), \tag{3.2}$$

which can be written as

$$u_m^{k+1} = (1 - 2DR)u_m^k + (DR)(u_{m+1}^k + u_{m-1}^k), \tag{3.3}$$

with

$$R \equiv \frac{\Delta t}{(\Delta x)^2} \,. \tag{3.4}$$

Note that the positivity condition

$$0 \leq u_m^k \Rightarrow 0 \leq u_m^{k+1} , \tag{3.5}$$

forces the restriction

$$1 - 2DR \geq 0 \Rightarrow \Delta t \leq \frac{(\Delta x)^2}{2D} \,. \tag{3.6}$$

This condition is usually obtained by the requirement that the solutions of Eq. (3.3) be stable, i.e., bounded $\{u_m^k\}$ imply bounded $\{u_m^{k+1}\}$; see Hildebrand [1] or Strikwerda [2] for the details of this argument which relies on the von Neumann stability test, i.e., the scheme is stable if solutions are non-increasing in time. One way of checking this form of stability is to assume that u_m^k takes the form

$$u_m^k = C_k e^{im\theta} \tag{3.7}$$

where θ is a real constant. Substituting this into the scheme under investigation gives a linear, first-order difference equation for C_k. If the characteristic equation has roots whose absolute value are less than or equal to one, then the scheme is stable.

We propose the following numerical integration scheme for the linear diffusion equation

$$\frac{u_m^{k+1} - \bar{u}_m^k}{\Delta t} = D\left[\frac{u_{m+1}^k - 2u_m^k + u_{m-1}^k}{(\Delta x)^2}\right] , \tag{3.8}$$

where

$$\bar{u}_m^k = \frac{u_{m+1}^k + 2u_m^k + u_{m-1}^k}{4} \,. \tag{3.9}$$

Solving for u_m^{k+1} and simplifying the resulting expression gives

$$u_m^{k+1} = \left(\frac{1}{4} + DR\right)(u_{m+1}^k + u_{m-1}^k) + \left(\frac{1}{2} - 2DR\right)u_m^k . \tag{3.10}$$

Examination of Eq. (3.10) indicates that positivity holds if

$$\frac{1}{2} - 2DR \geq 0 \quad \text{or} \quad \Delta t \leq \frac{(\Delta x)^2}{4D}. \tag{3.11}$$

Let

$$\frac{1}{2} - 2DR = \gamma DR, \qquad \gamma \geq 0 \tag{3.12}$$

where γ is a non-negative parameter; this gives for DR

$$DR = \frac{1}{2(2+\gamma)} \leq \frac{1}{4}, \tag{3.13}$$

and Eq. (3.10) becomes

$$u_m^{k+1} = \left[\frac{4+\gamma}{4(2+\gamma)}\right](u_{m+1}^k + u_{m-1}^k) + \left[\frac{\gamma}{2(2+\gamma)}\right] u_m^k. \tag{3.14}$$

For the von Neumann stability test, substitute Eq. (3.7) into Eq. (3.14) and obtain

$$\left|\frac{C_{k+1}}{C_k}\right| = \left|\frac{(4+\gamma)\cos\theta + \gamma}{2(2+\gamma)}\right|. \tag{3.15}$$

The extremes of this ratio occurs at $\theta = 0$ and π. An easy calculation shows that

$$\left|\frac{C_{k+1}}{C_k}\right| \leq 1, \qquad \text{for } 0 \leq \theta < 2\pi. \tag{3.16}$$

Our conclusion is that the scheme is stable based on the von Neumann test.

Finally, it should be observed that the new scheme can be derived from the standard forward Euler representation, Eq. (3.2), by replacing u_m^k in the discrete derivative with $\bar{u}_m^k$, i.e., u_m^k is replaced by a nonlocal weighted average of its values at the discrete space points x_{m+1}, x_m, and x_{m-1}.

4. DAMPED WAVE EQUATION

The Maxwell-Cattaneo thermal conduction law [12] is a modified form of the classical Fourier flux relation [9]. It results in a heat transport equation that mathematically has the same structure as the damped wave equation (DWE), i.e.,

$$u_t + \tau u_{tt} = u_{xx}, \tag{4.1}$$

where in normalized form the parameter τ is related to the thermal relaxation or lag time [12]. Since $u(x,t)$ is the absolute temperature at space location x and time t, its value must be non-negative [9, 11, 12]. Thus, any finite difference scheme for the DWE must also have this property, i.e., for $u(x,0) = f(x)$, we have

$$0 \leq f(x) \leq \bar{f} \Rightarrow 0 \leq u(x,t) \leq \bar{f}, \qquad t > 0, \tag{4.2}$$

where $\bar{f}$ is a positive constant. In Mickens and Jordan [13], we demonstrated how to construct a positivity preserving discrete model for the DWE. However, that scheme was first-order in the discrete time in contrast to the fact that the DWE is of second-order in the time derivative. We now show that a scheme can be constructed for which a second-order discrete time derivative explicitly appears in the final expression, i.e., u_m^{k+1} is given in terms of u_m^k and u_m^{k-1}.

To begin, consider the following discretization of the DWE [14]:

$$\begin{aligned}(1-\alpha)\left[\frac{u_m^{k+1}-u_m^k}{\Delta t}\right] &+ \alpha\left[\frac{u_m^{k+1}-u_m^{k-1}}{2\Delta t}\right] \\ +\tau\left[\frac{u_m^{k+1}-2u_m^k+u_m^{k-1}}{(\Delta t)^2}\right] &= \frac{u_{m+1}^k-(u_m^{k+1}+u_m^{k-1})+u_{m-1}^k}{(\Delta x)^2},\end{aligned} \tag{4.3}$$

where α is an unknown (at the moment) parameter; the term u_{tt} is replaced by a central difference approximation; the first-order time derivative is approximated by a linear combination of forward-Euler and central difference discrete derivatives, i.e.,

$$\begin{aligned}u_t &= (1-\alpha)u_t + \alpha u_t \\ &\approx (1-\alpha)\left[\frac{u_m^{k+1}-u_m^k}{\Delta t}\right] + \alpha\left[\frac{u_m^{k+1}-u_m^{k-1}}{2\Delta t}\right];\end{aligned} \tag{4.4}$$

and the $2u_m^k$ term appearing in the standard definition of the discrete second-order space derivative term, i.e.,

$$u_{xx} \to \frac{u_{m+1}^k - 2u_m^k + u_{m-1}^k}{(\Delta x)^2} \tag{4.5}$$

is replaced by

$$2u_m^k \to u_m^{k+1} + u_m^{k-1}. \tag{4.6}$$

Equation Eq. (4.3) can be rearranged to give

$$\begin{aligned}\left(1 - \frac{\alpha}{2} + R + \frac{1}{\beta}\right) u_m^{k+1} &= \left(\frac{\alpha}{2} - R - \frac{1}{\beta}\right) u_m^{k-1} \\ &\quad + \left(1 - \alpha + \frac{2}{\beta}\right) u_m^k + R(u_{m+1}^k + u_{m-1}^k),\end{aligned} \tag{4.7}$$

where

$$\beta = \frac{\Delta t}{\tau}, \qquad R = \frac{\Delta t}{(\Delta x)^2}. \tag{4.8}$$

The parameter α is determined if the coefficient of u_m^{k-1} is set equal to R; doing this gives

$$\alpha = 2\left(2R + \frac{1}{\beta}\right) > 0. \tag{4.9}$$

Thus, given $(\tau, \Delta t, R)$, the parameter α can be calculated. Substituting this value back into Eq. (4.7) and solving for u_m^{k+1} gives

$$u_m^{k+1} = \frac{Ru_m^{k-1} + (1-4R)u_m^k + R(u_{m+1}^k + u_{m-1}^k)}{1-R}. \tag{4.10}$$

Positivity required that

$$1 - 4R \geq 0 \quad \text{or} \quad \Delta t \leq \frac{(\Delta x)^2}{4}. \tag{4.11}$$

For the special case, $R = 1/4$, we have

$$u_m^{k+1} = \frac{u_m^{k-1} + u_{m+1}^k + u_{m-1}^k}{3}. \tag{4.12}$$

This explicit three level representation clearly indicates that if $\{u_m^k\}$ and $\{u_m^{k-1}\}$ are non-negative, then u_m^{k+1} is non-negative.

The stability of this scheme can be tested by substituting Eq. (3.7) into Eq. (4.12) to obtain the following linear, second-order difference equation

$$3C_{k+1} - (2\cos\theta)C_k - C_{k-1} = 0. \tag{4.13}$$

The associated characteristic equation [4] is

$$3r^2 - (2\cos\theta)r - 1 = 0. \tag{4.14}$$

In terms of the two roots, $r_+ = r_+(\theta)$ and $r_- = r_-(\theta)$, the solution to Eq. (4.13) is

$$C_k = A(r_+)^k + B(r_-)^k, \tag{4.15}$$

where (A, B) are arbitrary constants. It is easy to show that [14]

$$\frac{1}{3} \leq r_+ \leq 1, \qquad -1 \leq r_- \leq -\frac{1}{3}, \tag{4.16}$$

and thus conclude that all solutions to Eq. (4.13) are non-increasing. Thus, the NSFD scheme given by Eq. (4.10) is stable under the von Neumann test.

5. CANCER DYNAMICS

The modeling of phenomena involving cross-diffusion [15], tumor encapsulation [16], and malignant cancer invasion [17] give rise to coupled PDE's for which the diffusion coefficients may depend on all the dependent variables. Such types of mathematical models can also occur in studying insect dispersal and other ecological systems [18]. Our task in this section is to construct a NSFD scheme for the particular case of a model for malignant invasion as derived by Marchant et al. [17]. This model involves three variables u, c, and p, coupled together by interaction terms and nonlinear cross-diffusion. In dimensionless form, the equations are

$$\frac{\partial u}{\partial t} = u(1-u) - \frac{\partial}{\partial x}\left(u\frac{\partial c}{\partial x}\right), \tag{5.1}$$

$$\frac{\partial c}{\partial t} = -pc, \tag{5.2}$$

$$\frac{\partial p}{\partial t} = r(uc - p), \tag{5.3}$$

where r is a (large) parameter. Under certain conditions [17], these equations reduce to two, i.e.,

$$\frac{\partial u}{\partial t} = u(1-u) - \frac{\partial}{\partial x}\left(u\frac{\partial c}{\partial x}\right), \tag{5.4}$$

$$\frac{\partial c}{\partial t} = -uc^2. \tag{5.5}$$

The u and c variables denote quantities that are non-negative. Also, observe that the effective diffusion coefficient in Eq. (5.4) is negative and depends on the variable u, i.e., writing the diffusion term as

$$\frac{\partial}{\partial x}\left(D(u)\frac{\partial c}{\partial x}\right) = -\frac{\partial}{\partial x}\left(u\frac{\partial c}{\partial x}\right), \tag{5.6}$$

we have

$$D(u) = -u. \tag{5.7}$$

Further, it follows from Eq. (5.5) that, for fixed x, $c(x,t)$ is monotonic decreasing. Any NSFD scheme developed must also have all these features built into it.

Starting with the second of these equations, the following NSFD can be constructed for it:

$$\frac{c_m^{k+1} - c_m^k}{\Delta t} = -u_m^k c_m^k c_m^{k+1}. \tag{5.8}$$

Solving for c_m^{k+1} gives

$$c_m^{k+1} = \frac{c_m^k}{1 + (\Delta t)u_m^k c_m^k}. \tag{5.9}$$

It is clearly seen that c_m^k is a monotonic decreasing function of the discrete time for a fixed value of the discrete space variable.

For the discretization of Eq. (5.4), we make the replacements:

$$\frac{\partial u}{\partial t} \to \frac{u_m^{k+1} - \bar{u}_m^k}{\Delta t}, \tag{5.10}$$

where

$$\bar{u}_m^k = \frac{u_{m+1}^k + 2u_m^k + u_{m-1}^k}{4}; \tag{5.11}$$

$$u\,\frac{\partial c}{\partial x} \rightarrow \left(\frac{u_m^k + u_{m-1}^k}{2}\right)\left(\frac{c_m^k - c_{m-1}^k}{\Delta x}\right) \tag{5.12}$$

and

$$\begin{aligned}\frac{\partial}{\partial x}\left(u\,\frac{\partial c}{\partial x}\right) \longrightarrow \left(\frac{1}{\Delta x}\right)&\left[\left(\frac{u_{m+1}^k + u_m^k}{2}\right)\left(\frac{c_{m+1}^k - c_m^k}{\Delta x}\right)\right.\\ &\left.-\left(\frac{u_m^k + u_{m-1}^k}{2}\right)\left(\frac{c_m^k - c_{m-1}^k}{\Delta x}\right)\right],\end{aligned} \tag{5.13}$$

where forward- and backward-differences were used, respectively, in Eqs. (5.12) and (5.13); and

$$u - u^2 \rightarrow u_m^k - \bar{u}_m^k u_m^{k+1}. \tag{5.14}$$

Making these substitutions into (5.4) and solving for u_m^{k+1} gives

$$\begin{aligned}[1 + (\Delta t)\bar{u}_m^k]u_m^{k+1} = {}& 2R\bar{u}_m^k c_m^k + (\Delta t)u_m^k\\ &+ \left(\frac{1}{4}\right)\left[1 - 2Rc_{m+1}^k\right]u_{m+1}^k\\ &+ \left(\frac{1}{4}\right)\left[1 - 2Rc_{m-1}^k\right]u_{m-1}^k\\ &+ \left(\frac{1}{4}\right)\left[1 - 2R(c_{m+1}^k + c_{m-1}^k)\right]u_m^k.\end{aligned} \tag{5.15}$$

Note that this finite difference scheme is explicit since u_m^{k+1} is given in terms of the other variables evaluated at discrete time k.

Time step-size restrictions can be determined by first defining c^* as

$$\underset{\substack{k=0\\ \text{all } m}}{\text{Max}}\{c_m^0\} \equiv c^* > 0. \tag{5.16}$$

Positivity requires that the coefficients of $(u_{m+1}^k, u_m^k, u_{m-1}^k)$, in Eq. (5.15), be non-negative. Since $R = \Delta t/(\Delta x)^2$, it follows that

$$1 - 2Rc^* \geq 0 \;\Rightarrow\; \Delta t \leq \frac{(\Delta x)^2}{2c^*} = (\Delta t)^*. \tag{5.17}$$

In actual numerical simulations, we would use

$$\Delta t = \gamma(\Delta t)^*, \qquad 0 < \gamma < 1. \tag{5.18}$$

In summary, we have constructed NSFD schemes for Eqs. (5.4) and (5.5); they are given, respectively, by the expressions in Eqs. (5.15) and (5.9). These schemes are valid under the restriction on Δt given by the conditions of Eqs. (5.16) and (5.17).

To the best of our knowledge, this is the first time that an intrinsic positivity preserving finite difference scheme has been derived for systems of partial DE's where cross-diffusion terms appear.

6. DISCUSSION

The applications considered in this overview have the common feature that their dependent variables must be non-negative. Many systems of importance to technology, the natural sciences and applied mathematics have this property. Specific examples include heat transfer problems where the temperature can always be measured on the absolute Kelvin scale; chemical reactions where the concentrations are non-negative; and population dynamics for which no meaning can be associated with negative populations. For all these cases, the governing differential equations must have mathematical structures such that positive initial data gives rise to future states that are non-negative. It should be noted that theorems exist for systems of ordinary DE's that allow us to determine whether this condition holds [18]. The work presented here and elsewhere [19, 20] clearly show that in the construction of discrete numerical integration schemes for DE's modeling such systems, the corresponding finite difference equations must also be positivity preserving. If not, NI's are certain to exist. This means that in the creation of the related discrete equations priority must be given to making certain that they have structures for which positivity of their solutions is assured.

Since positivity is, in general, only one of several properties that the discrete equations must incorporate, DC can be applied to insure that these other features are also built into the discrete representations of the DE's. While these procedures will not produce a unique discretization, the possibilities are usually restricted to a small number of cases. Examples of how this "method" works are given in Mickens [4, 6, 13, 19, 20, 21].

In summary, we have introduced the concept of dynamic consistency and illustrated through examples how it can be used to aid in the construction of discrete models for DE's. These so-called nonstandard finite difference schemes have features not found in discrete models constructed by means of the usual procedures. We also derived a new result for the discrete modeling of a system having cross-diffusion. Currently, we are extending this result to several problems in mathematical biology for which this type of nonlinear diffusion naturally appears [15, 16, 17, 23].

Acknowledgments

The work reported in this paper was supported in part by research grants from DOE and the MBRS-SCORE Program at Clark Atlanta University. The author would also like to express his appreciate to Professor Abba B. Gumel, University of Manitoba, Dr. Pedro M. Jordan, Stennis Space Center, Professor Kale Oyedeji, Morehouse College, and Professor Sandra Rucker, Department of Mathematical Sciences, Clark Atlanta University, for their many helpful discussions regarding diffusion equations, numerical analysis, and the philosophy of science as it relates to theory construction.

References

[1] Hildebrand, F. B., (1968) *Finite Difference Equations and Simulations* (Prentice-Hall; Englewood Cliffs, NJ).

[2] Strikwerda, J. C. (1989) *Finite Difference Schemes and Partial Differential Equations* (Wadsworth & Brooks/Cole; Pacific Grove, CA).

[3] Mickens, R. E. (1989) "Exact solutions to a finite difference model of a nonlinear reaction-advection equation: Implication for numerical analysis," *Numerical Methods for Partial Differential Equations* **5**, 313–325.

[4] Mickens, R. E. (1994) *Nonstandard Finite Difference Models of Differential Equations* (World Scientific, Singapore).

[5] Anguelov, R., Lubuma, Jean M.-S., and Mahudu, S. M. (2003) "Qualitatively stable finite difference schemes for advection-reaction equations," *Journals of Computational and Applied Mathematics* **158**, 19–30.

[6] Mickens, R. E. (2005) "Dynamic Consistency: A fundamental principle for constructing nonstandard finite difference for differential equations," *Journal of Difference Equations and Applications* (accepted for publication).

[7] Oran, E. S. and Boris, J. P. (1987) *Numerical Simulation of Reactive Flow* (Elsevier, New York).

[8] Hairer, E., Lubich, C., and Wanner, G. (2002) *Geometric Numerical Integration: Structure-Preserving Algorithms for Ordinary Differential Equations* (Springer, Berlin)

[9] Crank, J. (1975) *The Mathematics of Diffusion*, 2nd Edition (Clarendon Press, Oxford).

[10] Williamson, R. E. (1997) *An Introduction to Differential Equations and Dynamical Systems* (McGraw-Hill, New York).

[11] Smith, G. D. (1978) *Numerical Solution of Partial Differential Equations*, 2nd Edition (Clarendon Press, Oxford).

[12] Joseph, D. D. and Preziosi, L. (1989) "Heat waves," *Reviews of Modern Physics* **61**, 41–73; (1990) **62**, 375–391.

[13] Mickens, R. E. and Jordan, P. M. (2004) "A positivity-preserving nonstandard finite difference scheme for the damped wave equation," *Numerical Methods for Partial Differential Equations* **20**, 639–649.

[14] Mickens, R. E. and Jordan, P. M. (2004) "A new positivity-preserving nonstandard finite difference scheme for the DWE," *Numerical Methods for Partial Differential Equations* (under review).

[15] Ni, W.-M. (1998) "Diffusion, cross-diffusion, and their spike-layer steady states," *Notices of the American Mathematical Society* **45**, 9–25.

[16] Sherratt, J. A. (2000) "Traveling wave solutions of a mathematical model for tumor encapsulation," *SIAM Journal of Applied Mathematics* **60**, 392–407.

[17] Marchant, B. P., Norbury, J. and Perumpani, A. J. (2000) "Traveling shock waves arising in a model of malignant invasion," *SIAM Journal of Applied Mathematics* **60**, 463–476.

[18] Thieme, H. R. (2003) *Mathematics in Population Biology* (Princeton University Press; Princeton, NJ); Appendix A.

[19] Mickens, R. E. (2000) "The role of positivity in the construction of NSFD schemes for PDE's," in Schultz, D. et al. (editors), *Proceedings of International Conference on Scientific Computing and Mathematical Modeling* (University of Wisconsin-Milwaukee, May 25–27).

[20] Mickens, R. E. (1999) "NSFD schemes for reaction-diffusion equations," *Numerical Methods for Partial Differential Equations* **15**, 201–204.

[21] Mickens, R. E., editor (2000) *Application of Nonstandard Finite Difference Schemes* (World Scientific, Singapore).

[22] See Special Issue of *Journal of Difference Equations and Applications* **9** (2003), #11 and #12, for a broad range of the application of NSFD methods to systems in the natural, biomedical, and engineering sciences.

[23] Murray, J. D. (1989) *Mathematical Biology* (Springer-Verlag, Berlin); section 9.4.

Enveloping Implies Global Stability

Paul Cull
Computer Science Dept.,
Oregon State University, Corvallis, OR 97331 USA
pc@cs.orst.edu

Abstract

Some of the simplest models of population growth are one dimensional nonlinear difference equations. While such models can display wild behavior including chaos, the standard biological models have the interesting property that they display global stability if they display local stability. Various researchers have sought a simple explanation for this agreement of local and global stability. Here, we show that **enveloping** by a linear fractional function is sufficient for global stability. We also show that for seven standard biological models local stability implies enveloping and hence global stability. We derive two methods to demonstrate enveloping and show that these methods can easily be applied to the seven example models.

1 Introduction

Simple population growth models have a pleasant property, they display global convergence if they have local convergence. This fact was established for a number of models by Fisher et al [9, 10] who constructed an explicit Lyapunov function for each model they studied. Since then a number of workers have created a variety of sufficient conditions to demonstrate global stability. [21, 19, 4, 2, 1, 3] Each of these methods suffer from the difficulty that either the method does not apply to one of the commonly used models or the method is computationally difficult to apply.

In this paper, we describe a simple condition which is satisfied by all the commonly used simple population models, and we show that for these models the computation for the method is not difficult. Our simple condition is that the population models are *enveloped* by *linear fractional functions.* No single linear fractional serves for all models. Instead the linear fractionals depend on a single parameter which must be adjusted for the particular model. In some cases, this parameter will also change depending on the parameters of the model. This parameter dependence may be why this simple condition has not been discovered before.

Our pleasure with this result is not solely mathematical. There is also a psychological component. We suspect that the original creators of these models were good biologists and not sophisticated mathematicians. If the similarity among these models required deep and complicated mathematics, we would feel that we had not captured the simple vision of the original modelers. We will argue that the usual way of writing these models suggests an implicit constraint that will force enveloping by a linear fractional.

2 Background and Definitions

In the most general sense, we want to study difference equations of the form

$$x_{t+1} = f(x_t)$$

but with this degree of generality, little can be said. If we require that f is a function which is defined for all values of x, then given an initial condition x_0, we can show that there is a unique solution to the difference equation, that is, x_t traces out a well-defined trajectory. To obtain stronger results, we will assume that f is continuous and has as many continuous derivatives as necessary. As we will see in the examples, we will assume even more structure for a population model. Intuitively, if there is no population now, there will be no population later. If the population is small, we expect it to be growing. If the population is large, we expect it to be decreasing. These ideas suggest that there should be an *equilibrium point* where the population size will remain constant. We expect the function f to be *single-humped*, that is, f should rise to a maximum and then decrease. For some models, f will go to 0 for some finite x, but for other models f will continually decrease toward 0.

We want to know what will happen to x_t for large values of t. Clearly we expect that if x_0 is near $\overline{x}$ then x_t will overshoot and undershoot $\overline{x}$. Possibly this oscillation will be sustained, or possibly x_t will settle down at $\overline{x}$. The next definitions codify these ideas. A population model is ***globally stable*** if and only if for all x_0 such that $f(x_0) > 0$ we have

$$\lim_{t\to\infty} x_t = \overline{x}$$

where $\overline{x}$ is the unique equilibrium point of $x_{t+1} = f(x_t)$. A population model is ***locally stable*** if and only if for every small enough neighborhood of $\overline{x}$ if x_0 is in this neighborhood, then x_t is in this neighborhood for all t, and

$$\lim_{t\to\infty} x_t = \overline{x}.$$

How can we decide if a model has one of these properties? The following well-known theorem gives one answer.

Theorem 1. *If $f(x)$ is differentiable then, a population model is locally stable if $|f'(\overline{x})| < 1$, and if the model is locally stable then $|f'(\overline{x})| \leq 1$.*

For global stability, a slight modification of a very general theorem of Sarkovskii [20] gives:

Theorem 2. *A continuous population model is globally stable iff it has no cycle of period 2. (That is, there is no point except* $\overline{x}$ *such that* $f(f(x)) = x$*.)*

This theorem has been noted by Cull[1] and Rosenkranz[19].

Unfortunately, this global stability condition may be difficult to test. Further, there is no obvious connection between the local and global stability conditions.

Various authors have demonstrated global stability for some population models. Fisher *et al* [9] and Goh [10] used Lyapunov functions[13] to show global stability. This technique suffers from the drawbacks that a different Lyapunov function is needed for each model and that there is no systematic method to find these functions. Singer [21] used the negativity of the Schwarzian to show global stability. This technique does not cover all the models we will consider, and it even requires modification to cover all the models it was claimed to cover. Rosenkranz [19] noted that no period 2 was implied by $|f'(x)f'(f(x))| < 1$ and showed that this condition held for a population genetics model. This condition seems to be difficult to test for the models we will consider. Cull [1, 2, 4, 3] developed two conditions **A** and **B** and showed that each of the models we will consider satisfied at least one of these conditions. These conditions used the the first through third derivatives and so were difficult to apply. Also, as Hwang [12] pointed out these conditions required continuous differentiability. All of these methods are relatively mathematically sophisticated, and so it is not clear how biological modelers could intuitively see that these conditions were satisfied.

If we return to the condition for local stability, we see that it says if for x slightly less than 1, $f(x)$ is below a straight line with slope -1, and if for x slightly greater than 1, $f(x)$ is above the same straight line, then the model is locally stable. If we consider the model

$$x_{t+1} = x_t e^{2(1-x_t)},$$

we can see that the local stability bounding line is $2-x$. Somewhat surprisingly, this line is an upper bound on $f(x)$ for all x in $[0,1)$ and a lower bound for all $x > 1$. Since $2-(2-x) = x$, the bounding by this line can be used to argue that for this model there are no points of period 2, and hence the model is globally stable. From this example, we abstract the following definition. A function $\phi(x)$ ***envelops*** a function $f(x)$ if and only if

- $\phi(x) > f(x)$ for $x \in (0,1)$
- $\phi(x) < f(x)$ for $x > 1$ such that $\phi(x) > 0$ and $f(x) > 0$

We will use the notation $\phi(x) \bowtie f(x)$ to symbolize this enveloping.

As we will see, our example population models have one or more parameters, and a model with one choice of parameters will envelop the same model with a different choice of parameters. For example, the function $xe^{2(1-x)}$ envelops all the functions of the form $xe^{r(1-x)}$ for $r \in (0,2)$.

While a straight line was sufficient to envelop $xe^{2(1-x)}$, a straight line fails to envelop the closely related function $x[1+2(1-x)]$. To get a more general enveloping function, we consider the ratio of two linear functions and assume that the ratio is 1 when $x = 1$ and the derivative of this function is -1 when $x = 1$, which gives the following definition.

A ***linear fractional function*** is a function of the form

$$\phi(x) = \frac{1-\alpha x}{\alpha - (2\alpha - 1)x} \qquad \text{where } \alpha \in [0,1) \ .$$

These functions have the properties

- $\phi(1) = 1$
- $\phi'(1) = -1$
- $\phi(\phi(x)) = x$
- $\phi'(x) < 0.$

The shape of our linear fractional functions changes markedly as α varies. For $\alpha = 0$, $\phi(x) = 1/x$, which has a pole at $x = 0$, and decreases with an always positive second derivative. For $\alpha \in (0, 1/2)$, $\phi(x)$ starts (for $x = 0$) at $1/\alpha$ and decreases with a positive second derivative. For $\alpha = 1/2$, $\phi(x) = 2 - x$, which starts at 2 and decreases to 0 with a zero second derivative. For $\alpha \in (1/2, 1)$, $\phi(x)$ starts at $1/\alpha$, decreases with a negative second derivative, and hits 0 at $1/\alpha$ which is greater than 1. We are only interested in these functions when $x > 0$ and $\phi(x) > 0$, so we do not care about the pole in these linear fractionals because the pole occurs outside the area of interest. The following figure shows the three different shapes of linear fractional functions.

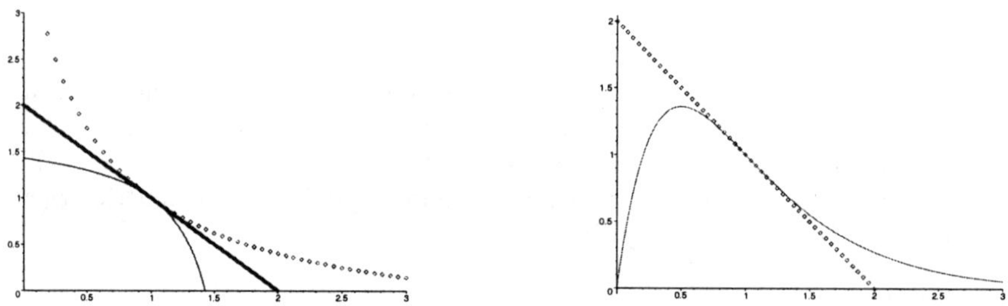

Figure 1: (a)Three types of linear fractionals. Dotted line $\alpha = 1/4$. Heavy line $\alpha = 1/2$. Light line $\alpha = .7$. (b) Model I is enveloped by the straight line $2 - x$ which is the linear fractional with $\alpha = 1/2$.

3 Theorems

We are now in a position to prove the necessary theorems. In what follows, we will assume that our model is $x_{t+1} = f(x_t)$, and that the model has been normalized so that the equilibrium point is 1, that is $f(1) = 1$. We will use the notation $f^{(k)}(x)$ to mean that the function f has been applied k times to x. This notation can be recursively defined by $f^{(0)}(x) = x$ and $f^{(i)}(x) = f(f^{(i-1)}(x))$ for $i \geq 1$.

Theorem 3. *Let $\phi(x)$ be a monotone decreasing function which is positive on $(0, x_-)$ and so that $\phi(\phi(x)) = x$. Assume that $f(x)$ is a continuous function such that:*

- $\phi(x) > f(x)$ *on* $(0, 1)$
- $\phi(x) < f(x)$ *on* $(1, x_-)$
- $f(x) > x$ *on* $(0, 1)$
- $f(x) < x$ *on* $(1, \infty)$
- $f(x) > 0$ *on* $(1, x_\infty)$

then for all $x \in (0, x_\infty)$, $\lim_{k\to\infty} f^{(k)}(x) = 1$.

Proof. From Sarkovskii's theorem, it suffices to show that $f(x)$ has no cycle of period 2. We show that $f(f(x)) > x$ for $x \in (0, 1)$. If $f(f(x)) > 1$ then $f(f(x)) > x$. If $f(f(x)) < 1$ and $f(x) < 1$ then $f(f(x)) > f(x) > x$. If $f(f(x)) < 1$ and $f(x) > 1$, $\phi(f(x)) < f(f(x))$ and $x_- > \phi(x) > f(x)$, and since $\phi(x)$ is decreasing and self inverse $x = \phi(\phi(x)) < \phi(f(x)) < f(f(x))$. A similar argument shows that $x > f(f(x))$ for $x > 1$. (Even if $f(x) > 1$, $f(x) < x_-$ because $x_- > \phi(x)$.) Even though Sarkovskii's theorem assumes a closed interval, we are showing that there are no cycles in an open interval, and so none within the closed intervals inside the open interval. Further our assumptions on $f(x)$ allow us to argue that there is a small ε so that $f^{(k)}(x)$ will eventually fall into the the closed interval $[\varepsilon, \phi(\varepsilon)]$. □

A slight recasting of the above argument gives:

Corollary 1. *If $f_1(x)$ is enveloped by $f_2(x)$, and $f_2(x)$ is globally stable, then $f_1(x)$ is globally stable.*

Corollary 2. *If $f(x)$ is enveloped by a linear fractional function then $f(x)$ is globally stable.*

A function $h(z)$ is ***doubly positive*** iff

1. $h(z)$ has a power series $\sum_{i=0}^{\infty} h_i z^i$
2. $h_0 = 1$, $h_1 = 2$
3. For all $n \geq 1 \quad h_n \geq h_{n+1}$
4. For all $n \geq 2 \quad h_n - 2h_{n+1} + h_{n+2} \geq 0$

Theorem 4. *Let* $x_{t+1} = f(x_t)$ *where* $f(x) = xh(1-x)$ *and* $h(z)$ *is doubly positive, then* $f(x)$ *is enveloped by the linear fractional function*

$$\phi(x) = \frac{1-\alpha x}{\alpha + (1-2\alpha)x}$$

where $\alpha = \frac{3-h_2}{4-h_2} \geq \frac{1}{2}$ *and the model* $x_{t+1} = f(x_t)$ *is globally stable.*

Proof. Recasting in terms of $z = 1-x$ we want to show that $\phi(z) - (1-z)h(z) > 0$ for $z \in (0,1)$ and $\phi(z) - (1-z)h(z) < 0$ for $z \in (-\frac{1-\alpha}{\alpha}, 0)$ where $\phi(z) = \frac{1+\beta z}{1-(1-\beta)z}$ and $\beta = \frac{\alpha}{1-\alpha}$. Assuming that $h(z)$ has a power series, the function we want to bound can be written as:

$$\begin{array}{lllll} 1 & +\beta z & & & \\ -h_0 & -h_1 z & -h_2 z^2 & -h_3 z^3 & -\cdots \\ & +(2-\beta)h_0 z & +(2-\beta)h_1 z^2 & +(2-\beta)h_2 z^3 & +\cdots \\ & & -(1-\beta)h_0 z^2 & -(1-\beta)h_1 z^3 & -\cdots \end{array}$$

By the assumption on h_0 and h_1, the coefficients of z^0 and z^1 vanish. By choosing $\beta = 3 - h_2$ the coefficient of z^2 vanishes. The succeeding coefficients can be written as

$$(\beta - 1)[h_n - h_{n+1}] + [h_{n+1} - h_{n+2}]$$

with $n \geq 1$. By assumption $\beta = 3-h_2 \geq 3-h_1 = 1$. So assuming that $h_n \geq h_{n+1}$ makes all these coefficients nonnegative, and for the power series to converge at least one of these inequalities must be strict, and hence $\phi(z) - (1-z)h(z) > 0$ for $z \in (0,1)$. We have shown that the function has the form $z^3 p(z)$, so to show that it is *negative* on $(-1/\beta, 0)$, which will follow if $p(z)$ is positive on $(-1/\beta, 0)$ and this will follow if $p_n - \frac{1}{\beta}p_{n+1} \geq 0$ where p_n and p_{n+1} are the nth and $n+1^{st}$ coefficients of $p(z)$. From above, this is

$$(\beta - 1)[h_n - h_{n+1}] + \frac{1}{\beta}[h_{n+1} - 2h_{n+2} + h_{n+3}] \geq 0$$

which will be nonnegative by the assumptions, and at least one inequality will be positive if the power series converges. □

While this doubly positive condition will be sufficient for a number of models, it is not sufficient for all the examples because, in particular, β will be less than 1 for some of the models. The following observation will be useful in many cases.

Observation 1. *Let* $\phi(x) = A(x)/B(x)$, $f(x) = C(x)/D(x)$ *and* $G(x) = A(x)D(x) - B(x)C(x)$. *If* $G(1) = 0$, $G'(1) = 0$, *and* $G''(x) > 0$ *on* $(0,1)$ *and* $G''(x) < 0$ *for* $x > 1$, *then* $\phi(x)$ *envelops* $f(x)$. *(We are implicitly assuming that* A, B, C, D *are all positive, and all functions are twice continuously differentiable.)*

Proof. Obviously, if $G'(1) = 0$ and $G''(x) > 0$ on $(0,1)$ then $G'(x) < 0$ on $(0,1)$. Also, if $G''(x) < 0$ for $x > 1$, $G'(x) < 0$ for $x > 1$. But then $G(x)$ is always decreasing, and since $G(1) = 0$, $G(x) > 0$ for $x < 1$ and $G(x) < 0$ for $x > 1$. Rewriting this result shows that $\phi(x)$ envelops $f(x)$. □

If convenient we can switch to the variable $z = 1-x$, and, of course, $G''(x) = G''(z)$. So if $G''(z) = zp(z)$ where $p(z)$ is strictly positive then $\phi(x)$ envelops $f(x)$.

4 Simple Models of Population Growth

4.1 Model I

The model $x_{t+1} = x_t e^{r(1-x_t)}$ is widely used (see, for example [14, 15, 18]). Our first observation is that $0 < r \leq 2$ is the necessary condition for local stability. It is easy to show that this model with $0 < r < 2$ is enveloped by this model with $r = 2$. As we showed earlier, this model with $r = 2$ is enveloped by $\phi(x) = 2-x$ and hence local and global stability coincide.

It is also easy to check that the doubly positive condition holds for this model. Specifically,

$$h(z) = e^{2z} = 1 + z + \frac{2^2 z^2}{2!} + \frac{2^3 z^3}{3!} + \cdots$$

and $h_0 = 1$, and $h_1 = 2$, and

$$h_n - h_{n+1} = \frac{2^n}{(n+1)!}[n+1-2] = \frac{2^n(n-1)}{(n+1)!} \geq 0$$

for $n \geq 1$ and

$$h_n - 2h_{n+1} + h_{n+2} = \frac{2^n}{(n+2)!}[n^2 - n - 2] \geq 0$$

for $n \geq 2$.

4.2 Model II

The model $x_{t+1} = x_t[1+r(1-x_t)]$ is widely used [22] and is sometimes considered to be a truncation of Model I. As for Model I the necessary condition for local stability is $0 < r \leq 2$, and like Model I it is easy to show that this model with $0 < r < 2$ is enveloped by this model with $r = 2$. Unlike Model I, this model is not enveloped by a straight line. But the doubly positive condition holds. Specifically, $h(z) = 1 + 2z$, so

$$h_n - h_{n+1} = \left\{ \begin{array}{ll} 2-0 & n=1 \\ 0-0 & n>1 \end{array} \right\} \geq 0$$

and

$$h_n - 2h_{n+1} + h_{n+2} = \left\{ \begin{array}{ll} 2 & n=1 \\ 0 & n>1 \end{array} \right\} \geq 0.$$

Since $h_2 = 0$, the enveloping function has $\alpha = \frac{3}{4}$ and is

$$\phi(x) = \frac{4-3x}{3-2x}.$$

In this simple example, it's easy to check that the enveloping condition is equivalent to $(1-x)^3$ having a single change of sign which occurs at $x = 1$.

4.3 Model III

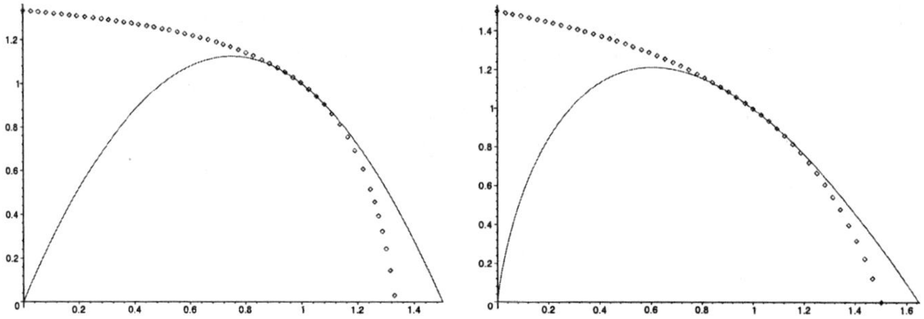

Figure 2: (a) The quadratic map (Model II) is enveloped by $(4-3x)/(3-2x)$. (b) Model III $f(x) = x[1 - 2\ln x]$ enveloped by $(3-2x)/(2-x)$.

The model $x_{t+1} = x_t[1 - r\ln x_t]$ is attributed to Gompertz and studied by Nobile *et al*[16]. As with the preceding two models $0 < r \leq 2$ is the necessary condition for local stability, the model with $r = 2$ envelops the model with $0 < r < 2$, and the doubly positive condition holds. Specifically,

$$h(z) = 1 - 2\ln(1-z) = 1 + 2z + \frac{2z^2}{2} + \cdots + \frac{2z^n}{n} + \cdots$$

and $h_n - h_{n+1} = \frac{2}{n(n+1)} > 0$ for $n \geq 1$ and $h_n - 2h_{n+1} + h_{n+2} = \frac{4}{n(n+1)(n+2)} > 0$ for $n \geq 1$. Since $h_2 = 1$, the enveloping function has $\alpha = 2/3$ and is $\phi(x) = \frac{3-2x}{2-x}$.

4.4 Model IV

Model IV is

$$x_{t+1} = x_t(\frac{1}{b + cx_t} - d)$$

from [24]. This model differs from the previous three in that there are two parameters, b and d, remaining after the carrying capacity has been normalized to 1.

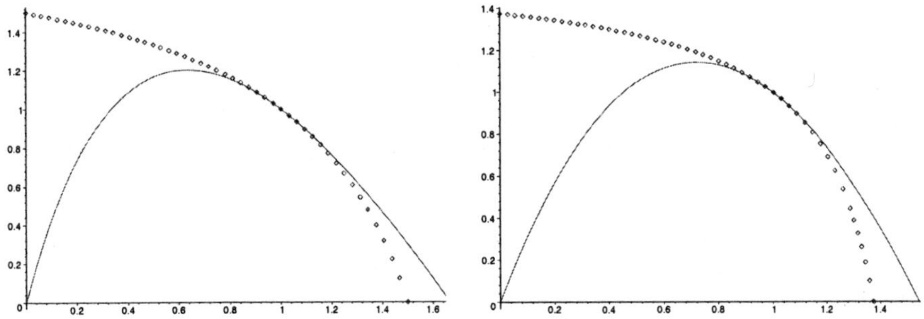

Figure 3: Two examples of model IV. The model with $d = 3$ is enveloped by $(3-2x)/(2-x)$. With $d = 11$ the model is enveloped by $(11 - 8x)/(8 - 5x)$.

The necessary condition for local stability gives

$$\frac{d-1}{(d+1)^2} \leq b < \frac{1}{d+1}.$$

To avoid a pole for $x > 0$, we also assume, $d > 1$. It is easy to check that this model with $b = \frac{d-1}{(d+1)^2}$ envelops this model with larger values of b. With these assumptions

$$f(x) = x[\frac{(d+1)^2}{d-1+2x} - d]$$

and

$$h(z) = \frac{d+1}{1 - \frac{2}{d+1}z} - d.$$

Since $d > 1$,

$$h(z) = 1 + 2z + \frac{2^2}{d+1}z^2 + \frac{2^3}{(d+1)^2}z^3 + \cdots$$

and

$$h_n = \frac{2^n}{(d+1)^{n-1}} \text{ for } n \geq 1.$$

So,

$$h_n - h_{n+1} = \frac{2^n}{(d+1)^n}(d-1) > 0$$

and

$$h_n - 2h_{n+1} + h_{n+2} = \frac{2^n(d-1)^2}{(d+1)^n} > 0.$$

The enveloping function is

$$\phi(x) = \frac{4d - (3d-1)x}{3d - 1 + 2(1-d)x}$$

and has

$$\alpha = \frac{3d-1}{4d} > \frac{1}{2}.$$

We note that $\phi(x)$ has a pole, but $\phi(x)$ goes to zero before the pole, so we can simply ignore the pole. Of course, we only need $\phi(x)$ to bound $f(x)$ on the interval $(0, \frac{4d}{3d-1})$ where $\phi(x)$ is positive.

4.5 Model V

Model V has

$$f(x) = \frac{(1+ae^b)x}{1+ae^{bx}}$$

and comes from Pennycuick *et al*[17]. This and the following two model are more complicated than the previous models because we have to consider different enveloping functions for different parameter ranges.

For $b \leq 2$, $xe^{b(1-x)}$ envelops $f(x)$ because $e^{b(1-x)} + ae^{bx} \bowtie 1 + ae^b$ since $e^{b(1-x)} \bowtie 1$ for $b > 0$. (Here we are using the notation $g(x) \bowtie h(x)$ to mean $g(x) > h(x)$ for $x \in (0,1)$ and $g(x) < h(x)$ for $x > 1$ and still in the range of interest.) But $xe^{b(1-x)}$ is just Model I, and as we showed it is enveloped by $2-x$. So Model V is globally stable for $b \leq 2$. Of course, the inequality still holds for $b > 2$, but since Model I is *not* stable for $b > 2$, the inequality does not help in establishing the stability of Model V.

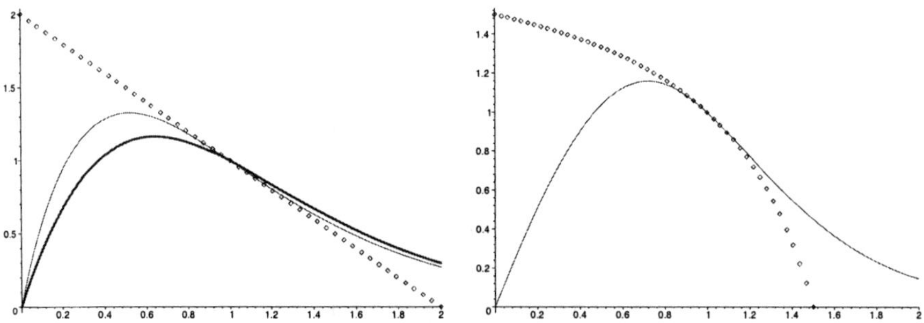

Figure 4: Two examples of model V. The model with $b \leq 2$ is enveloped by $(2-x)$. (The curve becomes steeper as a is increased.) With $b = 3$ the model is enveloped by $(3-2x)/(2-x)$.

For this model we assume that $a > 0$ and $b > 0$. The necessary condition for local stability gives $a(b-2)e^b \leq 2$. It is easy to show that this model with larger values of a envelops this model with smaller values of a. Letting $ae^b = \frac{2}{b-2}$ and using $z = 1 - x$ we have

$$f(z) = \frac{b(1-z)}{(b-2)+2e^{-bz}}.$$

The enveloping linear fractional is

$$\phi(x) = \frac{b-(b-1)x}{(b-1)-(b-2)x}$$

or converting to $z = 1 - x$,

$$\phi(z) = \frac{1+(b-1)z}{1+(b-2)z}.$$

Following the technique of the Observation, we have,

$$G(z) = (b-2)+2e^{-bz}+(b-1)(b-2)z+2(b-1)ze^{-bz}-b(1-z)-b(b-2)z(1-z).$$

It is easy to check that $G(0) = G'(0) = 0$. Finally,

$$G''(z) = z\{2b(b-2)\left(\frac{1-e^{-bz}}{z}\right)+2(b-1)b^2e^{-bz}\}.$$

Clearly, $\frac{1-e^{-bz}}{z}$ is positive for all $z \neq 0$ and since $b > 2$, the first term in {-brackets is positive. Of course, the second term is also positive. So by the Observation, Model V is enveloped as claimed.

4.6 Model VI

Model VI is from Hassel [11] and has

$$f(x) = \frac{(1+a)^b x}{(1+ax)^b} \qquad \text{with } a > 0,\ b > 0.$$

There are two cases to consider $0 < b \leq 2$ and $b > 2$. The enveloping function for $b \leq 2$ is $\phi(x) = 1/x$. Cross multiplication shows that we want $(1+ax)^b \bowtie (1+a)^b x^2$. Taking b^{th} roots and rearranging shows that we want $1-x+ax(1-x^{\frac{2-b}{b}}) \bowtie 0$. Clearly, each of the two terms is positive (nonnegative) below 1 and negative (nonpositive) above 1, and so enveloping is established.

The local stability condition implies that $a(b-2) \leq 2$. It is also is easy to show that this model with smaller values of a is enveloped by this model with larger values of a. So if $b > 2$, we can use $a = \frac{2}{b-2}$ or equivalently $b = \frac{2+2a}{2}$ to simplify formulas. Cross multiplication gives

$$G(x) = 2(b-1)(1+ax)^b - (b-2)(1+ax)^b - (b-2)(1+a)^b x - 2(1+a)^b x^2.$$

Simplification and multiplication by $a/2$ gives

$$\begin{aligned} G(x) &\equiv (2+a)(1+ax)^{b-1} - x(1+ax)^{b-1} - (1+a)^b x \qquad \text{and} \\ G''(x) &\equiv (2+a)^2 a(b-2)(1+ax)^{b-3} - a(b-1)(1+ax)^{b-2} \\ &\quad -(2+a)(1+ax)^{b-2} - (2+a)a(b-2)x(1+ax)^{b-3}. \end{aligned}$$

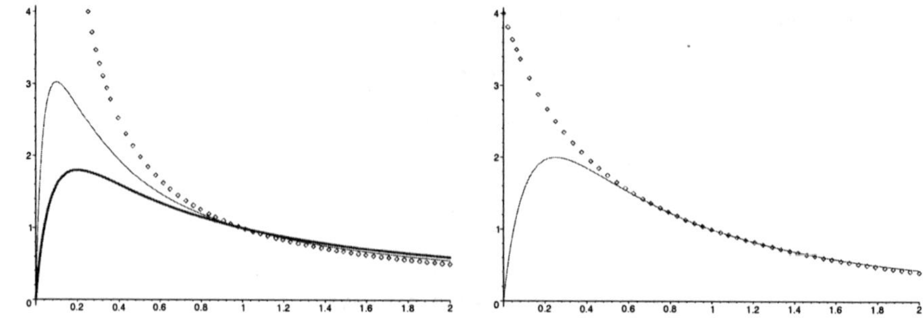

Figure 5: Two examples of model IV. The model with $b \le 2$ is enveloped by $1/x$. (The curve becomes steeper as a is increased.) With $b = 3$ the model is enveloped by $27x/(1+2x)^3$.

Dividing by $(1+ax)^{b-3}$ and simplifying gives

$$G''(x) \equiv (2+a)^2 2 - 2(2+a)(1+ax) - 2(2+a)x \equiv 2(2+a)(1+a)(1-x).$$

So $G''(x) \bowtie 0$ and enveloping is established. (In this argument we use $\equiv$ to indicate that two quantaties have the same sign, but not necessarily the same value.)

4.7 Model VII

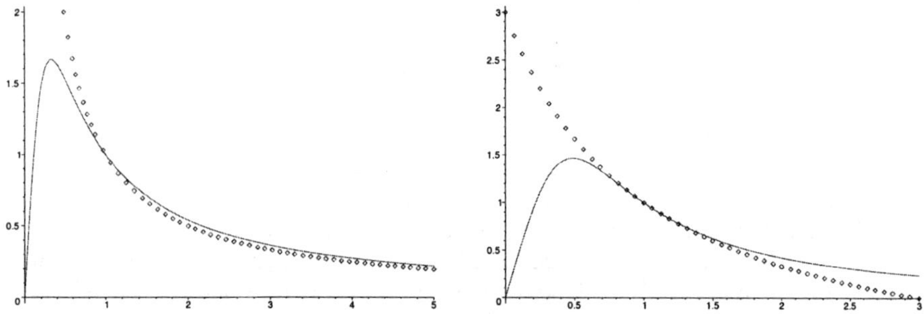

Figure 6: Different linear fractionals envelop Model VII with different parameter values. The fractional $1/x$ is used for $c = 2$. the fractional $(3-x)/(1+x)$ is used with $c = 2.5$, $r = 5$.

Model VII is due to Maynard Smith [23] and has

$$f(x) = \frac{rx}{1+(r-1)x^c}.$$

This seems to be the hardest to analyze model in our set of examples. For example, this model does not satisfy the Schwarzian derivative condition or Cull's condition **A**. Even for our enveloping analysis, we will need to consider this model as three subcases.

Similar to previous models, local stability implies $r(c-2) \leq c$, and it is easy to show that this model with smaller values of r is enveloped by this model with larger values of r.

We first consider the situation when $c \in (0, 2]$. Here, local stability does not place an upper bound on r. Of course, we assume $r > 1$ for this to be a population model. The enveloping function here is $\phi(x) = 1/x$, that is the linear fractional with $\alpha = 0$. Cross multiplication shows that we need

$$1 + (r-1)x^c - rx^2 \bowtie 0$$

for enveloping. Rewriting this gives $1 - x^c + rx^c(1 - x^{2-c}) \bowtie 0$. Clearly, $1 - x^c > 0$ and $1 - x^{2-c} \geq 0$ for $1 > x$ and $2 \geq c$, and $1 - x^c < 0$ and $1 - x^{2-c} \leq 0$ for $1 < x$ and $2 \geq c$, so enveloping is established.

For $c > 2$, we use $r = \frac{c}{c-2}$, and show that

$$\phi(x) = \frac{c - 1 - (c-2)x}{c - 2 - (c-3)x}$$

is the enveloping function. As before, we calculate

$$\begin{aligned}
G(x) &= 2x^c[(c-1) - (c-2)x] + (c-1)(c-2) - 2(c-1)(c-2)x + c(c-3)x^2 \\
G'(x) &= 2x^{c-1}[c(c-1) - (c+1)(c-2)x] - 2(c-1)(c-2) + 2c(c-3)x \\
G''(x) &= 2x^{c-2}[c(c-1)^2 - c(c+1)(c-2)x] + 2c(c-3) \\
&= 2c\{(c-1)^2 x^{c-2}[1-x] + (c-3)[1 - x^{c-1}]\}.
\end{aligned}$$

So for $c \geq 3$, $G''(x) \bowtie 0$ and $\phi(x)$ envelops $f(x)$.

We are left with the case when $c \in (2, 3)$.

$$\frac{G''(x)}{2c} = -(c+1)(c-2)x^{c-1} + (c-1)^2 x^{c-2} - (c-3)$$

and so, $G''(x)$ has two positive real roots. One of these is, of course, the root at $x = 1$. Now, taking another derivative, $G''(x)$ is clearly decreasing at $x = 1$, and hence the other root occurs for some $x < 1$. Since $G''(0) < 0$, G'' will start out negative, become positive, and then become negative for all $x > 1$. But now consider $G'(x)$. $G'(0) < 0$ and so while G'' is negative, G' will become more negative, and when G'' becomes positive, G' will increase from a negative value up to 0 at $x = 1$, and then since $G'' < 0$, G' will decrease and stay negative. Hence G which starts positive will decrease through 0 at $x = 1$ and continue decreasing. So, $G(x) \bowtie 0$ and $\phi(x)$ does envelop $f(x)$.

5 Enveloping is Only Sufficient

Here we want to give a simple model which has global stability, but cannot be enveloped by any linear fractional. Define $f(x)$ by

$$f(x) = \begin{cases} 6x & 0 \le x < 1/2 \\ 7 - 8x & 1/2 \le x < 3/4 \\ 1 & 3/4 \le x \end{cases}$$

then $x_{t+1} = f(x_t)$ has $x = 1$ as its globally stable equilibrium point because if $x_t \ge 1$ then $x_{t+1} = 1$, for $x_t \in [1/2, 1)$, $x_{t+1} > 1$ and $x_{t+2} = 1$, and for $x_t \in (0, 1/2)$, the subsequent iterates grow by multiples of 6 and eventuslly surpass $1/2$. This $f(x)$ cannot be enveloped by a linear fractional because $f(1/2) = 3$ which implies that the linear fractional would have $\alpha \le -1$ and hence have a pole in $(0, 1)$ and thus it could not envelop a positive function. We note that with $\alpha = -1$, $\phi(x)$ would have a pole at $1/3$ and could be used to show that $x = 1$ is globally stable for all $x > 1/3$.

6 Conclusion

Enveloping is a simple technique to demonstrate global stability for some one-dimensional difference equations. Enveloping was introduced by Cull and Chaffee [7, 6, 8]. We demonstrated that the usual population models can be enveloped by linear fractional functions. Such enveloping seems to capture the idea of *simple* function in that a "free-hand" drawing of a population model can usually be enveloped by a linear fractional. (Cull [5] gives a discussion of dynamical systems defined by linear fractionals.) As we showed by example, enveloping by a linear fractional is only a *sufficient* condition for global stability. The simplest population models which have local stability without global stability are discussed by Singer [21] and by Cull [4]. While the examples in this paper are all one-humped population models, **enveloping implies global stability** also holds for functions with multiple peaks, for discontinuous functions, and even for multi-functions.

References

[1] P. Cull. Global Stability of Population Models. *Bulletin of Mathematical Biology*, 43:47–58, 1981.

[2] P. Cull. Local and Global Stability for Population Models. *Biological Cybernetics*, 54: 141–149, 1986.

[3] P. Cull. Local and Global Stability of Discrete One-dimensional Population Models. In L. M. Ricciardi, editor, *Biomathematics and Related Computational Problems*, pages 271–278. Kluwer, Dordrecht, 1988.

[4] P. Cull. Stability of Discrete One-dimensional Population Models. *Bulletin of Mathematical Biology*, 50(1):67–75, 1988.

[5] P. Cull. Linear Fractionals - Simple Models with Chaotic-like Behavior. In D. M. Dubois, editor, *Computing Anticipatory Systems:CASYS 2001 - Fifth International Conference, Conference Proceedings 627*, pages 170–181. American Institue of Physics, Woodbury, N.Y., 2002.

[6] P. Cull and J. Chaffee. Stability in discrete population models. In D. M. Dubois, editor, *Computing Anticipatory Systems: CASYS'99*, pages 263–275. Conference Proceedings 517, American Institute of Physics, Woodbury, NY, 2000.

[7] P. Cull and J. Chaffee. Stability in simple population models. In R. Trappl, editor, *Cybernetics and Systems 2000*, pages 289–294. Austrian Society for Cybernetics Studies, 2000.

[8] Paul Cull. Stability in one-dimensional models. *Scientiae Mathematicae Japonicae*, 58:349–357, 2003.

[9] M.E. Fisher, B.S. Goh, and T.L. Vincent. Some Stability Conditions for Discrete-time Single Species Models. *Bulletin of Mathematical Biology*, 41:861–875, 1979.

[10] B. S. Goh. *Management and Analysis of Biological Populations.* Elsevier, New York, 1979.

[11] M.P. Hassel. Density Dependence in Single Species Populations. *Journal of Animal Ecology*, 44:283–296, 1974.

[12] Y.N. Huang. A Counterexample for P. Cull's Theorem. *Kexue Tongbao*, 31:1002–1003, 1986.

[13] J.P. LaSalle. *The Stability of Dynamical Systems.* SIAM, Philadelphia, 1976.

[14] R.M. May. Biological Populations with Nonoverlapping Generations: Stable Points, Stable Cycles, and Chaos. *Science*, 186:645–647, 1974.

[15] P.A.P. Moran. Some Remarks on Animal Population Dynamics. *Biometrics*, 6:250–258, 1950.

[16] A. Nobile, L.M. Ricciardi, and L. Sacerdote. On Gompertz Growth Model and Related Difference Equations. *Biological Cybernetics*, 42:221–229, 1982.

[17] C.J. Pennycuick, R.M. Compton, and L. Beckingham. A Computer Model for Simulating theGrowth of a Population, or of Two Interacting Populations. *Journal of Theoretical Biology*, 18:316–329, 1968.

[18] W.E. Ricker. Stock and Recruitment. *Journal of the Fisheries Research Board of Canada*, 11:559–623, 1954.

[19] G. Rosenkranz. On Global Stability of Discrete Population Models. *Mathematical Biosciences*, 64:227–231, 1983.

[20] A. Sarkovskii. Coexistence of Cycles of a Continuous Map of a Line to Itself. *Ukr. Mat. Z.*, 16:61–71, 1964.

[21] D. Singer. Stable Orbits and Bifurcation of Maps of the Interval. *SIAM Journal on Applied Mathematics*, 35(2):260–267, Sept 1978.

[22] J.M. Smith. *Mathematical Ideas in Biology.* Cambridge University Press, Cambridge, 1968.

[23] J.M. Smith. *Models in Ecology.* Cambridge University Press, Cambridge, 1974.

[24] S. Utida. Population Fluctuation, an Experimental and Theoretical Approach. *Cold Spring Harbor Symposium on Quantitative Biology*, 22:139–151, 1957.

Global Asymptotic Stability in the Jia Li Model for Genetically Altered mosquitoes

Robert J. Sacker*
Department of Mathematics, University of Southern California, Los Angeles, CA 90089-2532 USA,
e-mail: rsacker@usc.edu, http://rcf.usc.edu/~rsacker

Hubertus F. von Bremen
Department of Mathematics and Statistics, California State Polytechnic University, Pomona, CA 91768 USA
e-mail: hfvonbremen@csupomona.edu

Keywords: Difference equation, global stability, genetically altered mosquitoes
AMS 2000 Subject Classification: 39A11, 92D25

1 Introduction

Malaria remains a major killer with more than 1 million deaths each year in sub-Saharan Africa alone while yellow fever, dengue fever, West Nile virus, encephalitis and filariasis continue to have an impact on populations worldwide. The *Anopheles* strains of mosquitoes are largely responsible for the transmission of *Plasmodium* or malaria, the *Culex tarsalis* accounts largely for West Nile virus, encephalitis and filariasis and the *Aedes aegypti* is associated with yellow fever and dengue.

Much work has been done to genetically modify mosquitoes in the laboratory to hinder or block parasite transmission thus making the mosquitoes refractory. This is done by insertion of genes at appropriate sites to create a stable germline. The progress in this area is fairly recent. In 2000 Catteruccia, et al. [4] report: "Success has has only been achieved in the last five years, with the transformation of the Mediterranean fruitfly *Ceratitus capitata*, the yellow fever mosquito *Aedes aegypti* and the flour beetle *Tribolium castaneum*."

*Supported by University of Southern California, Letters Arts and Sciences Faculty Development Grant.
The authors thank the referee for many constructive remarks.

One of the many problems faced by researchers is a reduction of fitness caused by the mutations resulting from the gene insertions and the inbreeding while transformed lines are established, [4, 3]. More recent significant results were obtained by Moreira, Wang, Collins and Jacobs-Lorena, [15]. They produced transgenic *Anopheles stephensi* expressing either of two effector genes, a tetramer of the SM1 dodecapeptide or the phospholipase A2 gene (PLA2) from honey bee venom. Both were effective in impairing the transmission of *Plasmodium berghei* [14, 10, 9, 11]. However by measuring mosquito survival, fecundity and fertility they found that mosquitos transformed with the SM1 showed *no significant reduction* in these fitness parameters relative to the nontransgenic controls. On the other hand, the PLA2 transgenics had reduced fitness that seemed to be independent of the gene insertion site. This reduced fitness was also observed in the *Culex tarsalis* mosquito, [1] where they also report on a stable germline transformation of the *Culex quinquefasciatus* mosquito using a *Hermes* transposable element containing an enhanced green fluorescent protein marker. Similar results were also obtained by Jasinskiene, et al. [12] using *Hermes* to modify *Aedes aegypti.* Coates, et al. [5] show that the *mariner* transposable element functions as a heritable, stable and efficient mediator of gene insertion into *Aedes aegypti.*

Another factor affecting the mosquito's survival probability when attacking a defensive host is that sporozoite-infected mosquitoes probe more often and spend more time probing for a blood meal than their uninfected counterparts due to impaired salivary gland function, see Anderson, et al. [2] and references therein.

In this paper we consider the discrete-time mathematical model for populations consisting of wild and genetically altered mosquitoes proposed by Jia Li [13]. In that paper a two species model having a hybrid Ricatti/Ricker type nonlinearity and equal survival probabilities is developed and sufficient conditions are given guaranteeing the existence of a *locally* asymptotically stable fixed point. It is shown below that under less restrictive conditions the fixed point is actually *globally* asymptotically stable with respect to initial populations in which both species are present. We then investigate the cases in which the survival probabilities (fixed as well as density dependent) are different for the two species, and show that the model becomes sensitive to small changes in the survival parameters. In particular it is *very* sensitive to changes in the density-dependent survival parameters. For related results see Jim Cushing [6] where he presents linear and nonlinear matrix population models for structured species and for the interaction of several structured species and provides stability results for certain types of nonlinear models in which the nonlinearities appear as a common factor of each equation of the system.

2 Model for population of mosquitoes

The following description closely follows [13]. Let x_n be the number of wild mosquitoes present at generation n. The population dynamics of the wild mosquitoes is described by the difference equation

$$x_{n+1} = f(x_n)s(x_n)x_n, \tag{2.1}$$

where f is the birth function (per-capita rate of offspring production) and s is the survival probability (fraction of the off-spring that survive). The survival probability is assumed to have a Ricker-type nonlinearity $s(x_n) = e^{-d-kx_n}$.

Let y_n be the number of genetically altered mosquitoes present at generation n, and assume that before the wild and altered mosquitoes interact, the dynamics of the altered mosquito population is similar to that of the wild type. Once the altered mosquitoes are released into the wild mosquito habitat, the populations are governed by the system of difference equations

$$\begin{aligned} x_{n+1} &= f_1(x_n, y_n)x_n e^{-d-k(x_n+y_n)}, \\ y_{n+1} &= f_2(x_n, y_n)y_n e^{-d-k(x_n+y_n)}. \end{aligned} \tag{2.2}$$

It is assumed that both wild and altered mosquitoes have the same survival probability $e^{-d-k(x_n+y_n)}$. For $x_n > 0$, $y_n > 0$ the birth rate functions f_1 and f_2 are given by

$$\begin{aligned} f_1(x_n, y_n) &= c(N_n)\frac{\alpha_1 x_n + \beta_1 y_n}{x_n + y_n}, \\ f_2(x_n, y_n) &= c(N_n)\frac{\alpha_2 x_n + \beta_2 y_n}{x_n + y_n}, \end{aligned} \tag{2.3}$$

where $c(N_n)$ is the number of matings per individual, per unit time with $N_n = x_n + y_n$. At generation n the number of matings, per individual, with wild mosquitoes is $c(N_n)x_n/(x_n + y_n)$ and with altered mosquitoes, $c(N_n)y_n/(x_n + y_n)$. Let α_1 be the number of wild offspring that a wild mosquito produces through mating with a wild mosquito, and β_1 be the number of wild mosquitoes produced through mating with an altered mosquito. Similarly, α_2 and β_2 are the number of altered mosquitoes produced by the mating of altered mosquitoes with wild and altered mosquitoes respectively.

Combining (2.2) and (2.3) gives the following set of difference equations that govern the interacting populations of wild and altered mosquitoes

$$\begin{aligned} x_{n+1} &= c(N_n)\frac{\alpha_1 x_n + \beta_1 y_n}{x_n + y_n} x_n e^{-d-k(x_n+y_n)} \\ y_{n+1} &= c(N_n)\frac{\alpha_2 x_n + \beta_2 y_n}{x_n + y_n} y_n e^{-d-k(x_n+y_n)} . \end{aligned} \tag{2.4}$$

The mating rate depends on the population density. When the population is relatively small the mating rate will be assumed to be proportional to the total population, N_n, that is, $c(N_n) = c_0 N_n$. Once the population size exceeds a certain level, we expect the number of matings to saturate, and we assume the mating rate is constant, that is, $c(N_n) = c$.

In this paper we will focus on the constant mating rate case. Thus letting $a_i = c\alpha_i$ and $b_i = c\beta_i$, for $i = 1, 2$, (2.4) becomes

$$\begin{aligned} x_{n+1} &= \frac{a_1 x_n + b_1 y_n}{x_n + y_n} x_n e^{-d-k(x_n+y_n)} \\ y_{n+1} &= \frac{a_2 x_n + b_2 y_n}{x_n + y_n} y_n e^{-d-k(x_n+y_n)} . \end{aligned} \tag{2.5}$$

In (2.5) we assume that $x_n > 0,\ y_n > 0,\ n \geq 0$.

3 Global asymptotic stability

In this section we will study the ratio $z_n = x_n/y_n$ and we will show that under certain conditions the positive fixed point of (2.5) is globally asymptotically stable (GAS). In fact, we will show that this takes place under less stringent conditions than those imposed in [13] to obtain *local* asymptotic stability.

Lemma 3.1 *Suppose $a, b, c, d > 0,\ z \geq 0$ and consider the difference equation*

$$z_{n+1} = f(z_n), \quad \textit{with} \qquad (3.1)$$
$$f(z) = z\frac{az+b}{cz+d}.$$

If $b/d > 1$ and $c/a > 1$, then (3.1) has a unique positive, GAS fixed point.

Proof. From the hypothesis we have that $bc/ad > 1$ and so $bc - ad > 0$. Through direct computation we have that

$$f'(z) = \frac{acz^2 + 2adz + bd}{(cz+d)^2} > 0 \text{ for } z \geq 0. \qquad (3.2)$$

We have that $f'(0) = b/d > 1$ and

$$f'' = \frac{2d(ad - bc)}{(cz+d)^3}.$$

Therefore $f''(z) < 0$ for all $z > 0$, and (3.1) has the unique positive fixed point $z^* = (d-b)/(a-c)$. It is easily seen from the increasing and concave properties of f that the positive fixed point is unique and GAS (for $z > 0$). See [8] for more details. ■

Considering the ratio $z_n = x_n/y_n$ and using (2.5) we get

$$z_{n+1} = \frac{a_1 z_n + b_1}{a_2 z_n + b_2} z_n. \qquad (3.3)$$

Due to the decay survival probability term in (2.5) the populations can not grow indefinitely ($x_n + y_n \to \infty$ is not possible).

The nonzero fixed point of (3.3) is

$$\hat{z} = \frac{b_2 - b_1}{a_1 - a_2}. \qquad (3.4)$$

With $a = a_1$, $b = b_1$, $c = a_2$, and $d = b_2$ (3.3) is just (3.1). If $b_1/b_2 > 1$ and $a_2/a_1 > 1$ then the conditions of Lemma 3.1 apply and the positive fixed point of (3.3) given in (3.4) is GAS (for $z > 0$).

Note that a fixed *point* for (3.3) represents a fixed or invariant line in the (x, y) plane, i.e. the line $S = \{(x,y) : y/x = r\}$ is invariant where $r = \frac{a_2 - a_1}{b_1 - b_2}$. From the GAS of this fixed point we then have that $\frac{y_n}{x_n} \to \frac{1}{\hat{z}} = r$, i.e. the ω-limit set of any point (x, y) with $x > 0,\ y > 0$ lies in the line S. Then to study the solutions in S we can set $y_n = rx_n$ in the first equation of (2.5), and in the second equation we can set $x_n = (1/r)y_n$. Using these substitutions we get the following two uncoupled Ricker's equations (on S).

$$\begin{aligned} x_{n+1} &= \frac{a_1 + b_1 r}{1+r} x_n e^{-d-k(1+r)x_n}, \\ y_{n+1} &= \frac{a_2 + b_2 r}{1+r} y_n e^{-d-k(1+1/r)y_n}. \end{aligned} \tag{3.5}$$

The above two equations are of the general form of the Ricker's equation

$$w_{n+1} = R(w_n), \text{ where } \quad R(w) \doteq \rho w e^{p-\alpha w}. \tag{3.6}$$

The nonzero fixed point of (3.6) is $w^* = (p + \ln(\rho))/\alpha$, and w^* is GAS for $0 < p + \ln(\rho) < 2$, or $e^{-p} < \rho < e^{2-p}$, [7] with

$$\rho = \frac{a_1 + b_1 r}{1+r} = \frac{a_2 + b_2 r}{1+r} = \frac{a_1 b_2 - b_1 a_2}{a_1 - a_2 + b_2 - b_1}. \tag{3.7}$$

The fixed points of the decoupled system (3.5) are

$$\begin{aligned} \hat{x} &= (-d + \ln(\rho))/(k(1+r)), \\ \hat{y} &= r(-d + \ln(\rho))/(k(1+r)), \end{aligned} \tag{3.8}$$

and these fixed points are stable provided $0 < -d + \ln(\rho) < 2$.

This shows the following result.

Theorem 3.2 *The positive fixed point of (2.5) (given in (3.8)) is globally asymptotically stable in the first open quadrant provided that $b_1/b_2 > 1$, $a_2/a_1 > 1$ and $0 < -d + \ln(\rho) < 2$.*

Let

$$N = \frac{(a_1 - a_2)(b_2 - b_1)}{a_1 b_2 - b_1 a_2}, \quad \text{and} \quad P = -d + \ln(\rho). \tag{3.9}$$

Thus if $b_1/b_2 > 1$, $a_2/a_1 > 1$ and $0 < P < 2$, then (2.5) has a positive fixed point and it is GAS. In terms of N and P, Li's result (Theorem 3.2, [13]) requires $b_1/b_2 > 1$, $a_2/a_1 > 1$, $0 < P$, and

$$N(P-1) < P < 2 + (N/2)(P-2) \tag{3.10}$$

for *local* asymptotic stability of the positive fixed point. The right portion $(P < 2 + (N/2)(P-2))$ of the inequality in (3.10) is equivalent to $P < 2$. Note that $N < 0$, and thus if $P < 1$ then $N(P-1)$ is a finite positive quantity so that the left inequality in (3.10) is more restrictive than just $P > 0$ as the following example illustrates.

4 Numerical examples

Here we construct an example in which Theorem 3.2 applies but condition (3.10) is violated.
Example: Consider a system (2.5) with the following set of parameter values: $a_1 = 2$, $a_2 = 3$, $b_1 = 7$, $b_2 = 6.9$, $k = 0.2$, and $d = 1.87$. The parameter values satisfy $b_1/b_2 > 1$ and $a_2/a_1 > 1$. Using (3.4) the fixed point for the ratio system is $\hat{z} = 0.1$. Using (3.6) the fixed point of the decoupled system is $\hat{x} \approx 0.00398674828$ and $\hat{y} \approx 0.03986748281$. The condition $0 < -d + \ln(\rho) < 2$ is satisfied with $P = -d + \ln(\rho) = 0.008770846$, and thus by (Theorem 3.2) the point $(\hat{x}, \hat{y})$ of the coupled system (2.5) is GAS within the open first quadrant.

Using an initial condition of $(x_0, y_0) = (0.03, 0.07)$ direct simulations confirm the expected behavior of the system. Figure 1 shows the population level of the wild and altered mosquitos for the first 4000 generations. As expected, the population of the wild and altered mosquitos approach the fixed level $(\hat{x}, \hat{y})$ given above. The ratio dynamics for $z_n = x_n/y_n$ is shown in Figure 2 and as guaranteed by Theorem 3.2 the iterates approach the computed fixed value of $\hat{z} = 0.1$. The parameter values of this example do not satisfy condition (3.10). In fact $N(P-1) \approx 0.01376707158$.

5 The model is not robust

In the previous sections it was assumed that the survival probabilities of the wild and the altered mosquitoes are identical. In this section we will show that in spite of the global asymptotic stability exhibited by the system, only a slight deviation from identical survival probabilities yields dynamics which

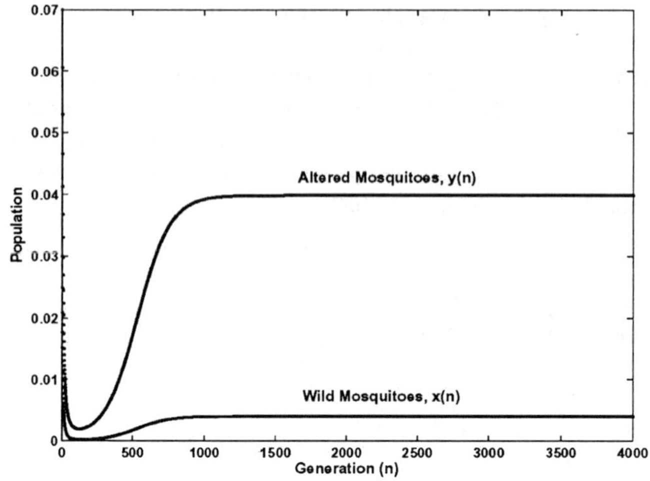

Figure 1: Wild and altered mosquitoes–4000 generations

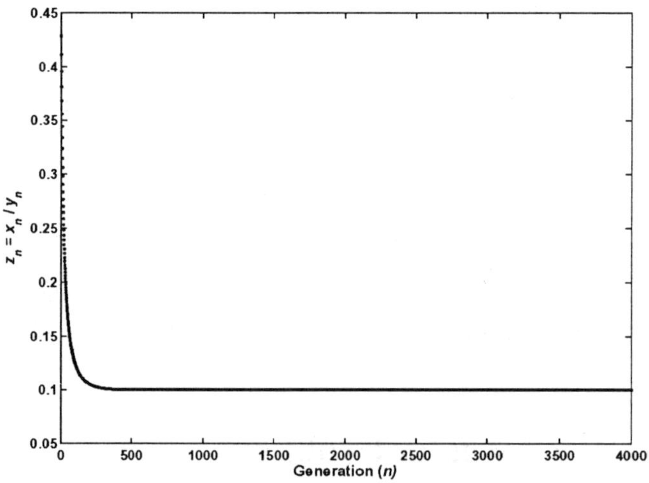

Figure 2: Ratio dynamics $z_n = x_n/y_n$ for 4000 generations

are quite different. In particular the frequency of the altered mosquitoes can decrease radically due to a small decrease in the survival probabilities, especially the density dependent survival probability k_y (below).

In this section we will focus on the case when the survival probability of the wild and altered mosquitoes are not equal. In this case (2.5) becomes

$$\begin{aligned} x_{n+1} &= \frac{a_1 x_n + b_1 y_n}{x_n + y_n} x_n e^{-d_x - k_x(x_n+y_n)} \\ y_{n+1} &= \frac{a_2 x_n + b_2 y_n}{x_n + y_n} y_n e^{-d_y - k_y(x_n+y_n)} . \end{aligned} \tag{5.1}$$

Two cases will be considered. In the first case $k_x = k_y$, and in the second case $k_x \neq k_y$. Consider the case when $k_x = k_y = k$. For this special case we have the advantage that we can easily follow the process given in section (3) to show global asymptotic stability of the positive fixed point of (5.1). Let $\Delta d = d_x - d_y$, then the ratio $z_n = x_n/y_n$ becomes

$$z_{n+1} = \frac{a_1 z_n + b_1}{a_2 z_n + b_2} z_n e^{-\Delta d}. \tag{5.2}$$

The ratio given in (5.2) is very similar to the one presented in (3.3) for the identical survival probability case. The only difference is that (5.2) contains the constant exponential term $e^{-\Delta d}$, and due to this term the stability of (5.2) requires that $b_1 e^{-\Delta d}/b_2 > 1$ and $a_2/(a_1 e^{-\Delta d}) > 1$. Comparing with the case $\Delta d = 0$, $\Delta d > 0$ increases the region of stability in the (a_1, a_2) plane and reduces the region of stability in the (b_1, b_2) plane. In contrast, $\Delta d < 0$ reduces the region of stability in the (a_1, a_2) plane and increases the region of stability in the (b_1, b_2) plane.

The nonzero fixed point of (5.2) is

$$\hat{z} = \frac{b_2 - b_1 e^{-\Delta d}}{a_1 e^{-\Delta d} - a_2}. \tag{5.3}$$

Similar to (3.5), the uncoupled equations of (5.1) become

$$\begin{aligned} x_{n+1} &= \frac{a_1 + b_1 r}{1 + r} x_n e^{-d_x - k(1+r)x_n}, \\ y_{n+1} &= \frac{a_2 + b_2 r}{1 + r} y_n e^{-d_y - k(1+1/r)y_n}, \end{aligned} \tag{5.4}$$

with now $r = \dfrac{a_2 - a_1 e^{-\Delta d}}{b_1 e^{-\Delta d} - b_2}$. The following conditions are required for the stability of the uncoupled Ricker's equations (5.4)

$$0 < -d_x + \ln\left(\frac{a_1 + b_1 r}{1+r}\right) < 2, \quad (5.5)$$
$$0 < -d_y + \ln\left(\frac{a_2 + b_2 r}{1+r}\right) < 2.$$

Note that the two above inequalities are equivalent, i.e.

$$-d_x + \ln\left(\frac{a_1 + b_1 r}{1+r}\right) = -d_y + \ln\left(\frac{a_2 + b_2 r}{1+r}\right).$$

Thus (5.5) imposes only one set of inequalities.

The fixed points of the decoupled system (5.4) are

$$\hat{x} = \frac{-d_x + \ln\left(\frac{a_1+b_1 r}{1+r}\right)}{k(1+r)}, \quad (5.6)$$
$$\hat{y} = r\left(\frac{-d_y + \ln\left(\frac{a_2+b_2 r}{1+r}\right)}{k(1+r)}\right).$$

Similarly to Theorem 3.2, the positive fixed point of (5.1) (given in (5.6)) is GAS in the first open quadrant provided that $b_1 e^{-\Delta d}/b_2 > 1$, $a_2/a_1 e^{-\Delta d} > 1$ and (5.5) holds.

One way to measure which type of mosquito is the dominant mosquito in the population at generation m is to determine the ratio (frequency) $R_m = y_m/(x_m + y_m)$. Here, as before, x_m and y_m are the number of wild and genetically altered mosquitoes present at generation m respectively. If $R_m > 0.5$, then the genetically altered mosquitoes are the dominant type of mosquito in the population.

The following example illustrates the sensitivity of the model to changes in the survival probability, specifically sensitivity to changes in d_x and d_y. Consider a system (5.1) with the following set of parameter values: $a_1 = 2$, $a_2 = 3$, $b_1 = 7$, $b_2 = 6.9$, $k_x = k_y = k = 0.2$, and d_x and d_y vary. In Figure 3 the values of R_m are shown as a function of d_x and d_y. The values of R_m were numerically computed by using (5.1) and $m = 5000$. It is clear that that by simply changing d_y or d_x there can be a switch in the dominant type of mosquito.

In the general case with $k_x \neq k_y$ the equation for the ratio dynamics can not be reduced to an equation in terms of z only, and thus we can not follow the above process. The following example illustrates the sensitivity

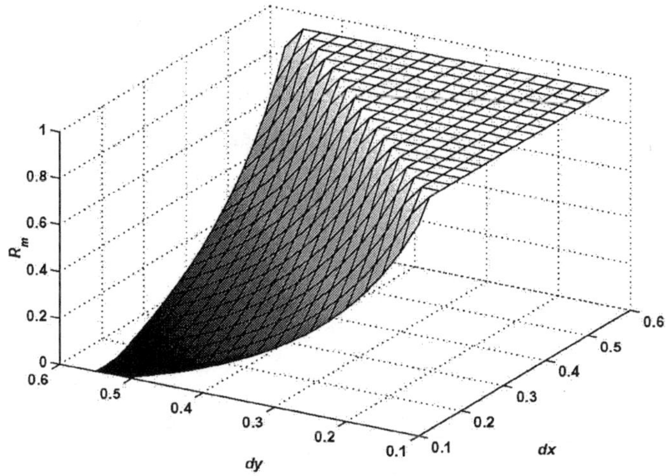

Figure 3: Ratio R_m as a function of d_x and d_y

of the model to changes in k_x and k_y. Consider a system (5.1) with the following set of parameter values: $a_1 = 2$, $a_2 = 3$, $b_1 = 7$, $b_2 = 6.9$, $d_x = d_y = d = 0.25$, and k_x and k_y vary.

In Figure 4 the values of R_m are shown for the cases when either k_x or k_y was kept fixed and either k_y or k_x was allowed to vary for two different values of a_1. The values of R_m were numerically computed by using (5.1) and $m = 5000$. This example show a very small change in k_y or k_x can induce a very large change in R_m and thereby induce a switch in the dominant type of mosquito. This effect is even stronger for the parameter value of $a_1 = 2.5$.

6 Conclusions

In this paper we explored some aspects of a discrete-time mathematical model for populations consisting of wild and genetically altered mosquitoes proposed by Jia Li [13]. We show that under certain conditions the fixed point of the system studied is globally asymptotically stable. The global

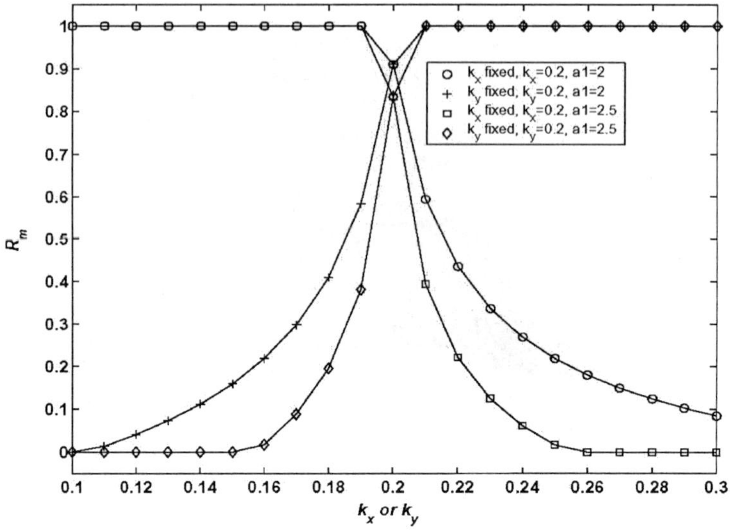

Figure 4: Ratio R_m as a function of k_x and k_y keeping one of k_x or k_y constant

stability is achieved under less stringent conditions than those imposed by [13] to obtain *local* asymptotic stability. A numerical example is presented to illustrate the result.

The model proposed by [13] assumes that the survival probabilities of the wild and the altered mosquitoes are identical. In this paper we show that with only a slight deviation from identical survival probabilities the model can yield dynamics which are quite different. We provide several examples where it is possible to change which population is the dominant one by just slightly changing the survival probability of one of the populations.

References

[1] Margaret L. Allen, David A. O'Brochta, Peter W. Atkinson, and Cynthia S. Levesque. Stable, germ-line transformations of *Culex quinquefasciatus* (Diperta: Culicidae). *J. Med. Entomol.*, 38(5):701–710, 2001.

[2] R. A. Anderson, B. G. J. Knols, and J.C. Koella. *Plasmodium falciparum* sporozoites increase feeding-associated mortality of their mosquito hosts *Anopheles gambia s.1. Parasitology*, 120:329–333, 2000.

[3] Flaminia Catteruccia, H. Charles J. Godfray, and Andrea Crisanti. Impact of genetic manipulation on the fitness of *Anopheles stephensi* mosquitoes. *Science*, 299:1225–1227, February 21, 2003.

[4] Flaminia Catteruccia, Tony Nolan, Thanasis G. Loukeris, Claudia Blass, Charalambos Savakis, Fotis C. Kafatos, and Andrea Crisanti. Stable germline transformation of the malaria mosquito *Anopheles stephensi. Nature*, 405:959–962, June 22, 2000.

[5] Craig J. Coates, Nijole Jasinskiene, Linda Mitashiro, and Anthony A. James. *Mariner* transposition and transformaton of the yellow fever mosquito, *Aedes aegypti. Proc. Natl Acad. Sci. USA*, 95:3748–3751, March 1998.

[6] J.M. Cushing. *An Introduction to Structured Population Dynamics.* SIAM, Philadelphia, PA, USA, 1998.

[7] Saber Elaydi and Robert J. Sacker. Basin of attraction of periodic orbits of maps on the real line. *J Difference Eq and Appl*, 10(10):881–888, 2004.

[8] Saber Elaydi and Robert J. Sacker. Global stability of periodic orbits of nonautonomous difference equations. *J Differential Eq*, 208(11):258–273, Jan. 2004.

[9] A. K. Ghosh, L. A. Moreira, and M. Jacobs-Lorena. *Plasmodium*-mosquito interactions, phage display libraries and transgenic mosquitoes impaired for malaria transmission. *Insect Biochem. and Molec. Bio.*, 32:1325–1331, 2002.

[10] Anil K. Ghosh, Paulo E.M. Ribolla, and Marcelo Jacobs-Lorena. Targeting *Plasmodium* ligands on mosquito salivary glands and midgut with a phage display peptide library. *Proc. Natl. Acad. Sci. USA*, 98(23):13278–13281, November 6, 2001.

[11] Junitsu Ito, Anil Ghosh, Luciano A. Moreira, Ernst A. Wimmer, and Marcelo Jacobs-Lorena. Transgenic anopheline moquitoes impaired in transmission of a malaria parasite. *Nature*, 417:452–455, May 23, 2002.

[12] Nijole Jasinskiene, Craig J. Coates, Mark Q. Benedict, Anthony J. Cornel, Cristina Salazar Rafferty, Anthony A. James, and Frank H.

Collins. Stable transformaton of the yellow fever mosquito, *Aedes aegypti*, with the *Hermes* element from the housefly. *Proc. Natl Acad. Sci. USA*, 95:3743–3747, March 1998.

[13] Jia Li. Simple mathematical models for mosquito populations with genetically altered mosquitos. *Math. Bioscience*, 189:39–59, 2004.

[14] Luciano A. Moreira, Junitsu Ito, Anil Ghosh, Martin Devenport, Helge Zieler, Eappen G. Abraham, Andrea Crisanti, Tony Nolan, Flaminia Catteruccia, and Marcelo Jacobs-Lorena. Bee venom phospholipase inhibits malaria parasite development in transgenic mosquitoes. *J. Bio. Chem.*, 277:40839–40843, October 25, 2002.

[15] Luciano A. Moreira, Jing Wang, Frank H. Collins, and Marcelo Jacobs-Lorena. Fitness of anopheline mosquitoes expressing transgenes that inhibit plasmodium development. *Genetics*, 166:1337–1341, March 2004.

Global Behavior of Solutions of a Nonlinear Second-Order Nonautonomous Difference Equation

V. L. Kocic

Department of Mathematics
Xavier University of Louisiana
New Orleans, LA 70125

Abstract. Our aim in this paper is to investigate the permanence and the extreme stability of the nonlinear second-order nonautonomous difference equation of the form

$$x_{n+1}= f(x_n, x_{n-1}, a_n), \qquad n = 0, 1, \ldots.$$

The results are applied to the difference equation

$$x_{n+1}=\frac{a_n x_n}{1 + bx_n+cx_{n-1}}, \qquad n = 0, 1, \ldots$$

where $\{a_n\}$ is a positive bounded sequence, $b, c > 0$. In the case when $\{a_n\}$ is a periodic sequence, sufficient conditions for the existence of a unique attracting periodic solution are obtained.

AMS Subject Classification 39A12

1 Introduction and Preliminaries

Cushing-Henson's conjecture [3] about the periodically forced Beverton-Holt equation

$$x_{n+1} = \frac{rK_n x_n}{K_n + (r-1)x_n}, \qquad n = 0, 1, ... \tag{1}$$

where $r > 1$, $\{K_n\}$ is a periodic positive sequence and $x_0 > 0$, recently led to a number of interesting results (including some rather general) about the dynamics of periodic difference equations [5, 6, 9, 11, 12]. In addition, some generalizations and variations of the above equation, such as

$$x_{n+1} = \frac{a_n x_n}{1 + x_{n-k}}, \qquad n = 0, 1, ...$$

where $\{a_n\}$ is a bounded positive sequence, are also studied in recently [9, 11]. Some earlier results about such equations are also known in literature [1, 2, 8, 10, 13, 17]. Both equations belong to a class of so-called "nonautonomous rational recursive sequences". Global asymptotic stability of some special second-order nonautonomous rational recursive sequences with period-two coefficients are studied in [7, 14, 15, 19, 20].

The study of stability properties and attractivity in nonautonomous systems is far more complex than in the case of autonomous systems due to the fact that nonautonomous systems, in most cases, do not posses an equilibrium. One way to overcome this difficulty is to study the so-called "extreme stability", the property of the system whose solutions all converge to each other. Extreme stability was originally introduced (for continuous systems) by LaSalle and Lefschetz [16]. Another approach includes the so-called "convergent systems" - systems whose solutions all tend to a particular bounded and asymptotically stable solution called a "limiting" solution (see for example [18] and references cited therein).

In our case, a difference equation (3) is said to be **extremely stable** if for any pair of (positive) solutions $\{x_n\}$ and $\{\bar{x}_n\}$

$$\lim_{n\to\infty} (x_n - \bar{x}_n) = 0 \qquad \text{or} \qquad x_n \sim \bar{x}_n. \tag{2}$$

Clearly, if $\{x_n\}$ and $\{\bar{x}_n\}$ are bounded from below and above by positive constants ($0 < c < x_n, \bar{x}_n < d < \infty$, $n = 0, 1, ...$) the condition (2) is equivalent to

$$\lim_{n\to\infty} \frac{x_n}{\bar{x}_n} = 1.$$

A difference equation (3) is said to be **permanent** if there exist numbers $0 < C \leq D < \infty$ such that for any initial conditions $x_{-1}, x_0 \in (0, \infty)$ there exists a positive integer N which depends on initial conditions such that

$$C \leq x_n \leq D \qquad \text{for} \qquad n \geq N.$$

In this paper, we first study the permanence and extreme stability in nonautonomous difference equation of the form

$$x_{n+1} = f(x_n, x_{n-1}, a_n), \; n = 0, 1, ... \tag{3}$$

where the function f and the sequence $\{a_n\}$ satisfy the following hypotheses:

(H_1) *$f \in C\left[[0,\infty)^3, [0,\infty)\right]$ and $f(u,v,w) > 0$, for $u > 0, v \geq 0, w > 0$;*

(H_2) *$f(u,v,w)$ is decreasing in v, and nondecreasing in w and $\dfrac{f(u,v,w)}{u}$ is nonincreasing in u;*

(H_3) *$\{a_n\}$ is a positive bounded sequence with $0 < A \leq a_n \leq B < \infty$;*

(H_4) *there exists $\bar{x} \in (0,\infty)$ such that $f(\bar{x}, \bar{x}, B) \leq \bar{x}$;*

(H_5) *there exists $\tilde{x} \in (0,\infty)$ such that $f(\tilde{x}, \tilde{x}, A) \geq \tilde{x}$;*

(H_6) *$(f(tu,tv,w) - f(u,v,w))(t-1) > 0$, for $t > 0, t \neq 1$, $u,v,w \in (0,\infty)$;*

(H_7) *for every $0 < p < 1 < q < \infty$, the function G, defined by*

$$G(t) = \begin{cases} \max\limits_{u,v\in[p,q],w\in[A,B]} H(t,u,v,w) & \text{if} \quad 0 \leq t < 1 \\ 1 & \text{if} \quad t = 1 \\ \min\limits_{u,v\in[p,q],w\in[A,B]} H(t,u,v,w) & \text{if} \quad t > 1 \end{cases} \tag{4}$$

has no periodic orbits of prime period 2, where the function H is given by

$$H(t,u,v,w) = \begin{cases} \dfrac{f(ut,vt,w)}{tf(u,v,w)} & \text{if} \quad t > 0, u,v,w > 0 \\ \lim\limits_{t\to 0+} \dfrac{f(ut,vt,w)}{tf(u,v,w)} & \text{if} \quad t = 0, u,v,w > 0 \end{cases} \tag{5}$$

provided the limit exists.

The results are applied to the following difference equation

$$x_{n+1} = \frac{a_n x_n}{1 + bx_n + cx_{n-1}} \tag{6}$$

where $\{a_n\}$ is a positive bounded and persistent sequence and b and c are positive constants. In the case when $\{a_n\}$ is periodic we obtain sufficient conditions for the existence and uniqueness of a globally attracting periodic solution.

A sequence $\{x_n\}$ is said to **oscillate about zero** or simply to **oscillate** if the terms x_n are neither eventually all positive nor eventually all negative. Otherwise the sequence is called **nonoscillatory**. A sequence $\{x_n\}$ is called **strictly oscillatory** if for every $n_0 \geq 0$, there exist $n_1 > n_2 \geq n_0$ such that $x_{n_1}x_{n_2} < 0$. A sequence $\{x_n\}$ is said to **oscillate about** the sequence $\{\bar{x}_n\}$ if the sequence $\{x_n - \bar{x}_n\}$ oscillates. The sequence $\{x_n\}$ is called **strictly oscillatory about** $\{\bar{x}_n\}$ if the sequence $\{x_n - \bar{x}_n\}$ is strictly oscillatory.

Let $\{x_n\}$ be a positive sequence which oscillates about the positive sequence $\{\bar{x}_n\}$. A **positive semicycle** of $\{x_n\}$ **relative to the sequence** $\{\bar{x}_n\}$ consists of a "string" of terms $C^+ = \{x_{l+1}, x_{l+2}, \ldots, x_m\}$ such that $x_i \geq \bar{x}_i$ for $i = l+1, ..., m$ with $l \geq -1$ and $m \leq \infty$ and such that either $l = -1$ or $l \geq 0$, $x_l < \bar{x}_l$ and either $m = \infty$ or $m < \infty$, $x_{m+1} < \bar{x}_{m+1}$. A term $x_p \in C^+$ is said to be a maximum of the positive semicycle C^+ relative to $\{\bar{x}_n\}$ if $\frac{x_p}{\bar{x}_p} = \max\{\frac{x_{l+1}}{\bar{x}_{l+1}}, ..., \frac{x_m}{\bar{x}_m}\}$. A **negative semicycle** of $\{x_n\}$ **relative to the sequence** $\{\bar{x}_n\}$ consists of a "string" of terms $C^- = \{x_{k+1}, x_{k+2}, \ldots, x_l\}$, such that $x_i \geq \bar{x}_i$ for $i = k+1, ..., l$, with $k \geq -1$ and $l \leq \infty$ and such that either $k = -1$ or $k \geq 0$ and $x_k < \bar{x}_k$ and either $l = \infty$ or $l < \infty$, $x_{l+1} \geq \bar{x}_{l+1}$. A term $x_q \in C^-$ is said to be a minimum of the negative semicycle C^- relative to $\{\bar{x}_n\}$ if $\frac{x_p}{\bar{x}_p} = \min\{\frac{x_{k+1}}{\bar{x}_{k+1}}, ..., \frac{x_l}{\bar{x}_l}\}$. A solution may have a finite number of semicycles or infinitely many.

The following theorem and two technical lemmas will be useful in the sequel.

Theorem A. (Brower Fixed Point Theorem [21]): *The continuous operator*

$$A : M \to M$$

has at least one fixed point when M is a compact, convex, nonempty set in a finite dimensional normed space over $\mathbb{K}$ ($\mathbb{K} = \mathbb{R}$ or $\mathbb{K} = \mathbb{C}$).

Lemma B. *Assume hypothesis (H_1), (H_2), (H_6), and (H_7) are satisfied. Then for every $0 < p < 1 < q < \infty$, the function G, defined by (4) is decreasing and continuous on $[0, \infty)$.*

Proof. Let $0 < t < 1$. From (H_2) it follows that the function H is decreasing in t, so $H(t, u, v, w) < H(0, u, v, w)$ and $G(t) < G(0)$. Furthermore,

$$H(t, u, v, w) = \frac{f(ut, vt, w)}{tf(u, v, w)} > \frac{f(u, v, w)}{(1)f(u, v, w)} = 1$$

and $G(t) > G(1) = 1$. Similarly, for $t > 1$ we have $G(t) < G(1) = 1$. Let

$0 < t_1 < t_2$. If $t_1 \le 1 \le t_2$, where at least one inequality is strict, then

$$G(t_1) \ge G(1) \ge G(t_2)$$

where, again, at least one inequality is strict.

Now consider the case when $0 < t_1 < t_2 < 1$. Choose $(u_2, v_2, w_2) \in [p,q]^2 \times [A,B]$ such that

$$G(t_2) = \max_{u,v\in[p,q],w\in[A,B]} H(t_2,u,v,w) = H(t_2,u_2,v_2,w_2).$$

Then

$$\begin{aligned} G(t_2) = H(t_2,u_2,v_2,w_2) &< H(t_1,u_2,v_2,w_2) \\ &\le \max_{u,v\in[p,q],w\in[A,B]} H(t_1,u,v,w) = G(t_1). \end{aligned}$$

Similarly, in the case when $1 < t_1 < t_2$ one can show $G(t_2) < G(t_1)$ and the proof that G is decreasing is complete.

Now we prove that G is continuous. Otherwise there exists $t_0 \ge 0$ where G is discontinuous. We will assume that $0 < t_0 \le 1$. The cases when $t_0 = 0$ and $t_0 > 1$ are similar and they will be omitted. First, if $t_0 = 1$, let $\{t_n\}$ be a sequence of points in $(0,1]$ with limit 1 such that

$$\lim_{n\to\infty} G(t_n) \ne 1 = G(1).$$

Let $\{(u_n, v_n, w_n)\}$ be sequence from $[p,q]^2 \times [A,B]$ such that

$$G(t_n) = H(t_n, u_n, v_n, w_n).$$

Since $\{(u_n, v_n, w_n)\}$ is bounded, there exists a convergent subsequence $\{(u_{n_i}, v_{n_i}, w_{n_i})\}$ with limit (u', v', w') as $i \to \infty$ such that

$$\lim_{i\to\infty} G(t_{n_i}) = \lim_{i\to\infty} H(t_{n_i}, u_{n_i}, v_{n_i}, w_{n_i}) = H(1,u',v',w') = \frac{f(u',v',w')}{(1)f(u',v',w')} = 1$$

which is a contradiction. Therefore, the function G is continuous from the left at $t = 1$.

Now consider the case where the discontinuity of G is at a point $0 < t_0 < 1$. Then there exists $\varepsilon_1 > 0$ such that for every $\delta_1 > 0$ there exists $t_1 > 0$ such that

$$|t_0 - t_1| < \delta_1 \qquad \text{and} \qquad |G(t_0) - G(t_1)| \ge \varepsilon_1. \tag{7}$$

On the other hand the function $H(t,u,v,w)$ is uniformly continuous on $[\varepsilon_3, 1] \times [p,q]^2 \times [A,B]$, for every $0 < \varepsilon_3 < 1$. Then for every $\varepsilon_2 > 0$ there exists $\delta_2 = \delta_2(\varepsilon_2) > 0$ such that

$$|t' - t''| < \delta_2, |u' - u''| < \delta_2, |v' - v''| < \delta_2, |w' - w''| < \delta_2$$
$$\Longrightarrow |H(t', u', v', w') - H(t'', u'', v'', w'')| < \varepsilon_2.$$

Take $\varepsilon_2 = \varepsilon_1/2$, $\delta_1 = \delta_2$, and let $(u_0, v_0, w_0) \in [p, q]^2 \times [A, B]$ be such that $G(t_0) = H(t_0, u_0, v_0, w_0)$. We will assume that $t_1 > t_0$ and $\varepsilon_3 < t_0$. The case when $t_1 < t_0$ is similar and will be omitted. Now choose $(u_1, v_1, w_1) \in [p, q]^2 \times [A, B]$ such that

$$|u_1 - u_0| < \delta_1, \frac{f(\bar{x}, 0, B)}{\bar{x}} \max_{u \in [0, \bar{x}]} f(u, 0, B) \, |v_1 - v_0| < \delta_1, |w_1 - w_0| < \delta_1.$$

Then

$$H(t_0, u_0, v_0, w_0) - \varepsilon_1/2 < H(t_1, u_1, v_1, w_1) < H(t_0, u_0, v_0, w_0) + \varepsilon_1/2.$$

Since

$$G(t_1) \geq H(t_1, u_1, v_1, w_1) > H(t_0, u_0, v_0, w_0) - \varepsilon_1/2 = G(t_0) - \varepsilon_1/2$$

and so

$$0 \leq G(t_0) - G(t_1) \leq \varepsilon_1/2$$

which contradicts (7) and the proof is complete. □

Lemma C. [10]*Let $F \in C[[0, \infty), (0, \infty)]$ be a nonincreasing function and let $\bar{x}$ denotes the (unique) fixed point of F. Then the following statements are equivalent:*

(a) $\bar{x}$ is the only fixed point of F^2 in $(0, \infty)$;

(b) if λ and μ are positive numbers such that

$$F(\mu) \leq \lambda \leq \bar{x} \leq \mu \leq F(\lambda)$$

then

$$\lambda = \bar{x} = \mu.$$

2 Permanence

Lemma 1 *Assume the hypothesis (H_1)-(H_3) are satisfied and there exists an integer N_0 such that*

$$f(x, x, a_n) < x \quad \textit{for} \quad n \geq N_0, \quad \textit{and} \quad x \in (0, \infty). \tag{8}$$

Then every positive solution $\{x_n\}$ of Eq.(3) is eventually decreasing and

$$\lim_{n \to \infty} x_n = 0.$$

Proof. For $n > N_0$ we have

$$x_{n+1} = x_n \frac{f(x_n, x_{n-1}, a_n)}{x_n} \le x_n \frac{f(\min\{x_n, x_{n-1}\}, \min\{x_n, x_{n-1}\}, a_n)}{\min\{x_n, x_{n-1}\}} < x_n$$

and the positive sequence $\{x_n\}$ is decreasing and therefore convergent, that is

$$\lim_{n\to\infty} x_n = \bar{x} \ge 0.$$

Assume, for the sake of contradiction, that $\bar{x} > 0$. Let $\lambda = \limsup_{n\to\infty} a_n$ and let $\{n_i\}$ be a sequence of positive integers such that $\lim_{i\to\infty} a_{n_i} = \limsup_{n\to\infty} a_n = \lambda$. Then

$$\bar{x} = \lim_{i\to\infty} x_{n_i+1} = \lim_{i\to\infty} f(x_{n_i}, x_{n_i-1}, a_{n_i}) = f(\bar{x}, \bar{x}, \lambda)$$

and since $\bar{x} > 0$ we find $f(\bar{x}, \bar{x}, \lambda) = \bar{x}$.

On the other hand, from (8) it follows

$$\limsup_{n\to\infty} f(x, x, a_n) = f(x, x, \lambda) \le x \quad \text{for every} \quad x \in (0, \infty).$$

Since $f(u, v, w)/u$ is nonincreasing in u and $f(u, v, w)$ is decreasing in v, then for $0 < x < \bar{x}$ we have

$$\frac{f(x, x, \lambda)}{x} > \frac{f(\bar{x}, \bar{x}, \lambda)}{\bar{x}} = 1$$

which is impossible. Therefore $\bar{x} = 0$ and the proof is complete. □

Without proof we state the following lemma dual to Lemma 1.

Lemma 2 *Assume the hypothesis (H_1)-(H_3) are satisfied and there exists an integer N_0 such that*

$$f(x, x, a_n) > x \quad \textit{for} \quad n \ge N_0, \quad \textit{and} \quad x \in (0, \infty).$$

Then every positive solution $\{x_n\}$ of Eq.(3) is eventually increasing and

$$\lim_{n\to\infty} x_n = \infty.$$

Lemma 3 *Assume that the hypotheses (H_1)-(H_3) are satisfied.*
(i) If (H_4) holds then $x_n, x_{n+1} > \bar{x}$ implies $x_{n+2} < x_{n+1}$.
(ii) If (H_5) holds then $x_n, x_{n+1} < \tilde{x}$ implies $x_{n+2} > x_{n+1}$.

Proof. We will only prove part (i). The proof of part (ii) is similar and will be omitted. Since $x_n, x_{n+1} > \bar{x}$ we have

$$x_{n+2} = x_{n+1} \frac{f(x_{n+1}, x_n, a_{n+1})}{x_{n+1}} < x_{n+1} \frac{f(\bar{x}, \bar{x}, B)}{\bar{x}} \le x_{n+1}.$$

Theorem 4 *Assume that the hypotheses (H_1)-(H_4) are satisfied. Then every solution of Eq.(3) is bounded from above by a positive constant.*

Proof. Consider the function $f(x, x, B)$ where $a_n \le B < \infty$. The following cases are possible:
Case 1: $f(x, x, B) < x$ for every $x \in (0, \infty)$. From Lemma 1 if follows $\lim_{n\to\infty} x_n = 0$, for every positive solution $\{x_n\}$ of Eq.(3). Therefore all solutions are bounded.
Case 2: There exists $\bar{x} \in (0, \infty)$ such that $f(\bar{x}, \bar{x}, B) = \bar{x}$. Let $\{x_n\}$ be a nontrivial solution of Eq.(3). Then we have the following three cases:
Case 2a: $x_n \le \bar{x}$, for $n \ge N_0$. Clearly $\{x_n\}$ is bounded.
Case 2b: $x_n > \bar{x}$, for $n \ge N_0$. Then, from Lemma 3(i) it follows that $\{x_n\}$ is monotone decreasing, and therefore convergent and bounded.
Case 2c: $\{x_n\}$ strictly oscillates about $\bar{x}$. Let $C_i^- = \{x_{q_i+1}, x_{q_i+2}, \dots, x_{p_i}\}$ be the i-th negative semicycle of $\{x_n\}$ about $\bar{x}$, followed by the positive semicycle $C_i^+ = \{x_{p_i+1}, x_{p_i+2}, \dots, x_{r_i}\}$. Denote by x_{M_i} the maximum in C_i^+. From Lemma 3(i) it follows that the maximum in the positive semicycle (relative to $\bar{x}$) occurs in the first or second term of the semicycle.

First, consider the case when maximum occurs in the first term. Then $x_{M_i-1} < \bar{x}$, and $f(\bar{x}, 0, B) > f(\bar{x}, \bar{x}, B) = \bar{x}$ so we have

$$\begin{aligned} x_{M_i} &= f(x_{M_i-1}, x_{M_i-2}, a_{M_i-1}) < f(x_{M_i-1}, 0, B) \le \max_{u\in[0,\bar{x}]} f(u, 0, B) \\ &= \frac{f(\bar{x}, \bar{x}, B)}{\bar{x}} \max_{u\in[0,\bar{x}]} f(u, 0, B) < \frac{f(\bar{x}, 0, B)}{\bar{x}} \max_{u\in[0,\bar{x}]} f(u, 0, B). \end{aligned}$$

Next, assume that the maximum occurs in the second term. Then $x_{M_i-2} < \bar{x}$, $x_{M_i} > x_{M_i-1} \ge \bar{x}$ and we have

$$\begin{aligned} x_{M_i} &= x_{M_i-1} \frac{f(x_{M_i-1}, x_{M_i-2}, a_{M_i-1})}{x_{M_i-1}} \\ &= f(x_{M_i-2}, x_{M_i-3}, a_{M_i-2}) \frac{f(x_{M_i-1}, x_{M_i-2}, a_{M_i-1})}{x_{M_i-1}} \\ &< \frac{f(\bar{x}, 0, B)}{\bar{x}} f(x_{M_i-2}, 0, B) \le \frac{f(\bar{x}, 0, B)}{\bar{x}} \max_{u\in[0,\bar{x}]} f(u, 0, B). \end{aligned}$$

Therefore,

$$\limsup_{n\to\infty} x_n = \limsup_{i\to\infty} x_{M_i} \le \frac{f(\bar{x}, 0, B)}{\bar{x}} \max_{u\in[0,\bar{x}]} f(u, 0, B), \tag{9}$$

so the solution $\{x_n\}$ is bounded and the proof is complete. □

Theorem 5 *Assume that the hypotheses (H_1)-(H_5) are satisfied. Then Eq.(3) is permanent.*

Proof. Let $\{x_n\}$ be a solution of Eq. (3). In the first part of the proof we will show that for sufficiently large n

$$x_n < \frac{f(\bar{x},0,B)}{\bar{x}} \max_{u\in[0,\bar{x}]} f(u,0,B) + \varepsilon = D \tag{10}$$

where $\varepsilon > 0$. The proof follows directly from the proof of Theorem 4. First, case 1 in the proof of Theorem 4 is not possible; we have $f(x,x,B) < x$, for every $x \in (0,\infty)$, so $f(x,x,A) \leq f(x,x,B) < x$ and that contradicts the hypothesis (H_5).

We will focus only on the case 2 in the proof of Theorem 4. Then there exists $\bar{x} \in (0,\infty)$ such that $f(\bar{x},\bar{x},B) = \bar{x}$. Clearly, in the case 2c, (10) follows from (9). It remains to consider the cases 2a and 2b. In the case 2a

$$\begin{aligned} 1 &= \frac{f(\bar{x},\bar{x},B)}{\bar{x}} < \frac{f(\bar{x},0,B)}{\bar{x}} \\ \bar{x} &= f(\bar{x},\bar{x},B) < f(\bar{x},0,B) \leq \max_{u\in[0,\bar{x}]} f(u,0,B) \end{aligned}$$

and we obtain

$$\bar{x} < \frac{f(\bar{x},0,B)}{\bar{x}} \max_{u\in[0,\bar{x}]} f(u,0,B) < D. \tag{11}$$

Hence inequality (10) holds in the case 2a.

The remaining case is 2b. In this case we have $\lim_{n\to\infty} x_n = x \geq \bar{x}$. Since $\{a_n\}$ is bounded, there exists a sequence of integers $\{n_i\}$ such that

$$\lim_{i\to\infty} a_{n_i} = \limsup_{n\to\infty} a_n = b \leq B.$$

Assume, $x > \bar{x}$. Then $x = f(x,x,b)$ and

$$1 = \frac{f(x,x,b)}{x} \leq \frac{f(x,x,B)}{x} < \frac{f(\bar{x},\bar{x},B)}{\bar{x}} = 1$$

and that is a contradiction. Therefore, $\lim_{n\to\infty} x_n = \bar{x} < D$, so inequality (10) holds in this case.

By following the same procedure as in the case 2 of Theorem 4 and above we obtain, for sufficiently large n,

$$x_n > \frac{f(\tilde{x},D,A)}{\tilde{x}} \min_{u\in[\tilde{x},D]} f(u,D,A) - \eta = C > 0 \tag{12}$$

where $\tilde{x} = f(\tilde{x},\tilde{x},A)$ and η is a sufficiently small positive number such that $C > 0$. The proof is complete. □

3 Extreme Stability

Let $\{\bar{x}_n\}$ be a positive bounded solution of Eq.(3).

Lemma 6 *Assume that the hypotheses (H_1)-(H_3) and (H_6) are satisfied and let $\{\bar{x}_n\}$ be a positive solution of Eq.(3). Then the following statements are true:*
(i) If, for some $k \geq 0, x_k = \bar{x}_k$ and $x_{k+1} = \bar{x}_{k+1}$ then $x_n = \bar{x}_n$ for every $n = 0, 1, \ldots$.
(ii) Every semicycle, except perhaps the first one, has at least two terms.
(iii) The extremum in every semicycle, except perhaps in the first one, occurs in the first or second term of the semicycle.
(iv) The extremum in every semicycle, except perhaps in the first one, can not be equal to the last term of the semicycle.

Proof. (i) It follows directly from the monotonic character of the function f.
(ii) Let $x_{n-1} < \bar{x}_{n-1}$ be the last term in a negative semicycle relative to $\{\bar{x}_n\}$ followed by the $x_n \geq \bar{x}_n$ - the first term in the following positive semicycle. Clearly,

$$\frac{x_{n-1}}{\bar{x}_{n-1}} < 1 \leq \frac{x_n}{\bar{x}_n}$$

and we have

$$\begin{aligned} x_{n+1} &= f(x_n, x_{n-1}, a_n) = f(\frac{x_n}{\bar{x}_n}\bar{x}_n, \frac{x_{n-1}}{\bar{x}_{n-1}}\bar{x}_{n-1}, a_n) \\ &> f(\frac{x_n}{\bar{x}_n}\bar{x}_n, \frac{x_n}{\bar{x}_n}\bar{x}_{n-1}, a_n) \geq f(\bar{x}_n, \bar{x}_{n-1}, a_n) = \bar{x}_{n+1}. \end{aligned}$$

(iii) Let $x_{n-1} \geq \bar{x}_{n-1}$, and $x_n \geq \bar{x}_n$, where at least one of the inequalities is strict, be the first two terms in a positive semicycle relative to $\{\bar{x}_n\}$. Then

$$\begin{aligned} x_{n+1} = f(x_n, x_{n-1}, a_n) &= x_n \frac{f(x_n, x_{n-1}, a_n)}{x_n} \\ &< x_n \frac{f(\bar{x}_n, \bar{x}_{n-1}, a_n)}{\bar{x}_n} = \frac{x_n}{\bar{x}_n}\bar{x}_{n+1}. \end{aligned}$$

Therefore,

$$\frac{x_{n+1}}{\bar{x}_{n+1}} < \frac{x_n}{\bar{x}_n}$$

so the maximum occurs in the first or second term of the semicycle.
(iv) Let $x_{n+1} \geq \bar{x}_{n+1}$, and $x_n \geq \bar{x}_n$, be the last two terms in the positive semicycle relative to $\{\bar{x}_n\}$. Assume, for the sake of contradiction, that the maximum occurs in the last term. Then

$$\frac{x_{n+1}}{\bar{x}_{n+1}} > \frac{x_n}{\bar{x}_n} \geq 1$$

and

$$\begin{aligned} x_{n+2} &= f(x_{n+1}, x_n, a_{n+1}) = f(\frac{x_{n+1}}{\bar{x}_{n+1}}\bar{x}_{n+1}, \frac{x_n}{\bar{x}_n}\bar{x}_n, a_{n+1}) \\ &> f(\frac{x_{n+1}}{\bar{x}_{n+1}}\bar{x}_{n+1}, \frac{x_{n+1}}{\bar{x}_{n+1}}\bar{x}_n, a_{n+1}) \geq f(\bar{x}_{n+1}, \bar{x}_n, a_{n+1}) = \bar{x}_{n+2} \end{aligned}$$

which is a contradiction. □

Lemma 7 *Assume that the hypotheses (H_1)-(H_6) are satisfied and let $\{\bar{x}_n\}$ be a positive solution of Eq.(3). Then for every positive solution $\{x_n\}$ of Eq.(3) that is nonoscillatory relative to $\{\bar{x}_n\}$,*

$$x_n \sim \bar{x}_n.$$

Proof. We will consider only the case when $x_n \geq \bar{x}_n$ for $n \geq N_0$. The proof in the case when $x_n < \bar{x}_n$ for $n \geq N_0$ is similar and it will be omitted. For $n > N_0$ we find

$$x_{n+1} = f(x_n, x_{n-1}, a_n) = x_n \frac{f(x_n, x_{n-1}, a_n)}{x_n} \leq x_n \frac{f(\bar{x}_n, \bar{x}_{n-1}, a_n)}{\bar{x}_n} = \bar{x}_{n+1}\frac{x_n}{\bar{x}_n}.$$

Therefore $1 \leq \dfrac{x_{n+1}}{\bar{x}_{n+1}} \leq \dfrac{x_n}{\bar{x}_n}$ which implies that the sequence $\{x_n/\bar{x}_n\}$ is convergent, that is

$$\lim_{n\to\infty} \frac{x_n}{\bar{x}_n} = r \geq 1. \tag{13}$$

To complete the proof it remains to show that $r = 1$. Assume, for the sake of contradiction, that $r > 1$. Since $\{x_n\}$, $\{\bar{x}_n\}$, and $\{a_n\}$ are bounded and persistent sequences, there exists an increasing sequence of integers $\{n_i\}$ such that the following subsequences

$$\{x_{n_i}\}, \{x_{n_i-1}\}, \{\bar{x}_{n_i}\}, \{\bar{x}_{n_i-1}\}, \text{ and } \{a_{n_i}\}$$

converge to respective positive limits. Let

$$\lim_{i\to\infty} \bar{x}_{n_i} = x', \quad \lim_{i\to\infty} \bar{x}_{n_i-1} = x'', \quad \text{and} \quad \lim_{i\to\infty} a_{n_i} = a.$$

From the above and (13) follows

$$\lim_{i\to\infty} x_{n_i} = rx' \quad \text{and} \quad \lim_{i\to\infty} x_{n_i-1} = rx''.$$

Finally

$$1 < r = \lim_{i\to\infty} \frac{x_{n_i+1}}{\bar{x}_{n_i+1}} = \lim_{i\to\infty} \frac{f(x_{n_i}, x_{n_i-1}, a_{n_i})}{f(\bar{x}_{n_i}, \bar{x}_{n_i-1}, a_{n_i})} = \frac{f(rx', rx'', a)}{f(x', x'', a)}.$$

On the other hand, since $f(u,v,w)/u$ is nonincreasing in u and $f(u,v,w)$ is decreasing in v, and $r>1$ we find

$$r=\frac{f(rx',rx'',a)}{f(x',x'',a)}=\frac{rx'}{f(x',x'',a)}\frac{f(rx',rx'',a)}{rx'}$$
$$<\frac{rx'}{f(x',x'',a)}\frac{f(x',x'',a)}{x'}=r$$

which is a contradiction. Therefore, $r=1$ and the proof is complete. □

Theorem 8 *Assume that the hypotheses (H_1)-(H_7) are satisfied. Then Eq.(3) is extremely stable, that is for all positive solutions $\{x_n\}$ and $\{\bar{x}_n\}$ of Eq.(3)*

$$x_n\sim\bar{x}_n \qquad or \qquad \lim_{n\to\infty}\frac{x_n}{\bar{x}_n}=1\ .$$

Proof. From Lemma 7 it follows $x_n\sim\bar{x}_n$ for any nonoscillatory solution $\{x_n\}$ of Eq. (3). Assume that the solution $\{x_n\}$ of Eq. (3) is oscillatory relative to $\{\bar{x}_n\}$. Let $\{x_{q_i+1},x_{q_i+2},...,x_{p_i}\}$ be the i-th negative semicycle followed by the i-th positive semicycle $\{x_{p_i+1},x_{p_i+2},...,x_{r_i}\}$. Denote by x_{M_i} the maximum in the i-th positive semicycle and x_{m_i} the minimum in the i-th negative semicycle. Let

$$\lambda=\liminf_{n\to\infty}\frac{x_n}{\bar{x}_n}=\liminf_{i\to\infty}\frac{x_{m_i}}{\bar{x}_{m_i}}\quad\text{and}\quad\mu=\limsup_{n\to\infty}\frac{x_n}{\bar{x}_n}=\limsup_{i\to\infty}\frac{x_{M_i}}{\bar{x}_{M_i}}.\qquad(14)$$

From Lemma 3(i) it follows that the maximum in the positive semicycle occurs in the first or second term of the semicycle. First, consider the case when the maximum occur in the first term of the positive semicycle. Then, for every $\varepsilon>0$, and i sufficiently large

$$\lambda-\varepsilon<\frac{x_{M_i-2}}{\bar{x}_{M_i-2}},\frac{x_{M_i-1}}{\bar{x}_{M_i-1}}<1,\quad\text{and}\quad\frac{x_{M_i}}{\bar{x}_{M_i}}\geq 1.$$

Since $\bar{x}_{M_i-1}\frac{x_{M_i-1}}{\bar{x}_{M_i-1}}>\bar{x}_{M_i-1}(\lambda-\varepsilon)$ and $\bar{x}_{M_i-2}\frac{x_{M_i-2}}{\bar{x}_{M_i-2}}>\bar{x}_{M_i-2}(\lambda-\varepsilon)$,

$$\begin{aligned}\frac{x_{M_i}}{\bar{x}_{M_i}}&=\frac{f(x_{M_i-1},x_{M_i-2},a_{M_i-1})}{f(\bar{x}_{M_i-1},\bar{x}_{M_i-2},a_{M_i-1})}\\&=\frac{f(\bar{x}_{M_i-1}\frac{x_{M_i-1}}{\bar{x}_{M_i-1}},\bar{x}_{M_i-2}\frac{x_{M_i-2}}{\bar{x}_{M_i-2}},a_{M_i-1})}{\bar{x}_{M_i-1}\frac{x_{M_i-1}}{\bar{x}_{M_i-1}}}\frac{x_{M_i-1}}{f(\bar{x}_{M_i-1},\bar{x}_{M_i-2},a_{M_i-1})}\\&<\frac{f(\bar{x}_{M_i-1}(\lambda-\varepsilon),\bar{x}_{M_i-2}(\lambda-\varepsilon),a_{M_i-1})}{\bar{x}_{M_i-1}(\lambda-\varepsilon)}\frac{x_{M_i-1}}{f(\bar{x}_{M_i-1},\bar{x}_{M_i-2},a_{M_i-1})}\\&<\frac{f(\bar{x}_{M_i-1}(\lambda-\varepsilon),\bar{x}_{M_i-2}(\lambda-\varepsilon),a_{M_i-1})}{(\lambda-\varepsilon)f(\bar{x}_{M_i-1},\bar{x}_{M_i-2},a_{M_i-1})}\leq G(\lambda-\varepsilon).\end{aligned}$$

Now consider the case when the maximum occurs in the second term of the positive semicycle relative to $\{\bar{x}_n\}$. Then for every $\varepsilon > 0$ and i sufficiently large,

$$\lambda - \varepsilon < \frac{x_{M_i-2}}{\bar{x}_{M_i-2}}, \frac{x_{M_i-3}}{\bar{x}_{M_i-3}} < 1, \quad \text{and} \quad \frac{x_{M_i}}{\bar{x}_{M_i}} \geq \frac{x_{M_i-1}}{\bar{x}_{M_i-1}} \geq 1.$$

Since $\bar{x}_{M_i-1}\frac{x_{M_i-2}}{\bar{x}_{M_i-2}} < x_{M_i-1}$ and $f(u,v,w)/u$ is nonincreasing in u and decreasing in v we obtain

$$\begin{aligned}
\frac{x_{M_i}}{\bar{x}_{M_i}} &= \frac{f(x_{M_i-1}, x_{M_i-2}, a_{M_i-1})}{f(\bar{x}_{M_i-1}, \bar{x}_{M_i-2}, a_{M_i-1})} \\
&= \frac{f(x_{M_i-1}, x_{M_i-2}, a_{M_i-1})}{x_{M_i-1}} \frac{x_{M_i-1}}{f(\bar{x}_{M_i-1}, \bar{x}_{M_i-2}, a_{M_i-1})} \\
&\leq \frac{f(\bar{x}_{M_i-1}\frac{x_{M_i-2}}{\bar{x}_{M_i-2}}, \bar{x}_{M_i-2}\frac{x_{M_i-2}}{\bar{x}_{M_i-2}}, a_{M_i-1})}{\bar{x}_{M_i-1}\frac{x_{M_i-2}}{\bar{x}_{M_i-2}}} \frac{f(x_{M_i-2}, x_{M_i-3}, a_{M_i-2})}{f(\bar{x}_{M_i-1}, \bar{x}_{M_i-2}, a_{M_i-1})}.
\end{aligned}$$

By applying (H_6) with $t = \frac{x_{M_i-2}}{\bar{x}_{M_i-2}} < 1$ to the above we find

$$\begin{aligned}
\frac{x_{M_i}}{\bar{x}_{M_i}} &\leq \frac{f(\bar{x}_{M_i-1}, \bar{x}_{M_i-2}, a_{M_i-1})}{\bar{x}_{M_i-1}\frac{x_{M_i-2}}{\bar{x}_{M_i-2}}} \frac{f(x_{M_i-2}, x_{M_i-3}, a_{M_i-2})}{f(\bar{x}_{M_i-1}, \bar{x}_{M_i-2}, a_{M_i-1})} \\
&= \frac{f(\bar{x}_{M_i-2}\frac{x_{M_i-2}}{\bar{x}_{M_i-2}}, \bar{x}_{M_i-3}\frac{x_{M_i-3}}{\bar{x}_{M_i-3}}, a_{M_i-2})}{\bar{x}_{M_i-2}\frac{x_{M_i-2}}{\bar{x}_{M_i-2}}} \frac{\bar{x}_{M_i-2}}{f(\bar{x}_{M_i-2}, \bar{x}_{M_i-3}, a_{M_i-2})} \\
&< \frac{f(\bar{x}_{M_i-2}(\lambda-\varepsilon), \bar{x}_{M_i-3}(\lambda-\varepsilon), a_{M_i-2})}{\bar{x}_{M_i-2}(\lambda-\varepsilon)} \frac{\bar{x}_{M_i-2}}{f(\bar{x}_{M_i-2}, \bar{x}_{M_i-3}, a_{M_i-2})} \\
&\leq G(\lambda-\varepsilon).
\end{aligned}$$

Therefore, as ε is arbitrary, $\frac{x_{M_i}}{\bar{x}_{M_i}} \leq G(\lambda-\varepsilon)$ and from (14) we find

$$\mu \leq G(\lambda).$$

In the similar way one can show that

$$\lambda \leq G(\mu).$$

Therefore $\lambda = \mu = 1$ and $\lim_{n\to\infty} x_n/\bar{x}_n = 1$ which completes the proof. □

4 Applications

Consider the equation

$$x_{n+1} = \frac{a_n x_n}{1 + b x_n + c x_{n-1}} \tag{15}$$

where $b, c > 0$ and $\{a_n\}$ is a positive sequence such that

$$0 < A \leq a_n \leq B < \infty. \tag{16}$$

In this case

$$f(u,v,w) = \frac{wu}{1+bu+cv} \tag{17}$$

and clearly (H_1)-(H_3) are satisfied.

Theorem 9 *Consider Eq. (15) where $\{a_n\}$ is a positive sequence such that (16) holds. Then the following statements are true:*
(a) If $B \leq 1$ then all solutions of Eq. (15) are eventually decreasing and converge to 0.
(b) If $0 < A \leq 1 < B$ then all solutions of Eq. (15) are bounded from above by a positive constant.
(c) If $A > 1$ then Eq. (15) is permanent.
(d) If $A > 1$ and $\{\bar{x}_n\}$ is a positive solution of Eq. (15) then Eq. (15) is extremely stable, that is for all pairs of its positive solutions $\{x_n\}$ and $\{\bar{x}_n\}$

$$x_n \sim \bar{x}_n \qquad \mathit{or} \qquad \lim_{n\to\infty} \frac{x_n}{\bar{x}_n} = 1\ .$$

Proof. (a) If $B \leq 1$ then

$$f(x,x,B) = \frac{Bx}{1+(b+c)x} < x$$

for every $x \in (0,\infty)$ and from Lemma1 result follows.
(b) By taking $\bar{x} = \frac{B-1}{b+c}$ we see that (H_4) is satisfied and from Theorem 4 result follows.
(c) Since $B \geq A > 1$, by taking $\tilde{x} = \frac{A-1}{b+c}$ and $\bar{x} = \frac{B-1}{b+c}$ we see that (H_4) and (H_5) are satisfied and the result follows from the Theorem 5.
(d) We will apply Theorem 8. In this case $f(u,v,w) = \frac{wu}{1+bu+cv}$ and for $t > 1$ we have

$$\begin{aligned}(f(tu,tv,w) - f(u,v,w))(t-1) &= \left(\frac{wtu}{1+btu+ctv} - \frac{wu}{1+bu+cv}\right)(t-1) \\ &= \frac{wu(t-1)^2}{(1+btu+ctv)(1+bu+cv)} > 0\end{aligned}$$

and (H_6) is satisfied.

Now we have

$$H(t,u,v,w) = \frac{f(ut,vt,w)}{tf(u,v,w)} = \frac{1+bu+cv}{1+but+cvt}, \qquad \text{for} \qquad t \geq 0.$$

Since $H_u(t,u,v,w) = b(1-t)/(1+but+cvt)^2$ and $H_v(t,u,v,w) = c(1-t)/(1+but+cvt)^2$, for $0 < t < 1$ we have $\max_{u,v\in[p,q],w\in[A,B]} H(t,u,v,w) = \frac{1+(b+c)q}{1+(b+c)qt}$ and for $t > 1$ we obtain $\min_{u,v\in[p,q],w\in[A,B]} H(t,u,v,w) = \frac{1+(b+c)q}{1+(b+c)qt}$. Therefore $G(t) = \frac{1+(b+c)q}{1+(b+c)qt}$. Since

$$G^2(t) = \frac{1+(b+c)q}{1+(b+c)q\frac{1+(b+c)q}{1+(b+c)qt}} = \frac{(1+(b+c)q)(1+(b+c)qt)}{1+(b+c)qt+(b+c)q+(b+c)^2q^2},$$

the only positive solution of the equation $G^2(t) = t$ is $t = 1$. Hence the condition (H_7) is also satisfied and the proof is complete. □

The next lemma establishes the existence of an invariant interval for Eq. (15).

Lemma 10 *Let $B > A > 1$, and*

$$\frac{b}{c} > \frac{B-1}{A-1}. \tag{18}$$

Then

$$I = \left[\frac{b(A-1)-c(B-1)}{b^2-c^2}, \frac{b(B-1)-c(A-1)}{b^2-c^2}\right] \tag{19}$$

is an invariant interval for Eq.(15), that is

$$u, v \in I, w \in [A,B] \qquad \Longrightarrow \qquad f(u,v,w) \in I$$

where f is given by (17).

Proof. Consider the function f given by (17). Since

$$\begin{aligned} f(\tfrac{b(B-1)-c(A-1)}{b^2-c^2}, \tfrac{b(A-1)-c(B-1)}{b^2-c^2}, B) &\\ = \frac{B\frac{b(B-1)-c(A-1)}{b^2-c^2}}{1+b\frac{b(B-1)-c(A-1)}{b^2-c^2}+c\frac{b(A-1)-c(B-1)}{b^2-c^2}} &\\ = \frac{b(B-1)-c(A-1)}{b^2-c^2}, &\\ f(\tfrac{b(A-1)-c(B-1)}{b^2-c^2}, \tfrac{b(B-1)-c(A-1)}{b^2-c^2}, A) &\\ = \frac{A\frac{b(A-1)-c(B-1)}{b^2-c^2}}{1+b\frac{b(A-1)-c(B-1)}{b^2-c^2}+c\frac{b(B-1)-c(A-1)}{b^2-c^2}} &\\ = \frac{b(A-1)-c(B-1)}{b^2-c^2}, & \end{aligned}$$

then for $u, v \in I$ and $w \in [A, B]$ we get

$$f(\tfrac{b(A-1)-c(B-1)}{b^2-c^2}, \tfrac{b(B-1)-c(A-1)}{b^2-c^2}, A) \leq f(u, v, w) \leq f(\tfrac{b(B-1)-c(A-1)}{b^2-c^2}, \tfrac{b(A-1)-c(B-1)}{b^2-c^2}, B).$$

So the interval I is invariant interval for Eq. (15). □

Now we will examine the case when $\{a_n\}$ is periodic with prime period k:

$$a_{n+k} = a_n, \qquad \text{for every } n.$$

Lemma 11 *A necessary condition for the existence of a periodic solution $\{x_n\}$ of Eq.(15) with prime period k is that $\{a_n\}$ is periodic with period k.*

Proof. Assume that $\{x_n\}$ is periodic with prime period k, that is $x_{n+k} = x_n$, for $n = -1, 2, \ldots$.Then

$$a_{n+k} = \frac{x_{n+1+k}}{x_{n+k}}(1 + bx_{n+k} + cx_{n+k-1}) = \frac{x_{n+1}}{x_n}(1 + bx_n + cx_{n-1}) = a_n$$

and the proof is complete. □

Theorem 12 *Assume that $\{a_n\}$ is periodic with prime period k, and let*

$$1 < A = \min_{0\leq i\leq k} \{a_i\} < \max_{1\leq i\leq k} \{a_i\} = B, \qquad \textit{and} \qquad \frac{b}{c} > \frac{B-1}{A-1}. \tag{20}$$

Then the following statements are true:
(i) There exists a positive periodic solution $\{\bar{x}_n\}$ of Eq.(15) with prime period k.
(ii) The periodic solution $\{\bar{x}_n\}$ is unique and attracts all positive solutions of Eq.(15), that is,

$$\lim_{n\to\infty} \frac{x_n}{\bar{x}_n} = 1$$

for all positive solutions $\{x_n\}$ of Eq.(15).

Proof. (i) To prove that Eq. (15) has a periodic solution with period k, we must show that the following system has a positive solution:

$$\begin{aligned}
x_1 &= \frac{a_k x_k}{1 + bx_k + cx_{k-1}} \\
x_2 &= \frac{a_1 x_1}{1 + bx_1 + cx_k} \\
x_3 &= \frac{a_2 x_2}{1 + bx_2 + cx_1} \\
\vdots &= \\
x_k &= \frac{a_{k-1} x_{k-1}}{1 + bx_{k-1} + cx_{k-2}}.
\end{aligned}$$

Consider the function $F:\mathbb{R}_+^k\to\mathbb{R}_+^k$ defined by

$$F(u_1,\ldots,u_k)$$
$$=(\frac{a_ku_k}{1+bu_k+cu_{k-1}},\frac{a_1u_1}{1+bu_1+cu_k},\frac{a_2u_2}{1+bu_2+cu_1},...,\frac{a_{k-1}u_{k-1}}{1+bu_{k-1}+cu_{k-2}})$$
$$=(f(u_k,u_{k-1},a_k),f(u_1,u_k,a_1),f(u_2,u_1,a_2),...,f(u_{k-1},u_{k-2},a_{k-1})).$$

Let $u_1,...,u_k\in I$. Since $a_1,...,a_k\in[A,B]$, from Lemma 10 it follows that

$$f(u_k,u_{k-1},a_k),f(u_1,u_k,a_1),f(u_2,u_1,a_2),...,f(u_{k-1},u_{k-2},a_{k-1})\in I.$$

Therefore, $F:I^k\to I^k$. Clearly F is continuous on I^k, and I^k is a compact and convex set. Consequently, F has a fixed point in I^k , by the Brower Fixed Point Theorem.

Let $(\bar{u}_{1,...,}\bar{u}_k)\in I^k$ be a fixed point of F. Then the sequence $\{\bar{x}_n\}$ defined by

$$\bar{x}_{-1}=\bar{u}_{k-1},\ \bar{x}_0=\bar{u}_k,\text{ and }\bar{x}_{mk+i}=\bar{u}_i,\text{ for }i=1,2,\ldots k,\ m=0,1,\ldots.$$

satisfies Eq. (15) and is periodic with period k; this completes the proof of the part (i).

(ii) Theorem 9 implies that

$$\lim_{n\to\infty}\frac{x_n}{\bar{x}_n}=1$$

holds for any solution $\{x_n\}$ of Eq. (15).

Clearly the periodic solution $\{\bar{x}_n\}$ is unique. Otherwise, let $\{x'_n\}$ be another periodic solution of Eq. (15) with period k and different from $\{\bar{x}_n\}$. Then

$$x'_{n+k}=x'_n\quad\text{for}\quad n=-1,0,1,\ldots,$$

and there exists i such that

$$\frac{x'_{nk+i}}{\bar{x}_{nk+i}}=\frac{x'_i}{\bar{x}_i}\neq 1.$$

This contradicts the fact that

$$\lim_{n\to\infty}\frac{x'_n}{\bar{x}_n}=1,$$

and the proof is complete. □

Acknowledgement *Partial support for the work on this paper was provided by the Louisiana Board of Regents grant # LEQSF(2004-07)-RD-A40.*

Special thanks goes to D. Stutson and to the anonymous referee for providing valuable suggestions.

References

[1] M. E. Clark and L. J. Gross, Periodic solutions to nonautonomous difference equation. *Math. Biosciences*, 102, (1990), 105-119.

[2] J. M. Cushing, Periodically forced nonlinear systems of difference equations, *J. Diff. Eqn. Appl.* 3 (1998), 547-561.

[3] J. M. Cushing and S. M. Henson, A periodically forced Beverton-Holt equation. *J. Diff. Eqn. Appl.* 8 (2002), 1119-1120.

[4] R. DeVault, V. L. Kocic, and D. Stutson, Global behavior of solutions of the nonlinear difference equation $x_{n+1} = p_n + x_{n-1}/x_n$,*J. Diff. Eqn. Appl.* (2005), to appear.

[5] S. Elaydi and R. J. Sacker, Global stability of periodic orbits of nonautonomous difference equations in population biology and the Cushing-Henson conjectures, *Proc. 8th Internat. Conf. Difference Equat. Appl.*, Brno, Czech Republic, 2003 (to appear)

[6] S. Elaydi and R. J. Sacker, Global stability of periodic orbits of nonautonomous difference equations and population biology, *J. Diff. Eqn. Appl.* (to appear).

[7] C. H. Gibbons and C. B. Overdeep, On the trichotomy character of $x_{n+1} = (\alpha_n + \gamma_n x_{n-1})/(A_n + B_n x_n)$ with period-two coefficients, *J. Diff. Eqn. Appl.*, (to appear)

[8] J. R. Graef, C. Qian, and P. W. Spikes, Oscillation and global attractivity in a discrete periodic logistic equation. *Dynam. Systems. Appl.* 5 (1996), no. 2, 165-173.

[9] V. L. Kocic, A note on the nonautonomous Beverton-Holt model, *J. Diff. Eqn. Appl.* (2005), to appear.

[10] V. L. Kocic and G. Ladas, *Global Behavior of Nonlinear Difference Equations of Higher Order with Applications,* Kluwer Academic Publishers, Dordrecht, 1993.

[11] V. L. Kocic, D. Stutson, and G. Arora, Global behavior of solutions of a nonautonomous delay logistic difference equation, *J. Diff. Eq. Appl.* 10, (2004), 1267-1279.

[12] R. Kon, A note on attenuant cycles of population models with periodic carrying capacity, *J. Diff. Eqn. Appl.* 10, (2004), 791-793.

[13] Q. Kong, Oscillatory and asymptotic behavior of a discrete logistic model. *Rocky Mountain. J. Math.* 25 (1995), no. 1, 339-349.

[14] M. R. S. Kulenovic, G. Ladas, and C. B. Overdeep, On the dynamics of $x_{n+1} = p_n + x_{n-1}/x_n$. *J. Diff. Eqn. Appl.*, 9, (2003), 1053-1056.

[15] M. R. S. Kulenovic, G. Ladas, and C. B. Overdeep, On the dynamics of $x_{n+1} = p_n + x_{n-1}/x_n$ with period-two coefficient. *J. Diff. Eqn. Appl.*, 10, (2004), 905-914.

[16] J.P. LaSalle and S. Lefschetz. *Stability by Liapunov's Direct Method with Applications.* Academic Press, New York, 1961

[17] A. H. Nasr, Permanence of a nonlinear delayed difference equation with variable coefficients. *J. Diff. Eqn. Appl.* 3 (1997), no.2, 95-100.

[18] A. V. Pavlov, The output regulation problem: a convergent dynamics approach, PhD Thesis, Technische Universiteit Eindhoven, Eindhoven, 2004.

[19] S. Stevic, On the recursive sequence $x_{n+1} = \alpha_n + x_{n-1}/x_n$, *Int. J. Math. Sci.*, 2 (2003), 237-243.

[20] S. Stevic, On the recursive sequence $x_{n+1} = \alpha_n + x_{n-1}/x_n$, II. *Dyn.. Contin. Discrete Impuls. Syst. Ser. A Math. Anal.*, 10 (2003), 911-916.

[21] E. Zeidler, *Applied Functional Analysis, Applications to Mathematical Physics,* Springer-Verlag, New York, 1991.

How can three species coexist in a periodic chemostat? : Mathematical and Numerical Study

Shinji Nakaoka and Yasuhiro Takeuchi *

Graduate school of Science and Technology, Shizuoka University, Johoku 3-5-1, Hamamatsu, Shizuoka 432-8561, Japan

Address for manuscript correspondence:
Shinji Nakaoka and Yasuhiro Takeuchi
Graduate school of Science and Technology, Shizuoka University,
Johoku 3-5-1, Hamamatsu, Shizuoka 432-8561, Japan
E-mail r5445020@ipc.shizuoka.ac.jp, takeuchi@sys.eng.shizuoka.ac.jp
Tel.: 81 53 478 1200   Fax: 81 53 478 1200

ABSTRACT

A competition model of three species for one resource in a chemostat with a periodic washout rate is considered. Coexistence is indicated in [7] by numerical bifurcation analysis and in [12] by mathematical analysis. By introducing average competition functions, we obtain a necessary condition for the coexistence of a positive periodic solution and show that the condition restricts possible parameter value set to be relatively small. Further we show that the coexistence is enhanced when the period of the washout rate becomes large.

Key word: Chemostat equations, periodic washout rate, coexistence, Michaelis-Menten functional response, Conservation principle, Average competition function

1 Introduction

Chemostat equations have been used to study population dynamics of microorganisms in experimental apparatuses or aquatic ecosystems such as lakes. The Competitive Exclusion Principle states that among several species competing for common resources, the number of coexistent species does not exceed the number of available resources (see Grover [3] for example). The mathematical results on the standard chemostat equations of competition for a single limiting resource show that only the species with the lowest break even concentration

*The research was partly supported by the Ministry of Education, Culture, Sports, Science and Technology in Japan, under Grand-in-Aid for Scientific Research (A) 13304006.

survives (see Armstrong and McGehee [1], Smith and Waltman [10, Chapter 1, Chapter 2]). On the other hand, the competitive exclusion principle is not valid for chemostat equations if a fluctuating environment is under consideration. In fact, many studies have revealed that the coexistence of two species competing for one resource is possible if a nutrient input varies periodically (see Hsu [5] and Smith [9], for example). Butler *et al.* [2] showed that the coexistence of two species is also possible in the case where the washout rate varies periodically. In [2], coexistence is expected if the washout rate varies in such a way that each competitor has its own competitive advantage depending on the concentration of the resource (see also Lenas & Pavlou [6] and Pilyugin & Waltman [8]).

It is a fundamental interest and problem on chemostat equations whether fluctuating environments can support the coexistence of more than three species under only one resource. Lenas and Pavlou [7] showed by numerical bifurcation analysis that the coexistence of three species is possible. Wolkowicz and Zhou [12] gave sufficient conditions for the uniform persistence of competing arbitrary N-species on a periodic chemostat. In [12], for the three species competition case, they considered the following system of equations:

$$\begin{aligned} \frac{dS(t)}{dt} &= (S^0(t) - S(t))D_0(t) - \sum_{i=1}^{3} P_i(t, S(t))x_i(t), \\ \frac{dx_i(t)}{dt} &= x_i(t)(P_i(t, S(t)) - D_0(t)), \quad (i = 1, 2, 3). \end{aligned} \tag{1.1}$$

Here $S(t)$ denotes the concentration of the limiting nutrient, $x_i(t)$ $(i = 1, 2, 3)$ denotes the measure of i-th species at time t. $P_i(t, S)$ $(i = 1, 2, 3)$ represents the specific per capita nutrient uptake function of i-th species, $S^0(t)$ and $D_0(t)$ are the input nutrient concentration and the washout rate, respectively. $S^0(t)$ and $D_0(t)$ are continuous, ω-periodic and positive functions, and each $P_i(t, S)$ satisfies

(i) $P_i(t, S)$ is locally Lipschitz in S,

(ii) $P_i(t, 0) = 0$ for $t \geq 0$ and for any $t \geq 0$, $P_i(t, S)$ is strictly increasing for $S \in \mathbf{R}_+$.

They showed the existence condition for a positive ω-periodic solution $(S(t), x_1(t), x_2(t), x_3(t))$ of (1.1) with $S(0) \geq 0$ and $x_i(0) > 0$ $(i = 1, 2, 3)$. The detail is as follows:

Let $V_0^*(t)$ be the unique, globally attracting, positive ω-periodic solution of

$$\frac{dV}{dt} = (S^0(t) - V(t))D_0(t).$$

For each $1 \leq i \leq 3$, there is a corresponding single-species periodic equation

$$\frac{dx_i}{dt} = x_i(P_i(t, V_0^*(t) - x_i) - D_0(t)). \tag{E_i^*}$$

There is also, for each $1 \le i \le 3$, a corresponding two-species periodic competition system

$$\frac{dx_j}{dt} = x_j \left(P_j(t, V_0^*(t) - \sum_{k=1,k\neq i}^{3} x_k) - D_0(t) \right), \quad 1 \le j \le 3,\ j \neq i. \qquad (\bar{\mathrm{E}}_i)$$

Theorem 1.1. *[12, Theorem 4.2.] Assume that*

(i) $\mu_i \equiv \int_0^\omega (P_i(t, V_0^*(t)) - D_0(t))dt > 0,\ 1 \le i \le 3;$

(ii) $\mu_{ji} \equiv \int_0^\omega (P_i(t, V_0^*(t) - x_j^*(t)) - D_0(t))dt > 0,\ 1 \le i,j \le 3,\ i \neq j,\ j \neq 2,$ *and* $\mu_{21} \equiv \int_0^\omega (P_1(t, V_0^*(t) - x_2^*(t)) - D_0(t))dt > 0;$

(iii) $\bar{\mu}^i \equiv \int_0^\omega \left(P_i(t, V_0^*(t) - \sum_{j=1,j\neq i}^{3} \bar{x}_j^i(t)) - D_0(t) \right) dt > 0,\ 2 \le i \le 3,$

where $x_i^*(t)$ *is the unique positive* ω*-periodic solution of* (E_i^*) $(1 \le i \le 3)$ *and* $(\bar{x}_1^2(t), \bar{x}_3^2(t))$ *and* $(\bar{x}_1^3(t), \bar{x}_2^3(t))$ *are the unique positive,* ω*-periodic solution of* $(\bar{\mathrm{E}}_2)$ *and* $(\bar{\mathrm{E}}_3)$*. Then system* (1.1) *admits a positive* ω*-periodic solution* $(S(t), x_1(t), x_2(t), x_3(t))$ *with* $S(0) \ge 0$ *and* $x_i(0) > 0$ $(i = 1, 2, 3)$.

An interesting example of Theorem 1.1 is given in [12], which is also adopted to show the coexistence of three species in [7]: Let $S^0(t) = 11$ and $D_0(t) = u_0 + a\cos 2\pi/\omega t$ where $u_0 = 0.4675$, $a = 0.3$ and $\omega = 31.4$. Nutrient uptake functions are Michaelis-Menten type functional responses $P_i(t, S(t)) = m_i S(t)/(a_i + S(t))$, where $m_1 = 1$, $a_1 = 1$, $m_2 = 0.7$, $a_2 = 0.3$, $m_3 = 0.64$ and $a_3 = 0.2$. Each integral in Theorem 1.1 is evaluated numerically as $\mu_1 = 14.1111$, $\mu_2 = 6.7205$, $\mu_3 = 5.0603$, $\mu_{12} = 0.6780$, $\mu_{13} = 1.87861$, $\mu_{21} = 0.7909$, $\mu_{31} = 0.2990$, $\mu_{32} = 0.8122$, $\bar{\mu}^2 = 0.0012$ and $\bar{\mu}^3 = 0.009$. This example suggests that competition mediates coexistence in the following sense: If x_1 is absent and x_2 & x_3 compete, then x_2 drives x_3 to extinction. However, this extinction of x_3 is avoided simply by introducing x_1. Once x_1 is introduced, all three species persist in sustained oscillation (see [12, pp. 486–487]).

The purpose of our study is to investigate how three species can coexist in a periodic chemostat. There are still unknown aspects on periodic chemostat models. First, as in [7], the region of parameters which ensures the coexistence of three species actually exists, but it is narrow. An adequate interpretation why the coexistence region is narrow should be proposed. Second, it is unknown and therefore should be figured out what factors realize the coexistence of three species.

In this paper, let us consider periodic chemostat equations of the form

$$\begin{cases} S' = (S^0 - S)D(t) - \sum_{j=1}^{3} f_j(S)x_j, \\ x_i' = x_i\left(f_i(S) - D(t)\right), \quad (i = 1, 2, 3). \end{cases} \qquad (\mathrm{E})$$

Here S^0 is a positive constant and $D : [0, \infty) \to [0, \infty)$ is a positive, ω-periodic function. We assume that $D(t)$ is not constant, since the coexistence of three competitors is impossible under constant environment. The mean value of the periodic function $D(t)$ is denoted by $\langle D \rangle$:

$$\langle D \rangle = \frac{1}{\omega} \int_0^\omega D(s)\, ds.$$

We assume the following for the functional response f_i of i-th species.

F-(i) $f_i : \mathbf{R}_+ \to \mathbf{R}_+$ is continuously differentiable,

F-(ii) $f_i(0) = 0,\ f_i'(S) > 0$.

A typical example of f_i is Michaelis-Menten functional response of the form :

$$f_i(S) = \frac{m_i S}{a_i + S}, \quad (i = 1, 2, 3). \tag{1.2}$$

Here a_i and m_i $(i = 1, 2, 3)$ are positive constants.

In the next section, system (E) is reduced to the limiting system. In Section 3, an average competition function is introduced which is exploited to measure the degree of competition and to give sufficient conditions for competitive exclusion and a necessary condition for periodic coexistence. Section 4 gives some numerical simulation results which demonstrate how the coexistence of three species is realized as the period of the washout rate increases. Section 5 gives conclusions of our study.

2 Reduction to the limiting system

By measuring all variables in unit of S^0 and time in unit of $\langle D \rangle^{-1}$, that is,

$$\frac{S}{S^0} \mapsto S\ ,\ \frac{x_i}{S^0} \mapsto x_i \text{ and } \langle D \rangle t \mapsto t,$$

system (E) takes the form:

$$\begin{cases} S' = (1 - S)D(t) - \displaystyle\sum_{j=1}^{3} f_j(S) x_j, \\ x_i' = x_i \left(f_i(S) - D(t) \right), \quad (i = 1, 2, 3). \end{cases} \tag{2.1}$$

Here we relabeled $f_i(S)$ and $D(t)$ in the equations (2.1), each of which is actually $\langle D \rangle^{-1} f_i(S^0 S)$ and $\langle D \rangle^{-1} D(t / \langle D \rangle)$ in (E), respectively. Note that this scaling affects both the period and the mean value of D. The former becomes $\langle D \rangle \omega$, which we relabel ω and the latter becomes the unity: $\langle D \rangle = 1$.

Set $\Sigma = S + \sum_{j=1}^{3} x_j - 1$. Adding the equations (2.1) gives the periodic linear system

$$\Sigma'(t) = -D(t)\Sigma(t). \tag{2.2}$$

Then (2.1) corresponds to

$$\begin{cases} \Sigma' = -D(t)\Sigma, \\ x_i' = x_i \left(f_i \left(\Sigma - \sum_{j=1}^{3} x_j + 1 \right) - D(t) \right), \quad (i = 1, 2, 3). \end{cases} \tag{2.3}$$

Since $\langle D \rangle = 1$, solving (2.2) gives

$$\Sigma(t) = \Sigma(0) \exp \left[- \int_0^t (D(s) - 1) ds \right] e^{-t}.$$

Hence we have

$$\lim_{t \to \infty} \Sigma(t) = 0.$$

Hereafter let us consider the system (2.3) restricted to the invariant hyperplane $\Sigma = 0$, to which all solutions are attracted at an exponential rate. Therefore setting $\Sigma = 0$, or equivalently, $S = 1 - \sum_{j=1}^{3} x_j$ yields the limiting system:

$$x_i' = x_i \left(f_i \left(1 - \sum_{j=1}^{3} x_j \right) - D(t) \right), \quad i = 1, 2, 3. \tag{L}$$

Biologically relevant initial data for (L) belong to

$$\Omega = \left\{ (x_1, x_2, x_3)^T \in \mathbf{R}_+^3 : \sum_{j=1}^{3} x_j \leq 1 \right\},$$

where

$$\mathbf{R}_+^3 = \left\{ (x_1, x_2, x_3)^T \in \mathbf{R}^3 : x_i \geq 0, (i = 1, 2, 3) \right\}.$$

It is shown that Ω is positively invariant for (L). Convergence theorem obtained by Thieme [11] motivates us to consider the limiting system.Throughout the remainder of this paper, we consider system (L).

3 Average competition

In this section, let us introduce an average competition function. The definition of this function is motivated by Hutson [4].

Let $P_{kl} : [0, \infty) \times (0, \infty) \to [0, \infty)$ be continuously differentiable function $(k, l = 1, 2, 3,\ k \neq l)$. Average competition functions P_{kl} are defined by

$$P_{kl}(x_k, x_l) = x_k x_l^{-1}. \tag{3.1}$$

By the definition, it follows that

P-(i): $P_{kl} \cdot P_{lk} = 1$,

P-(ii): $P_{kl} = 0$ iff $x_k = 0$.

The derivative of P_{kl} along the solution of (L) is denoted by $\dot{P}_{kl}(x_k(t), x_l(t))$. Direct calculation gives

$$\frac{\dot{P}_{kl}(x_k(t), x_l(t))}{P_{kl}(x_k(t), x_l(t))} = f_k\left(1 - \sum_{j=1}^{3} x_j\right) - f_l\left(1 - \sum_{j=1}^{3} x_j\right). \tag{3.2}$$

Proposition 3.1. *Let $(\bar{x}_1(t), \bar{x}_2(t), \bar{x}_3(t))$ be a positive ω-periodic solution of* (L). *Then*

$$\left\langle \frac{\dot{P}_{12}(\bar{x}_1, \bar{x}_2)}{P_{12}(\bar{x}_1, \bar{x}_2)} \right\rangle = \left\langle \frac{\dot{P}_{13}(\bar{x}_1, \bar{x}_3)}{P_{13}(\bar{x}_1, \bar{x}_3)} \right\rangle = \left\langle \frac{\dot{P}_{23}(\bar{x}_2, \bar{x}_3)}{P_{23}(\bar{x}_2, \bar{x}_3)} \right\rangle = 0. \tag{BC}$$

Proof. Since $\bar{x}_1(t)$ is a positive ω-periodic solution of (L),

$$\bar{x}_1(0) = \bar{x}_1(\omega) = \bar{x}_1(0) \exp\left[\int_0^{\omega} (f_1(1 - \bar{x}_1(s) - \bar{x}_2(s) - \bar{x}_3(s)) - D(s))\, ds\right].$$

Since $\langle D \rangle = 1$, $\langle f_1(1 - \bar{x}_1 - \bar{x}_2 - \bar{x}_3)\rangle = 1$. In the same way, $\langle f_2(1 - \bar{x}_1 - \bar{x}_2 - \bar{x}_3)\rangle = 1$ and $\langle f_3(1 - \bar{x}_1 - \bar{x}_2 - \bar{x}_3)\rangle = 1$. Since $\langle \dot{P}_{kl}/P_{kl}\rangle = -\langle \dot{P}_{lk}/P_{lk}\rangle$, (BC) holds. This completes the proof. □

Suppose that there exist positive constants S^*_{kl} $(k, l = 1, 2, 3,\ k < l)$ such that $f_k(S^*_{kl}) = f_l(S^*_{kl})$. Without loss of generality, we can assume that $S^*_{12} < S^*_{13} < S^*_{23}$. Further we assume $f_3(S) < f_2(S) < f_1(S)$ for $S \in [0, S^*_{12})$, $f_3(S) < f_1(S) < f_2(S)$ for $S \in (S^*_{12}, S^*_{13})$, $f_1(S) < f_3(S) < f_2(S)$ for $S \in (S^*_{13}, S^*_{23})$ and $f_1(S) < f_2(S) < f_3(S)$ for $S \in (S^*_{23}, \infty)$ (see, for example, Fig.2).

Let us denote the nutrient $S(t)$ by

$$S(t) := 1 - \sum_{j=1}^{3} x_j(t). \tag{3.3}$$

In particular, $\bar{S}(t)$ denotes by the periodic nutrient when the solution of (L) is periodic. In addition, let us set $g_{kl}(S) \equiv f_k(S) - f_l(S)$. The graph of g_{kl} is illustrated on Fig. 1 in the case where f_i takes the form of Michaelis-Menten functional response. Note that g_{kl} corresponds to the right hand side of (3.2). Further, $g_{kl}(S) > 0$ for $S < S^*_{kl}$ and $g_{kl}(S) < 0$ for $S > S^*_{kl}$.

Remark 3.1. (BC) *represents that average competition among all species is balanced if the solution of* (L) *is periodic.* $\langle \dot{P}_{kl}(\bar{x}_k, \bar{x}_l)/P_{kl}(\bar{x}_k, \bar{x}_l)\rangle$ *represents the integral of $g_{kl}(\bar{S})$ on the range of the periodic nutrient (gray region on Fig. 1). Hence minima S^- and maxima S^+ of the periodic nutrient are determined in order that the integral equals to zero (that is, $\langle g_{kl}(\bar{S})\rangle = 0$).*

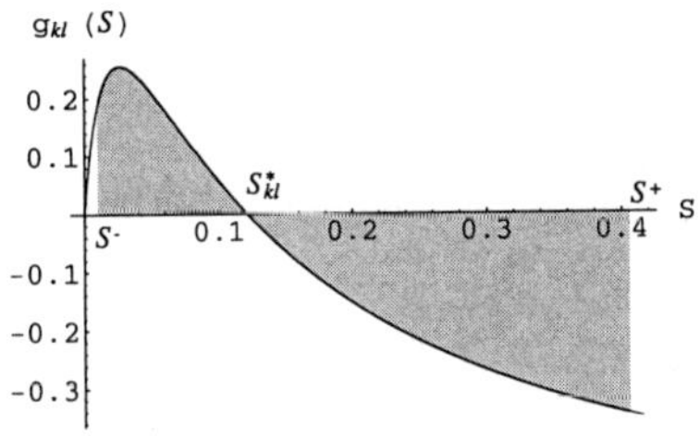

Figure 1: The graph of $g_{kl}(\bar{S})$, $(k < l)$

Note that since the integrand of $\langle \dot{P}_{kl}(\bar{x}_k, \bar{x}_l)/P_{kl}(\bar{x}_k, \bar{x}_l)\rangle$ is a composition function of $g_{kl}(S)$ and $S(t)$, it is possible that

$$\frac{1}{\omega}\int_{S^-}^{S^+} g_{kl}(r)dr > 0 \quad \text{but} \quad \frac{1}{\omega}\int_0^{\omega} g_{kl}(\bar{S}(t))dt = 0 \tag{3.4}$$

Proposition 3.2. (*competitive exclusion*) *Let $(x_1(t), x_2(t), x_3(t))$ be a positive solution of* (L).

C-(i) *If there exists $T_1 > 0$ such that $S(t) < S^*_{12}$ for all $t \geq T_1$, then $x_2(t) \to 0$ and $x_3(t) \to 0$ at some exponential rate as $t \to \infty$.*

C-(ii) *If there exists $T_2 > 0$ such that $S^*_{12} < S(t) < S^*_{23}$ for all $t \geq T_2$, then $x_1(t) \to 0$ and $x_3(t) \to 0$ at some exponential rate as $t \to \infty$.*

C-(iii) *If there exists $T_3 > 0$ such that $S^*_{23} < S(t)$ for all $t \geq T_3$, then $x_1(t) \to 0$ and $x_2(t) \to 0$ at some exponential rate as $t \to \infty$.*

Proof. It is sufficient to show case (i). The other cases are proved in the same manner. The derivative of P_{21} along the solution of (L) with a positive initial value $x(0) \in \Omega$ is

$$\frac{\dot{P}_{21}(x_2(t), x_1(t))}{P_{21}(x_2(t), x_1(t))} = f_2(S(t)) - f_1(S(t)) = g_{21}(S(t)).$$

Hereafter simply we write $P_{kl}(x_k(t), x_l(t))$ as $P_{kl}(t)$. Since we assume $S(t) < S^*_{12}$ for all $t \geq T_1$, $f_2(S(t)) < f_1(S(t))$ and hence $\dot{P}_{21}(t)/P_{21}(t) < 0$ for all $t \geq T_1$. Recall that $P_{21}(t) \geq 0$. Then there exists $P^*_{21} \geq 0$ such that $P_{21}(t) \to P^*_{21}$ as $t \to \infty$. In the same way, there exists $P^*_{31} \geq 0$ such that $P_{31}(t) \to P^*_{31}$ as $t \to \infty$. To complete the proof, it is sufficient to show that $P^*_{21} = P^*_{31} = 0$ since $P_{kl} = 0$ iff $x_k = 0$ by P-(ii). We claim that $S(t) \to S^*_{12}$ as $t \to \infty$ when $P^*_{21} > 0$. In fact, if not, there exist monotone increasing sequences $\{t_n\}_{n=1}^{\infty} > T_1$ and a sufficiently small positive constant δ such that $g_{21}(S(t_n)) < g_{21}(S^*_{12} - \delta) < 0$ for $n \in \mathbb{N}$. Then immediately we have $\dot{P}_{21}(t_n)/P_{21}(t_n) = g_{21}(S(t_n)) < g_{21}(S^*_{12} - \delta)$ for each $n \in \mathbb{N}$. Since $g_{21}(S^*_{12} - \delta)$ is a strictly negative constant, $P_{21}(t_n) \to 0$ as $n \to \infty$. This implies that $P_{21} \to P^*_{21} = 0$ as $t \to \infty$, but which is a contradiction. Hence $S(t) \to S_{12}^*$ as $t \to \infty$ when $P_{21}^* > 0$. Now we suppose $P_{21}^* > 0$ or $P_{31}^* > 0$. Since $x_1(t)$ is bounded, $x_2(t) \to P_{21}^* x_1(t)$ and $x_3(t) \to P_{31}^* x_1(t)$ as $t \to \infty$. Then $x_1(t) \to x_1^* \equiv (1 - S_{12}^*)/(1 + P_{21}^* + P_{31}^*) > 0$ as $t \to \infty$. This implies that $(x_1^*, P_{21}^* x_1^*, P_{31}^* x_1^*)$ is an equilibrium point of (L). Then the solution with a nonnegative initial value $(x_1^*, P_{21}^* x_1^*, P_{31}^* x_1^*)$ must satisfy

$$x_1^* = x_1^* \exp\left[\int_0^t (f_1(S_{12}^*) - D(s))ds\right]$$

for any positive t, or equivalently, $f_1(S_{12}^*) \equiv D(t)$. This is a contradiction since $D(t)$ is not constant. Hence $P_{21}^* = P_{31}^* = 0$. This completes the proof. □

Note on Proposition 3.2
Assume that $S(t) < S_{23}^*$ for all sufficiently large t. Here we do not necessarily assume that $S_{12}^* < S(t)$. Then either C-(i) or C-(ii) holds for each sufficiently large t. By the assumption, we can show that there exists $P_{32}^* \geq 0$ such that $P_{32}(t) \to P_{32}^*$ as $t \to \infty$. Moreover we can show that $S(t) \to S_{23}^*$ and $x_3(t) \to P_{32}^* x_2(t)$ as $t \to \infty$ when $P_{32}^* > 0$. Although $x_3(t) \to 0$ as $t \to \infty$ both in C-(i) and C-(ii), in this situation, there might be a *positive* solution $(x_1(t), x_2(t), x_3(t))$ of (L) such that $x_1(t) + x_2(t) + x_3(t) \to 1 - S_{23}^*$ and $x_3(t) \to P_{32}^* x_2(t)$ as $t \to \infty$. It is clear that species x_3 never enjoys competitive advantage as long as $S(t) < S_{23}^*$. However species x_3 still has the possibility to persist. The problem would be more difficult to figure out whether x_3 persists or not in this situation. It leaves for our future consideration.

4 Numerical simulation

Let us show some numerical simulation results which are carried out by using *Mathematica*. Assume that nutrient uptake functions of competing three species take the form of Michaelis-Menten functional response (1.2). The washout rate $D(t)$ is given by

$$D(t) = 1 + d\cos(2\pi t/\omega), \tag{4.1}$$

where d is a positive constant satisfying $0 < d < 1$. Throughout the remainder of this section, parameters and initial values are fixed at the following respective values:

$$\begin{cases} a_1 = 0.018181\ ,\ a_2 = 0.272727\ ,\ a_3 = 0.090909, \\ m_1 = 1.36898\ ,\ m_2 = 1.49733\ ,\ m_3 = 2.13904\ ,\ d = 0.64171, \\ x_1(0) = 0.5\ ,\ x_2(0) = 0.2\ ,\ x_3(0) = 0.3. \end{cases} \tag{P}$$

Note that these values are taken almost equal to the parameters adopted in [7, p.122] and [12, pp. 486–487]. The graph of respective functional response of competing three species with (P) is illustrated on Fig. 2. Note that every species can take competitive advantage since the washout rate varies between

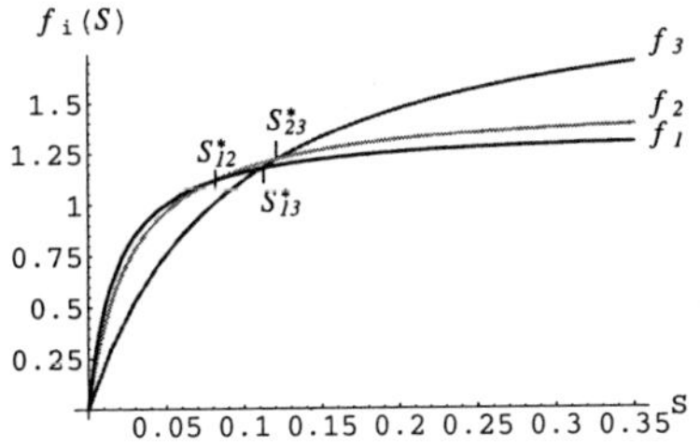

Figure 2: Functional responses of three species for (P)

the range in which every competitor has the chance to be superior to the other competitors in terms of nutrient uptake (note that $d = 0.64171$). The intersection points S^*_{12}, S^*_{13} and S^*_{23} are numerically calculated as $S^*_{12} = 0.078788$, $S^*_{13} = 0.111111$ and $S^*_{23} = 0.121212$.

Let us show some figures which illustrate trajectories of the solution of (L) with (P) for different values of period ω. Figs 3–8 illustrate the time series and the projections into $x_1 - x_2 - x_3$ phase space of trajectories for $\omega = 4$, 6, 8, 12.5, 20 and 50, respectively. Let us denote the end time of numerical simulation by t_{max}. Here $t_{max} = 6000$. The time series of trajectories are shown for $2900 \leq t \leq 3000 < t_{max}$. In the case $\omega = 4$, only x_2 can survive (see Fig. 3). We can confirm that $x_1(t_{max}) \sim 1.6 \times 10^{-21}$ and $x_3(t_{max}) \sim 3.7 \times 10^{-6}$. In the case $\omega = 6$, that is, on Fig. 4, it is observed that x_2 and x_3 survive, while x_1 goes extinct ($x_1(t_{max}) \sim 5.8 \times 10^{-6}$). In the case $\omega = 8$, three species coexist (see Fig. 5). On Figs 6–8, three species still coexist.

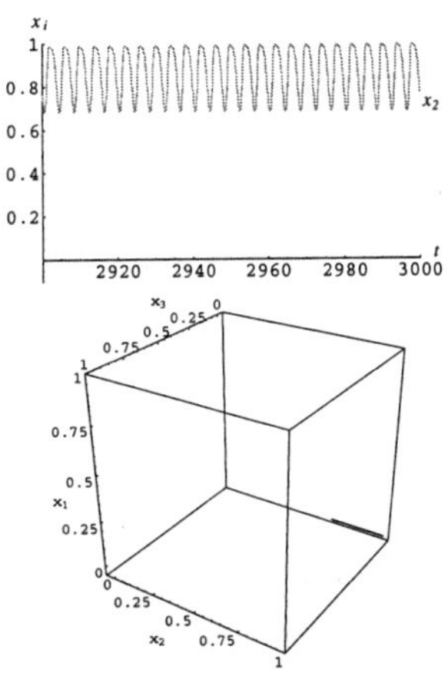

Figure 3: $\omega = 4$.
Only x_2 survives.

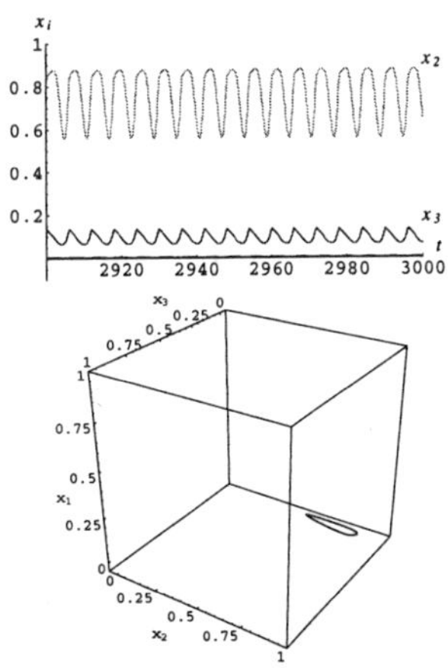

Figure 4: $\omega = 6$.
x_2 and x_3 coexist.

The mechanism of coexistence is intuitively interpreted as follows: Assume

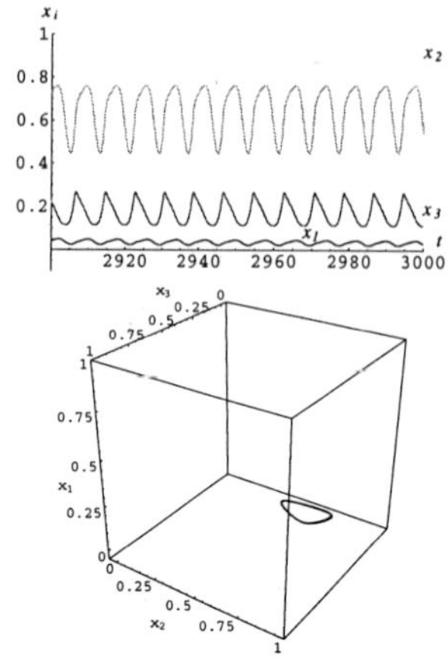

Figure 5: $\omega = 8$.
Three species coexist.

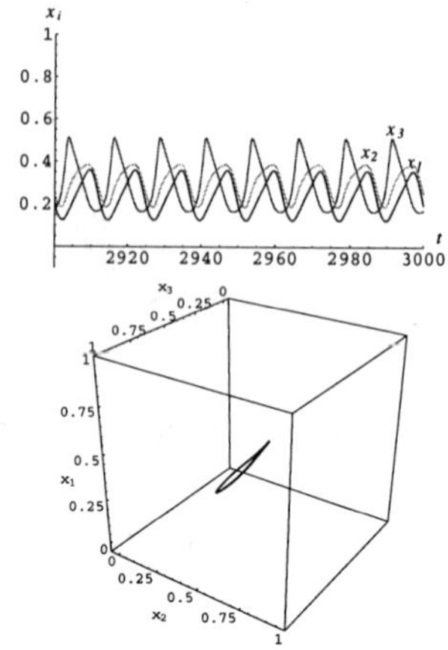

Figure 6: $\omega = 12.5$.
Three species coexist.

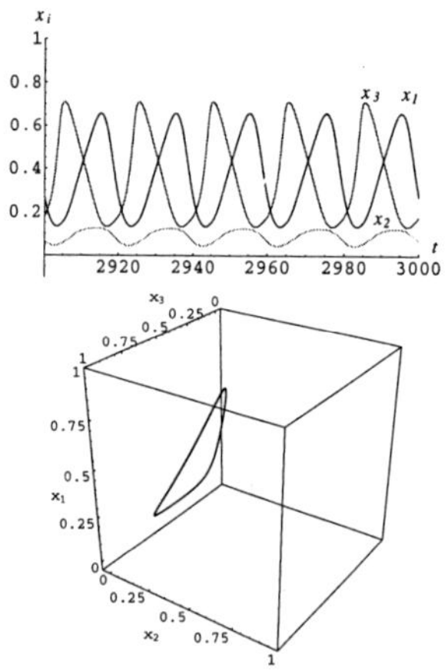

Figure 7: $\omega = 20$.
Amplitude grows large.

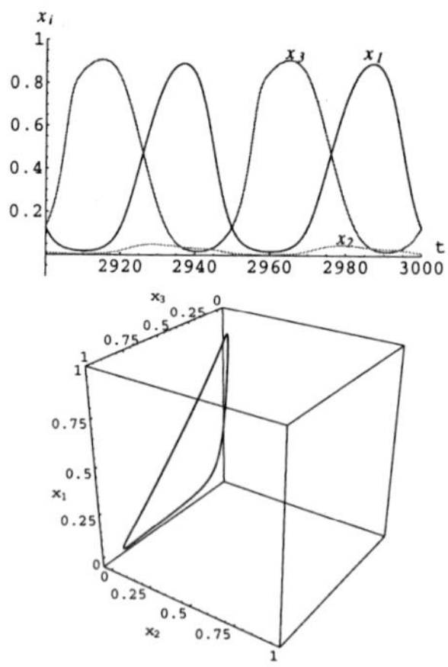

Figure 8: $\omega = 50$.
Three species still coexist.

that $f_3(1) > D_{max} = 1 + d$. First consider the situation that the nutrient $S(t)$ satisfies $S^*_{23} < S(t_1)$ and $f_3(S(t_1)) < D(t_1)$ at some t_1. Then all of x_1, x_2 and x_3 decrease at their respective exponential rates as long as t satisfies the relation $f_3(S^*_{23}) < f_3(S(t)) < D(t)$. If this relation holds true for all $t > t_1$, $S(t) \to 1$ as $t \to \infty$, but which contradicts with the assumption $f_3(1) > D_{max}$. Hence within a finite time $t_2 > t_1$, we have $f_3(S^*_{23}) < D(t_2) < f_3(S(t_2))$. Then according to the proof of Proposition 3.2 – (iii), $x_1(t_2)$ and $x_2(t_2)$ still decreases at their respective exponential rates, while $x_3(t_2)$ increases at some exponential rate. Note that $S(t)$ decreases as $x_3(t)$ increases. If the increase of $x_3(t)$ leads the inequality $S^*_{12} < S(t_3) < S^*_{23}$ for some $t_3 > t_2$, Proposition 3.2 – (ii) implies that $x_1(t_3)$ and $x_3(t_3)$ decreases, while $x_2(t_3)$ increases. Further if there exists $t_4 > t_3$ such that $S(t_4) < S^*_{12}$, then $x_2(t_4)$ and $x_3(t_4)$ decreases, while $x_1(t_4)$ increases. Consequently, if $S(t)$ moves in such a way that all species can grow in each dominant interval, the coexistence of three species is possible.

We can see on Figs 3–8, the amplitude of the nutrient (3.3) becomes large as ω increases. In fact, minima of $\bar{S}(t)$ for different values of ω are approximately equal to 0.01 (see Fig. 9), while maxima of $\bar{S}(t)$ increases as ω increases (see Fig. 10). Hence S^*_{12}, S^*_{13} and S^*_{23} belong to the range of $\bar{S}(t)$ and the assumptions of Proposition 3.2 don't hold.

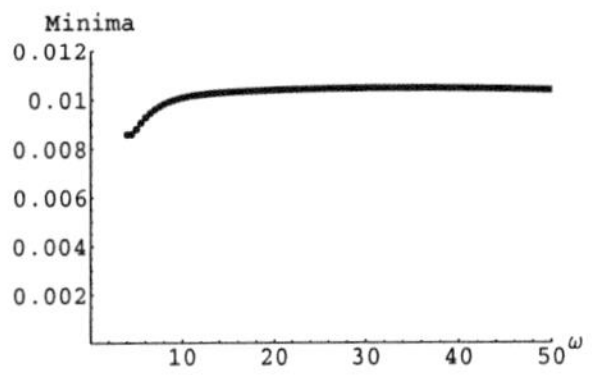

Figure 9: ω vs minima of $\bar{S}(t)$

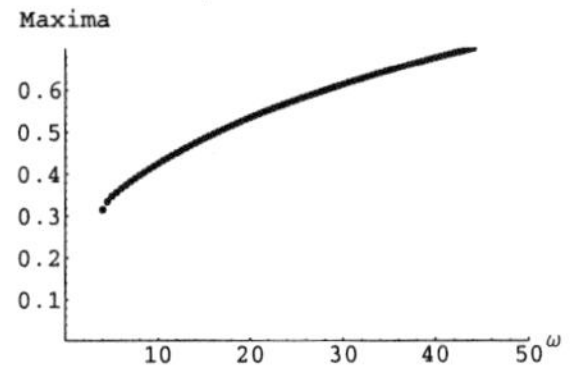

Figure 10: ω vs maxima of $\bar{S}(t)$

Finally let us propose an intuitive interpretation why maxima of $\bar{S}(t)$ increase as ω increases. Since the presence of species inhibits the increase of the nutrient, the timing when $\bar{S}(t)$ attains its maxima (minima) almost corresponds to that of $D(t)$. Note that when $D(t)$ changes slowly, that is, when ω is large, every species can enjoy competitive advantage for a long term. In particular, the long term dominance of x_3 makes the nutrient decrease intensively since x_3 consumes the nutrient with a high rate. Then x_1 takes competitive advantage before $D(t)$ attains its minima and begins to grow as $D(t)$ decreases. Since x_1 dominates the other competitors, x_2 and x_3 cannot grow rather decrease by Proposition 3.2. Soon $D(t)$ becomes large and then x_1 decreases. As all of nutrient, x_2 and x_3 are still low density, the nutrient can increase intensively. The slow change of $D(t)$ also promotes the increase of the nutrient (see Fig. 11).

Let us summarize numerical simulation results:

Remark 4.1. *Three species coexistence occurs as ω increases if $\bar{S}(t)$ has small minima and large maxima. Maxima of $\bar{S}(t)$ are likely to become large as ω in-*

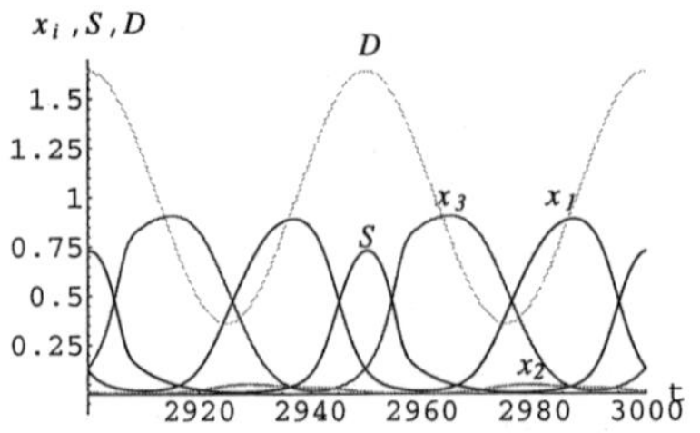

Figure 11: The time series of x_1, x_2, x_3, $S(t)$ and $D(t)$ for (P) with $\omega = 50$.

creases. This suggests that a long term period of periodic washout rate promotes the coexistence of three species competing for a single resource.

5 Conclusions

In this paper, we considered chemostat equations with a periodic washout rate in which three species compete for one limiting nutrient. We introduced an average competition function P_{kl} by which it was shown that positive ω-periodic solutions of system (L) must satisfy the condition (BC). As we remarked in Remark 3.1, (BC) highly restricts the range of amplitude of $\bar{S}(t)$ since $\langle \dot{P}_{kl}/P_{kl} \rangle = 0$ $(1 \le k, l \le 3,\ k < l)$. Hence (BC) would restrict the parameter sets of the equations to be narrow to ensure three species coexistence. In Section 4, it was demonstrated that the number of survivors increases as the period of the washout rate becomes large. In other words, a long term period enhances the coexistence of three species. Since the result obtained in this paper is just analyzed by mathematics partially, further mathematical analysis is necessary. This leaves for our future consideration.

Acknowledgements
The authors thank the referee for his careful and considerable comments which lead to a significant improvement of our manuscript.

References

[1] R. A. Armstrong and R. McGehee, Competitive exclusion, *Am. Nat.* **115** (1980), 151–170.

[2] G. J. Butler, S. B. Hsu and P. Waltman, A mathematical model of the chemostat with periodic washout rate, *SIAM Journal on Applied Mathematics*, **45** (1985), 435–49.

[3] J.P. Grover, "Resource Competition", Population and Community Biology Series, vol. 19, Chapman and Hall, New York, 1997.

[4] V. Hutson, A theorem on average Ljapunov functions, *Monatsh. Math.*, **98** (1984), 267–275.

[5] S. B. Hsu, A competition model for a seasonally fluctuating nutrient, *J. Math. Biol.*, **9** (1980), 115–132.

[6] P. Lenas and S. Pavlou, Periodic, quasi-periodic and chaotic coexistence of two competing microbial populations in a periodically operated chemostat, *Math. Biosci.* **121** (1994), 61–110.

[7] P. Lenas and S. Pavlou, Coexistence of three competing microbial populations in a chemostat with periodically varying dilution rate, *Math. Biosci.* **129** (1995), 111–142.

[8] S. Pilyugin and P. Waltman, Competition in the unstirred chemostat with periodic input and washout, *SIAM Journal on Applied Mathematics*, **59** (1999), 1157–1177.

[9] H. L. Smith, Competitive coexistence in an oscillating chemostat, *SIAM Journal on Applied Mathematics*, **40** (1981), 498–522.

[10] H. L. Smith and P. Waltman, "The Theory of the Chemostat", Cambridge University Press, Cambridge, 1995.

[11] H. R. Thieme, Convergence results and a Poincaré-Bendixson trichotomy for asymptotically autonomous differential equations, *J. Math. Biol.*, **30** (1992), 755–763.

[12] G. S.K. Wolkowicz and X.Q. Zhao, *N*-species competition in a periodic chemostat, *Differential Integral Equations* **11** (1998), 465–491.

Information-theoretic measures of discrete orthogonal polynomials

J.S. Dehesa∗ † ¶, R.J. Yáñez‡¶, R. Álvarez-Nodarse§¶ and P. Sánchez-Moreno†¶

The spreading of the four main families of classical orthogonal polynomials of a discrete variable (Hahn, Meixner, Kravchuk and Charlier), which are exact solutions of the second-order hypergeometric difference equation, is studied by means of some information-theoretic measures of global (variance, Shannon entropy power) and local (Fisher information) character. The variance is calculated in a closed an compact form by means of the three-term recurrence relation of the polynomials. Then, the Cramer-Rao and Heisenberg-Shannon inequalities are used to find rigorous bounds for the other two measures.

Keywords: Orthogonal polynomial in a discrete variable; Information theory; Hahn polynomials; Meixner polynomials; Information measures; Fisher information.

2000 Mathematics Subject Classifications: 33C45, 42C05, 33D45, 94A17.

1 Introduction

In the last few years it is emerging an information theory of the special functions of applied mathematics and mathematical physics [1–6], which underlies the information theory of the quantum-mechanical systems (see e.g. [1, 7–9]) with a solvable or quasi-solvable Schrödinger equation of motion [10, 11]. A main goal of this theory is the study of the spreading of the special functions all over their domain of definition by means of the variance, the Fisher information and the Shannon entropy of their associated Rakhmanov probability density, which are complementary information-theoretic measures. Up until now most efforts have been devoted to the classical orthogonal polynomials of a continuous variable, as it can be seen in the survey [4], updated to 2001, and [6].

∗ Corresponding author E-mail: dehesa@ugr.es

† Departamento de Física Moderna, Universidad de Granada, Granada, Spain.

‡ Departamento de Matemática Aplicada, Universidad de Granada, Granada, Spain.

§ Departamento de Análisis Matemático, Universidad de Sevilla, Sevilla, Spain.

¶ Instituto "Carlos I" de Física Teórica y Computacional, Universidad de Granada, Granada, Spain.

In this paper we want to contribute to extend this theory to the hypergeometric orthogonal polynomials of a discrete variable [12]. These mathematical objects play a central role in the theory of difference analogues of special functions [12]. Moreover, they have been extensively used in numerous branches of mathematics [12] and for mathematical modelling of a great deal of simple [13–16] and complex [17, 18] quantum systems. See also the recent survey [19].

The spreading of the classical discrete orthogonal polynomials (Hahn, Meixner, Kravchuk, Charlier) on their interval of orthogonality (a, b) and the distribution of the associated probability density $\rho(x) = p_n^2(x)\omega(x)$, where $\omega(x)$ is the corresponding weight function, can be most conveniently studied by means of information-theoretic measures [20–22]. The density $\rho_n(x)$ is relevant from both mathematical and physical standpoints. It governs the behavior of the ratio $p_{n+1}(x)/p_n(x)$ when n goes to infinity as E.A. Rakhmanov has shown [23]. Moreover, it characterizes the quantum-mechanical probability density of ground and excited states of numerous physical systems (see e.g. [24]).

Apparently, nothing is known about the information-theoretic properties of the classical discrete orthogonal polynomials apart from the recent work of L. Larsson-Cohn [25] on the asymptotics of Shannon information entropy of the Charlier polynomials, orthogonal with respect to a Poisson distribution. This author used a natural but non-trivial extension of the L^p-norm-based methodology previously developed for the asymptotics of the Shannon information entropy of the polynomials orthogonal with respect to continuous measures [1–3]. Here we plan to calculate the variance of the classical discrete orthogonal polynomials in a closed and compact form by use of their three-term recursion relation, and to bound the Fisher information and the Shannon entropy power by means of some information laws which mutually connect these three information-theoretic measures.

This paper is organized as follows. First, in Section 2 the required information-theoretic measures of a random variable are defined and some of their properties are pointed out. Then, in Section 3, the basic data of the classical discrete orthogonal polynomials and the associated Rakhmanov probability density are described. The results are formulated, proven and discussed in Section 4. Finally, some conclusions and open problems are given.

2 The information measures of a discrete random variable

Let X be a discrete random variable (i.e. a random variable which takes on a finite or countably infinite number N of values) characterized by the probability density $\rho(x_i)$, $i \in \mathbb{N}$ and $x_i \in (a, b) \subseteq \mathbb{R}$, assumed to be normalized to unity so that $\sum_{i=1}^{N} \rho(x_i) = 1$. Its distribution over the interval (a, b) can be studied by means of the following complementary spreading and information-theoretic measures: the variance, the Fisher information [20] and the Shannon entropy power [21].

The variance of the random variable is given by

$$V_X = \sum_i (x_i - \langle x \rangle)^2 \rho(x_i) = \langle x^2 \rangle - \langle x \rangle^2, \tag{1}$$

where $\langle x^m \rangle$ denotes the expectation value of x^m. The Fisher information [20, 26] and the Shannon entropy [21] of X are defined by

$$I_X = \sum_i \frac{[\rho(x_{i+1}) - \rho(x_i)]^2}{\rho(x_i)}, \tag{2}$$

and

$$H_X = -\sum_i \rho(x_i) \log \rho(x_i),$$

respectively. To avoid dimensionality and negativeness problems [22, 27], instead of H_X it is more convenient to use the Shannon entropy power of X defined as

$$P_X = \frac{1}{2\pi e} \exp(2H_X). \tag{3}$$

These three quantities, which have a qualitative different character, quantitatively measure the spreading of the random variable X in different and complementary ways. The variance, commonly known to measure the distribution of the probability mass around the centroid, is a global measure in a sense stronger than the Shannon entropy. This is because the variance gives a large weight to the tails of the density than the logarithmic Shannon functional. The strong dependence of the variance on the tails is not relevant, of course, when they fall off exponentially as for Gaussian or quasi-Gaussian distributions. In contrast to the variance and the Shannon entropy, the Fisher information is very sensitive to the difference of the density at adjacent points of the variable. Indeed, when the density $\rho(x_n)$ undergoes a rearrangement of points x_n, although the shape of the density may drastically change, the value of the entropy power remains constant according to Eq. (3); but the local slope values $\rho(x_{n+1}) - \rho(x_n)$ change drastically and so the sum (2), which defines the Fisher information, will also change strongly.

In addition to the non-negativeness, the variance, the Fisher information and the entropy power have a number of interesting properties [22,27]. Amid them we would like to highlight the scaling law of the variance and the Fisher information, and some information laws or information inequalities which mutually connect two of the aforementioned information measures. First, the variance and the Fisher information transform as

$$I_{aX} = |a|^{-2} I_X, \quad V_{aX} = |a|^2 V_X,$$

when the variable is scaled by a scalar factor $a \in \mathbb{C}$. Second, the variance and Fisher information satisfy the so-called Cramer-Rao inequality [22, 27]

$$I_X V_X \geq \alpha, \tag{4}$$

where $\alpha = 1$ and 0 when the support interval of the random continuous variable is unbounded and bounded respectively. For discrete variables one can only state the non-negativity of the Cramer-Rao product $I_X V_X$. Third, the variance and the entropy power verify the so-called Heisenberg-Shannon inequality [21, 22, 27]

$$P_X^{-1} V_X \geq 1, \tag{5}$$

which shows that the entropy power of a random continuous variable is bounded from above by its variance. Moreover, the finiteness of the variance implies the finiteness of the entropy power; but the converse does not hold. For discrete variables one cannot go further than the non-negativity of the Heisenberg-Shannon ratio.

3 The classical orthogonal polynomials of a discrete variable

The polynomials $\{p_n(x)\}$ of hypergeometric type of a discrete real variable are the polynomial solutions [12] of the second-order difference equation.

$$\sigma(x)\Delta\nabla p_n(x) + \tau(x)\Delta p_n(x) + \lambda_n p_n(x) = 0, \tag{6}$$

where $\sigma(x)$ and $\tau(x)$ are arbitrary polynomials of, at most, second and first degree, respectively; λ_n is the eigenvalue and

$$\Delta f(x) = f(x+1) - f(x); \quad \nabla f(x) = f(x) - f(x-1),$$

are the forward and backward operators, respectively.

It is well-known that these polynomials satisfy a number of interesting properties [12]. First, the orthonormality condition

$$\sum_{x_i=a}^{b-1} p_n(x_i)p_m(x_i)\omega(x_i) = \delta_{m,n}, \tag{7}$$

where the weight function $\omega(x)$ and the interval of orthogonality $(a, b-1)$ are given by

$$\Delta[\sigma(x)\omega(x)] = \tau(x)\omega(x), \quad x^k\sigma(x)\omega(x)|_{a,b} = 0,\ k = 0, 1, \ldots$$

Second, these polynomials fulfil the following three-term recurrence relation

$$xp_n(x) = \alpha_n p_{n+1}(x) + \beta_n p_n(x) + \alpha_{n-1}p_{n-1}(x), \tag{8}$$

where the parameters α_n and β_n can be expressed in terms of the coefficients σ and τ of the difference equation (6). See the excellent monograph of A.F. Nikiforov, S.K. Suslov and V.B. Uvarov [12], where one can find these expressions and a complete list of further difference and symmetry properties of the classical orthogonal polynomials of a discrete variable. Therein we can also find that there are four canonical families of classical orthogonal polynomials of a discrete variable: the Hahn, Meixner, Kravchuk and Charlier polynomials. In Table 1 we have gathered the basic data about these four families, namely the coefficients $\{\sigma, \tau, \lambda_n\}$ of their second-order difference equation, their interval of orthogonality (a, b), their weight function $\omega(x)$, and the parameters $\{\alpha_n, \beta_n\}$ of their recurrence relation (8). See [12] for further details.

There are two important sequences of probability densities which one can naturally associate to the polynomials $\{p_n(x)\}$. A spectral density, the normalized zero counting density, which is closely connected with the nth root asymptotics of p_n. This density has been recently studied [28] from the recurrence relation of the polynomials, by means of its moments around of the origin, which have been determined in a closed form in terms of the parameters and the degree n. Moreover, the asymptotic distribution density of these polynomials have been explicitly calculated [29] by use of the Nevai-Dehesa theorem [30]. There is another density, the structural density

$$\rho_n(x) = p_n^2(x)\omega(x), \tag{9}$$

which we call Rakhmanov density, which controls the asymptotic behavior of the ratio p_{n+1}/p_n [23, 31]. We will analyze the spreading of this density over the interval of orthogonality for all the four families of classical discrete orthogonal polynomials.

4 Information measures of the classical orthogonal polynomials of a discrete variable

Let $\{p_n(x)\}$ with $\deg p_n(x) = n$ denote a sequence of polynomials of a discrete variable orthogonal with respect to the weight function $\omega(x)$ on the real variable (a, b). So, it satisfies the orthogonality condition (7). Let us also assume, as usual, the additional condition $\omega(x) > 0$ for $a < x_i < b - 1$. Then the Rakhmanov probability functions $\rho_n(x) = p_n^2(x)\omega(x)$ are normalized density functions for the discrete random variable X.

These polynomials are usually called hypergeometric-type polynomials, and can be reduced by means of linear changes of the variable to one of the four classical families: Hahn, Meixner, Kravchuk and Charlier. The information measures of the polynomials $p_n(x)$ are defined as the information measures of its associated Rakhmanov density $\rho_n(x)$, which provide a quantitative measure of the spreading/concentration of the probability mass (so, of the polynomials themselves) on the orthogonality interval.

4.1 *The variance*

The variance of the classical orthogonal polynomials is, according to Eq. (1), the sum

$$V(p_n) = \sum_{x_i=a}^{b-1} (x_i - \langle x \rangle_n)^2 \rho_n(x_i) = \langle x^2 \rangle_n - \langle x \rangle_n^2, \tag{10}$$

where

$$\langle x^k \rangle_n := \sum_{x_i=a}^{b-1} x_i^k p_n^2(x_i)\omega(x_i), \text{ for } k = 1 \text{ and } 2.$$

Taking into account the three-term recurrence relation (8) it is not difficult to find that

$$\langle x \rangle_n = \beta_n,$$

and

$$\langle x^2 \rangle_n = \alpha_n^2 + \beta_n^2 + \alpha_{n-1}^2,$$

Table 1. Main data of classical discrete orthogonal polynomials.

$P_n(x)$	Hahn $h_n^{\alpha,\beta}(x;N)$	Meixner $M_n^{\gamma,\mu}(x)$	Kravchuk $K_n^p(x;N)$	Charlier $C_n^\mu(x)$
$\sigma(x)$	$x(N+\alpha-x)$	x	x	x
$\tau(x)$	$(\beta+1)(N-1)-(\alpha+\beta+2)x$	$(\mu-1)x+\mu\gamma$	$\dfrac{Np-x}{1-p}$	$\mu-x$
λ_n	$n(n+\alpha+\beta+1)$	$(1-\mu)n$	$\frac{n}{1-p}$	n
$(a,b-1)$	$[0,N-1]$	$[0,\infty)$	$[0,N]$	$[0,\infty)$
$\omega(x)$	$\dfrac{\Gamma(N+\alpha-x)\Gamma(\beta+x+1)}{\Gamma(N-x)\Gamma(x+1)}$	$\dfrac{\mu^x\Gamma(\gamma+x)}{\Gamma(\gamma)\Gamma(x+1)}$	$\dfrac{N!p^x(1-p)^{N-x}}{\Gamma(N+1-x)\Gamma(x+1)}$	$\dfrac{e^{-\mu}\mu^x}{\Gamma(x+1)}$
	$\alpha,\beta\geq-1\ ,\ n\leq N-1$	$\gamma>0, 0<\mu<1$	$0<p<1,\, n\leq N-1$	$\mu>0$
α_n	$\left(\frac{(n+1)(n+\alpha+1)(n+\beta+1)(n+\alpha+\beta+1)(N-n-1)(N+n+\alpha+\beta+1)}{(2n+\alpha+\beta+1)(2n+\alpha+\beta+2)^2(2n+\alpha+\beta+3)}\right)^{1/2}$	$\frac{(\mu(n+1)(n+\gamma))^{1/2}}{(1-\mu)}$	$((n+1)(N-1)(1-p)p)^{1/2}$	$(\mu(n+1))^{1/2}$
β_n	$\frac{(\beta+1)(N-1)(\alpha+\beta)+n(2N+\alpha-\beta-2)(\alpha+\beta+n+1)}{(\alpha+\beta+2n)(\alpha+\beta+2n+2)}$	$\frac{n(1+\mu)+\mu\gamma}{1-\mu}$	$Np+(1-2p)n$	$n+\mu$

so that the variance $V(p_n)$ is given in terms of the recursion coefficients by

$$V(p_n) = \alpha_n^2 + \alpha_{n-1}^2. \tag{11}$$

This expression together with the explicit values of the coefficients α_n, as given in Table 1, allows one to calculate the following expressions for the variance of the four classical families of orthogonal polynomials of a discrete variable in terms of the degree and the parameters which characterize them.

Charlier polynomials $C_n^{\mu}(x); \mu > 0; [0, \infty)$.

These polynomials, which are orthogonal with respect to the Poisson distribution, have the variance

$$\begin{aligned} V(C_n^{\mu}) &= 2\mu n + \mu \\ &= 2\mu n + O(1), \end{aligned} \tag{12}$$

which shows that the variance depends linearly on both the parameter μ and the degree of the polynomial under consideration.

Meixner polynomials $M_n^{\gamma,\mu}(x); \gamma > 0, 0 < \mu < 1; [0, \infty)$.

These polynomials, which are orthogonal with respect to the Pascal distribution, have the variance

$$\begin{aligned} V(M_n^{\gamma,\mu}) &= \frac{\mu(2n^2 + 2\gamma n + \gamma)}{(\mu - 1)^2} \\ &= \frac{2\mu}{(\mu - 1)^2} n^2 + O(n), \end{aligned}$$

which gives the behavior of the Meixner variance with respect to the parameters (γ, μ) and the degree n of the polynomials. This behavior is graphically shown in Figures 1, 2 and 3, respectively. In Figure 1 we see the linear dependence on γ for μ and n given. The dependence on μ for γ given and various n's is shown in Figure 2. It shows that the Rakhmanov density goes from an extremely narrow function around the centroid to a very smooth quasi-uniform function when μ is increasing along its full range of allowed values for fixed γ and n. This transition from peaked to uniform shapes is very fast when the degree n of the polynomials increases. Finally, in Figure 3 we have plotted the behavior of the variance as a function of n for fixed γ and μ. We observe that this behavior is slowly increasing for low values of μ and γ, indicating that the probability mass around the centroid smoothly increases with the degree; but for large values of μ, the

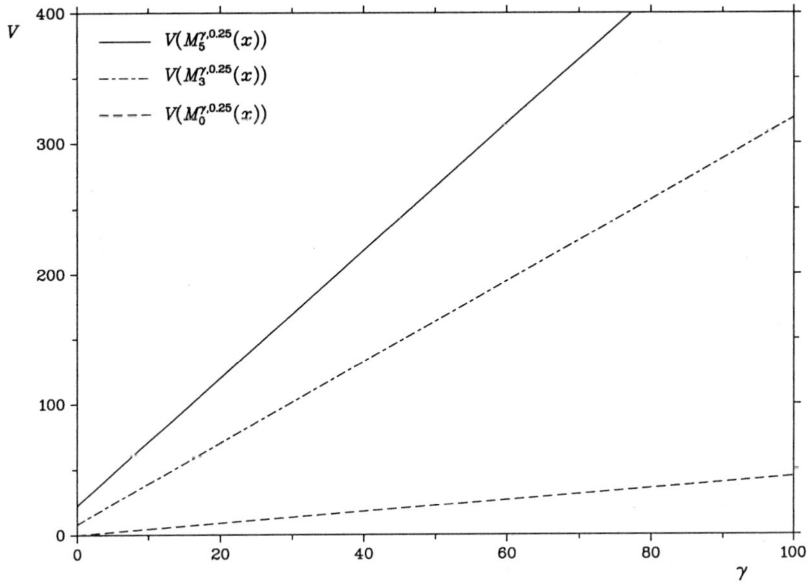

Figure 1. Variance of the Meixner polynomials $M_n^{\gamma,\mu}(x)$ in terms of γ, with $n = 0, 3$ and 5, and $\mu = 0.25$.

enhancement quadratic effect of the variance with the degree is much stronger.

Kravchuk polynomials $K_n^p(x, N); 0 < p < 1; [0, N]$.

The finite sequence of polynomials $\{K_n^p; n = 0, 1, \dots, N-1\}$, which is orthogonal with respect to the binomial distribution, has the variance

$$V(K_n^p) = (N + 2n\,(N - n))\,(1 - p)\,p. \tag{13}$$

Notice that the variance of the polynomials has a concave form centered around $N/2$ as a function of its degree n, so that the probability mass is most smoothly distributed around the centroid for the polynomials in the middle of the sequence. This is shown in Figure 4, where it is observed the behavior of the variance as a function of n for the Kravchuk polynomials with $p = 0.25$ and $N = 50, 150$ and 200. It is interesting to remark that for large values of n and N but with the constant ratio $t = n/N$, $0 \le t < 1$, the variance behaves asymptotically as

$$V(K_n^p) = \frac{2p\,(1-p)\,(1-t)}{t} n^2 + O(n).$$

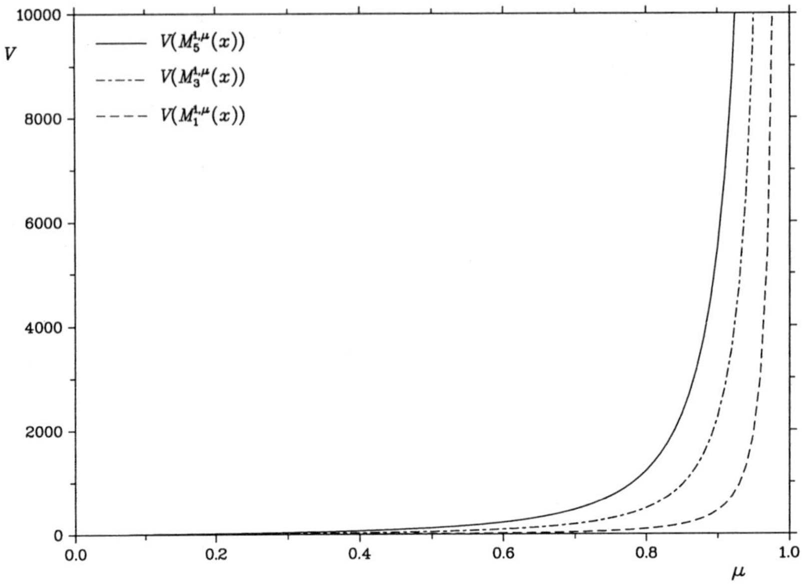

Figure 2. Variance of the Meixner polynomials $M_n^{\gamma,\mu}(x)$ in terms of μ, with $n = 1, 3$ and 5, and $\gamma = 1$.

On the other hand, the shape of the variance (13) as a function of the parameter has again a concave form centered around $p = 1/2$. This is shown if Figure 5 for $N = 50$ and $n = 1, 2$ and 10. We observe that the variance of the polynomials varies from zero at the two extremes of the Kravchuk sequence to the value $[N + 2n(N - n)]/4$ which occurs for $p = 1/2$, indicating that the spreading of the polynomials $K_n^{1/2}(x; N)$ is larger than that of the rest of the sequence.

Hahn polynomials $h_n^{\alpha,\beta}(x); \alpha > -1, \beta > -1; [0, N-1]$.

The finite sequence of the Hahn polynomials $\{h_n^{\alpha,\beta}(x; N); n = 0, 1, \ldots, N-1\}$ is orthogonal with respect to the distribution

$$\omega(x) = \frac{\Gamma(N + \alpha - x)\Gamma(\beta + x + 1)}{\Gamma(N - x)\Gamma(x + 1)},$$

on $[0, N - 1]$ when $\alpha > -1$ and $\beta > -1$. According to Eq. (11) and Table 1, the variance of the Hahn polynomial $h_n^{\alpha,\beta}(x; N)$ is given by

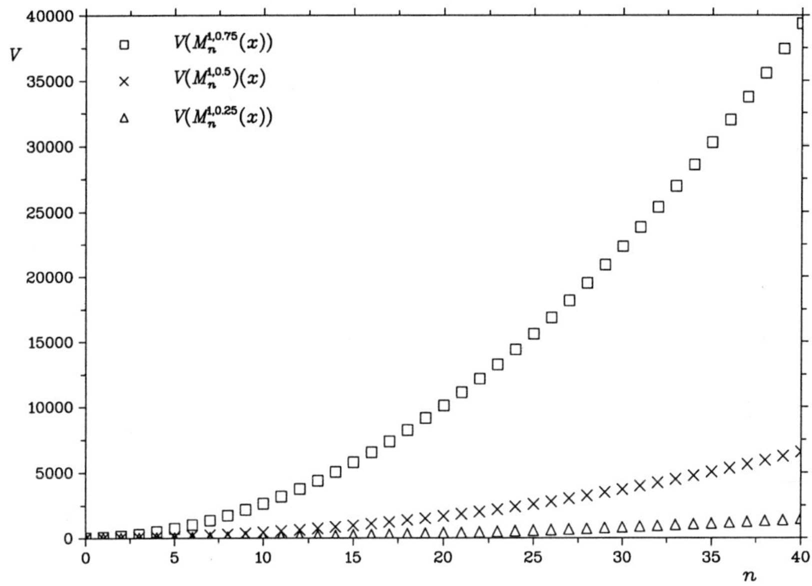

Figure 3. Variance of the Meixner polynomials $M_n^{\gamma,\mu}(x)$ in terms of its degree n, with $\gamma = 1$, and $\mu = 0.25, 0.5$ and 0.75.

$$V(h_n^{\alpha,\beta}) = \frac{n(n+\alpha)(n+\beta)(n+\alpha+\beta)(N-n)(n+N+\alpha+\beta)}{(2n+\alpha+\beta-1)(2n+\alpha+\beta)^2(2n+\alpha+\beta+1)} \tag{14}$$
$$+ \frac{(n+1)(n+\alpha+1)(n+\beta+1)(n+\alpha+\beta+1)(N-n-1)(n+N+\alpha+\beta+1)}{(2n+\alpha+\beta+1)(2n+\alpha+\beta+2)^2(2n+\alpha+\beta+3)}.$$

For the polynomials at the two extremes of the sequence (i.e. when $n = 0$ and $N-1$), the variance have the values

$$\begin{aligned} V(h_0^{\alpha,\beta}) &= \frac{(\alpha+1)(\beta+1)(N-1)(N+\alpha+\beta+1)}{(\alpha+\beta+2)^2(\alpha+\beta+3)} \\ &= \frac{(\alpha+1)(\beta+1)N^2}{(\alpha+\beta+2)^2(\alpha+\beta+3)} + O(N), \end{aligned}$$

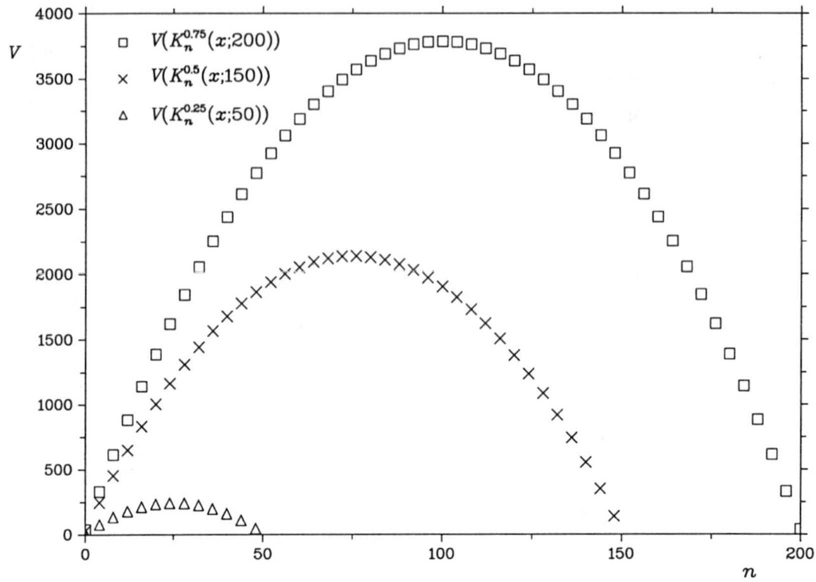

Figure 4. Variance of the Kravchuk polynomials $K_n^p(x;N)$ in terms of its degree n, with $N = 50, 150$ and 200, and $p = 0.25$.

and

$$V(h_{N-1}^{\alpha,\beta}) = \frac{(N-1)(N+\alpha-1)(N+\beta-1)(N+\alpha+\beta-1)}{(2N+\alpha+\beta-3)(2N+\alpha+\beta-2)^2}$$
$$= \frac{N}{8} + O(1).$$

respectively. Moreover, in the asymptotic case when n and N take large values but the ratio $t = n/N$ remains constant the variance of the Hahn polynomials behaves as

$$V(h_n^{\alpha,\beta}) = \frac{1-t^2}{8t^2}n^2 + O(n).$$

A very important subsequence of the Hahn sequence is formed by the Chebyshev polynomials $t_n(x;N)$, which corresponds to the case $\alpha = \beta = 0$; i.e.

$$h_n^{0,0}(x;N) = t_n(x;N).$$

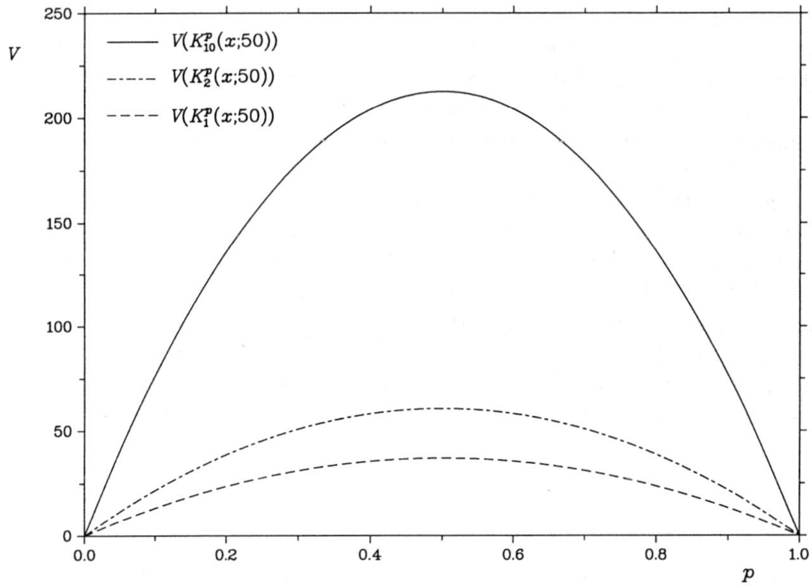

Figure 5. Variance of the Kravchuk polynomials $K_n^p(x;N)$ in terms of its parameter p, with $N=50$, and $n=1,2$ and 10.

Then, according to Eq. (14), we have that the variance of the Chebyshev polynomials is given by

$$V(h_n) = \frac{2n(1+n)(N^2 - n(1+n)) - N^2 + 1}{4(4n(1+n) - 3)}.$$

Moreover, notice that the Chebyshev polynomials with lowest and highest degrees have the following values for the variance

$$V(t_0) = \frac{N^2 - 1}{12} \text{ and } V(t_{N-1}) = \frac{(N-1)^2}{8N - 12},$$

respectively.

Furthermore, to better understand the expression (14) we have studied the behavior of the variance with respect to the parameters α, β and N, and the degree n of the involved polynomials. This is done in Figures 6-8.

Figure 6 shows the variance as a function of the degree n of the polynomial for fixed values of the parameters α, β and N; we have taken $N = 50$ and 100, and $\alpha = \beta = 1$. Then, in Figure 7 we observe the behavior with respect to n of the variance for fixed N (namely, $N = 100$ and pairs $(\alpha, \beta) = (0,0), (1,1)$ and $(2,3)$. Finally, in Figure 8 we plot the dependence of the variance on the parameter α for fixed N (namely, $N = 100$) and various pairs $(n, \beta) = (0,0)$, $(3,0)$, $(3,3)$ and $(5,0)$. The variation with β is similar to the variation with α because of the (α, β) symmetry of the variance. From these figures several observations follow. First, for n, α and β fixed the variance rapidly grows when N is increasing (see Figure 6). Second, for N fixed, the variance as a function of the degree n of the polynomial has an unimodal shape, and its maximum shifts to higher values of n when the values of the pair (α, β) are increasing (see Figure 7). Third, the function $V = V(\alpha)$ shown in Figure 8 illustrates that for polynomials with N fixed, both for β fixed and n increasing and for n fixed and β increasing, the location of the maximum shifts to higher values of α.

4.2 *The Fisher information and the entropy power*

The Fisher information (entropy power) of the classical discrete orthogonal polynomials $p_n(x)$ is the Fisher information (Shannon entropy power) of the associated Rakhmanov probability density $\rho_n(x)$. Then, taking into account Eqs. (2), (3) and (9), one has

$$\begin{aligned} I(p_n) &= \sum_{x_i=a}^{b-1} \frac{[\rho_n(x_{i+1}) - \rho_n(x_i)]^2}{\rho_n(x_i)} \\ &= \sum_{x_i=a}^{b-1} \frac{[p_n^2(x_{i+1})\omega(x_{i+1}) - p_n^2(x_i)\omega(x_i)]^2}{p_n^2(x_i)\omega(x_i)}, \end{aligned}$$

for the Fisher information, and

$$\begin{aligned} P(p_n) &= \frac{1}{2\pi e} \exp\left[-2\sum_{x_i=a}^{b-1} \rho_n(x_i)\log\rho_n(x_i)\right] \\ &= \frac{1}{2\pi e} \exp\left\{-2\sum_{x_i=a}^{b-1} p_n^2(x_i)\omega(x_i)\log\left[p_n^2(x_i)\omega(x_i)\right]\right\}, \end{aligned}$$

for the entropy power. These two quantities have not been calculated up until now. Here we will limit ourselves to bound them by use of two information inequalities, which are two well-known laws in information theory. Indeed, the

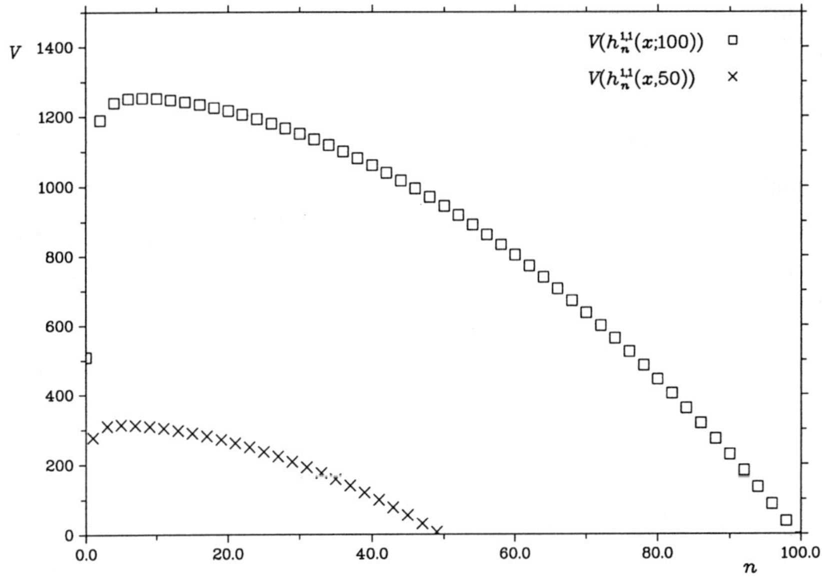

Figure 6. Variance of the Hahn polynomials $h_n^{1,1}(x;N)$ as a function of its degree n when the parameter $N = 50$ and 100.

Cramer-Rao product (4) of the Rakhmanov probability density $\rho_n(x)$ of the classical discrete polynomials orthogonal on an unbounded support interval (e.g. Meixner and Charlier) turns out to be bounded as

$$I(p_n)V(p_n) \geq 2;\ n \geq 2\,.$$

So, the Fisher information is bounded from below as

$$I(M_n^{\gamma,\mu}) \geq \frac{2(\mu-1)^2}{\mu(2n^2+2\gamma n+\gamma)};\ n \geq 2,$$

for Meixner polynomials, and

$$I(C_n^{\mu}) \geq \frac{2}{2\mu n+\mu};\ n \geq 2,$$

for Charlier polynomials.

As well, one obtains that the entropy power of the classical discrete orthog-

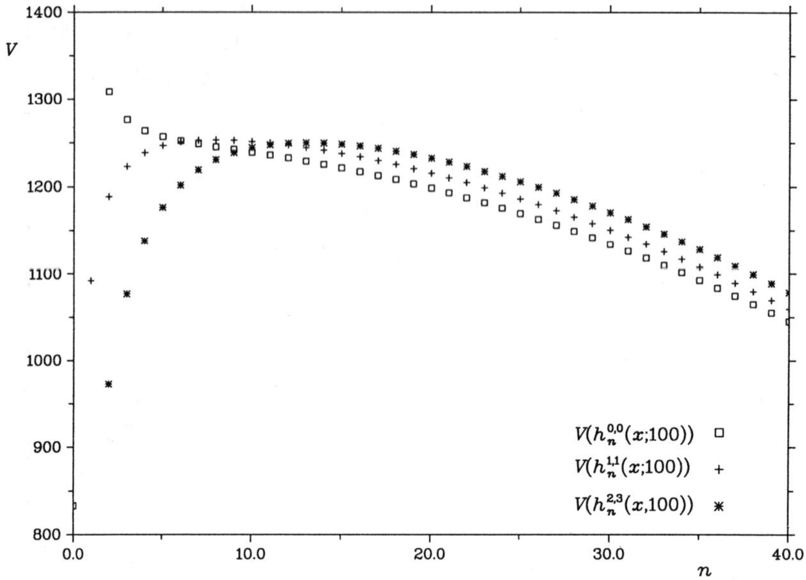

Figure 7. Variance of the Hahn polynomials $h_n^{\alpha,\beta}(x;100)$ as a function of its degree n when $(\alpha,\beta)=(0,0),(1,1)$ and $(2,3)$. Remember that the case $\alpha=\beta=0$ correspond to the Chebyshev polynomials $t_n(x;100)$.

onal polynomials is bounded from above by means of the variance as

$$P(p_n) \leq V(p_n).$$

when one applies the Heisenberg-Shannon inequality (5) to the associated Rakhmanov density (9). So, it is straightforward to find this bound for the four main classes of classical discrete orthogonal polynomials since we have previously computed their variance in this section. For the sake of brevity, let us only give this bound for the Charlier polynomials as

$$P(C_n^{\mu}) \leq 2\mu n + \mu,$$

which is in agreement with the only rigorous information-theoretic result known up to now, the asymptotical value of the entropy power

$$P(C_n^{\mu}) = \frac{2\pi\mu n}{e^3} + O(1),$$

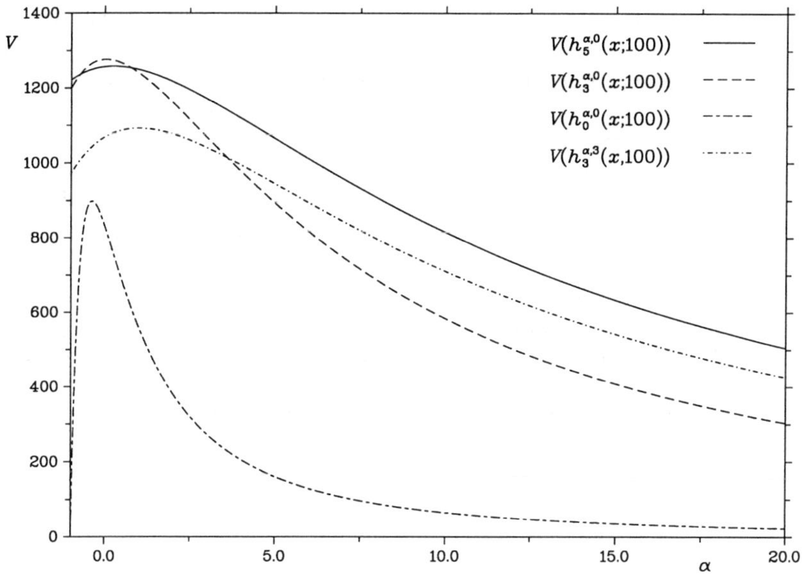

Figure 8. Variance of the Hahn polynomials $h_n^{\alpha,\beta}(x;100)$ as a function of the parameter α when $(n,\beta) = (0,0)$, $(3,0)$, $(5,0)$ and $(3,3)$.

recently found [25].

5 Conclusions

During the last decade there has been an intensive development in the theory of orthogonal polynomials in a discrete variable, whose foundations were laid in 1855 by P.L. Chebyshev in connection with certain problems in mathematical statistics and later on extended by other great names such as A.A. Markov, Stieltjes, Hahn and Rakhmanov among others, to numerous branches of mathematical analysis and applied mathematics. Most efforts have been devoted to two associated density functions, the zero counting density (which governs the nth root asymptotics of the polynomials p_n) and the Rakhmanov probability density (9) (which controls the asymptotics of the ratio p_{n+1}/p_n).

It is our goal to carry out an information-theoretic analysis of the classical orthogonal polynomials in a discrete variable. This analysis requires the determination of the information-theoretic measures (variance, Fisher information and Shannon entropy power) of the Rakhmanov density function for all the four main families of Hahn, Meixner, Kravchuk and Charlier polynomials. These measures quantitatively estimate the spreading of the polynomials both locally (Fisher information) and globally (variance, entropy power) in terms of the parameters and the degree n of the polynomials. So, they are complementary in the sense that they grasp the spreading in different manners.

In this paper we have found the explicit expressions of the variance of all the classical orthogonal polynomials in a discrete variable by means of their three-term recurrence relation. Moreover, we have bounded the Fisher information of the polynomials with an unbounded orthogonality interval (i.e. Meixner and Charlier) and the entropy power for the four main families of discrete orthogonal polynomials in terms of the variance by use of some information inequalities.

Among the related open problems we should mention the explicit computation of the Fisher information (or, at least, to find bounds for these measures in the Hahn and Kravchuk cases), and the entropy power of the four classical families and its asymptotics for the Hahn, Meixner and Kravchuk polynomials (since the Charlier case has been recently done as already mentioned).

Acknowledgements

This work has been partially supported by the MCYT Projects Nos. FIS2005-00973 (JSD, PSM and RJY) and BFM2003-6335-C03-01 (RAN), and by the European Research Network on Constructive Approximation (NeCCA) INTAS-03-51-6637. We belong to the P.A.I. Groups FQM-207 (JSD, PSM and RJY) and FQM-262 (RAN) of the Junta de Andalucía (Spain).

References

[1] J.S. Dehesa, W. van Assche, and R.J. Yáñez. Position and momentum information entropies of the D-dimensional harmonic oscillator and hydrogen atom. *Phys. Rev. A*, 50:3065–3079, 1994.
[2] A.I. Aptekarev, V.S. Buyarov, and J.S. Dehesa. Asymptotic behavior of the L^p-norms and the entropy for general orthogonal polynomials. *Russian Acad. Sci. Sb. Math.*, 82:373–395, 1995.
[3] J.S. Dehesa, W. van Assche, and R.J. Yáñez. Information entropy of classical orthogonal polynomials and their application to the harmonic oscillator and Coulomb potentials. *Methods Appl. Anal.*, 4:91–110, 1997.
[4] J.S. Dehesa, A. Martínez-Finkelshtein, and J. Sánchez-Ruiz. Quantum information entropies and orthogonal polynomials. *J. Comput. Appl. Math.*, 133:23–46, 2001.
[5] J.S. Dehesa, P. Sánchez-Moreno, and R.J. Yáñez. Cramer-Rao information plane of orthogonal hypergeometric polynomials. *J. Comput. Appl. Math.*, to appear, 2005.
[6] J. Sánchez-Ruiz and J.S. Dehesa. Fisher information of orthogonal hypergeometric polynomials. *J. Comput. Appl. Math.*, 182:150–214, 2005.
[7] J.S. Dehesa, A. Martínez Finkelshtein, and V.N. Sorokin. Short-wave asymptotics of the information entropy of a circular membrane. *Int. J. Bifurcation & Chaos*, 12:2387–2392, 2002.
[8] J.S. Dehesa, A. Martínez Finkelshtein, and V.N. Sorokin. Quantum-information entropies for highly-excited sates of single-particle systems with power-type potentials. *Phys. Rev. A*, 66:062109–(1–7), 2002.

[9] J.S. Dehesa, A. Martínez Finkelshtein, and V.N. Sorokin. Asymptotics of information entropies of some Toda-like potentials. *J. Mathem. Phys.*, 44:36–47, 2003.
[10] V.G. Bagrov and D.M. Gitman. *Exact Solutions of Relativistic Wavefunctions*. Kluwer Acad. Publ., Dordrecht, 1990.
[11] A.G. Ushveridze. *Quasi-exactly Solvable Models in Quantum Mechanics*. IOP, Bristol, 1994.
[12] A.F. Nikiforov, S.K. Suslov, and V.B. Uvarov. *Classical Orthogonal Polynomials of a Discrete Variable*. Springer, Berlin, 1991.
[13] S.K. Suslov. The Hahn polynomials in the Coulomb problem. *Sov. J. Nucl. Phys*, 40:79–82, 1984.
[14] M. Lorente. Continuous vs. discrete models for the quantum harmonic oscillator and the hydrogen atom. *Phys. Lett. A*, 285:119–126, 2001.
[15] N.M. Atakishiyev, E.I. Jafarov, S.M. Nagiyev, and K.B. Wolf. Meixner oscillators. *Revista Mexicana de Física*, 44:235–244, 1998.
[16] N.M. Atakisyiyev and S.K. Suslov. Difference analogs of the harmonic oscillator. *Theor. Math. Phys.*, 85:442–444, 1991.
[17] V.A. Savva, V.I. Zelenkov, and A.S. Mazurenko. Orthogonal polynomials in analytical method of solving differential equations describing dynamics of multilevel systems. *Int. Transforms & Special Functions*, 10:299–308, 2000.
[18] D. de Fazio, S. Cavalli, and V. Aquilanti. Orthogonal polynomials of a discrete variable as expansion bases sets in quantum mechanics: hyperquantization algorithm. *Int. J. Quantum Chem.*, 93:91–111, 2003.
[19] R. Álvarez-Nodarse, N.M. Atakishiyev, and R.S. Costas-Santos. Factorization of the hypergeometric-type difference equation on the uniform lattice. *ETNA*, to appear, 2004.
[20] R.A. Fisher. Theory of statistical estimation. *Proc. Cambridge Phil. Soc.*, 22:700–725, 1925.
[21] C.E. Shannon. A mathematical theory of communication. *Bell Syst. Tech. J.*, 27:379–423 and 623–656, 1948.
[22] T.M. Cover and J.A. Thomas. *Elements of Information Theory*. Wiley, N.Y., 1991.
[23] E.A. Rakhmanov. On the asymptotics of the ratio of orthogonal polynomials. *Math. USSR Sb.*, 32:199–213, 1977.
[24] A. Galindo and P. Pascual. *Quantum Mechanics*. Springer, Berlin, 1990.
[25] L. Larsson-Cohn. L^p-norms and information entropies of Charlier polynomials. *J. Approx. Theory*, 117:152–178, 2002.
[26] B.R. Frieden. *Science from Fisher information*. Cambridge University Press, Cambridge, 2004.
[27] A. Dembo, T.M. Cover, and J.A. Thomas. Information theoretic inequalities. *IEEE Trans. Infom. Theory*, 37:1501–1528, 1991.
[28] R. Álvarez-Nodarse and J.S. Dehesa. Distributions of zeros of discrete and continuous polynomials from their recurrence relation. *Appl. Math. Comput.*, 128:167–190, 2002.
[29] K.V. Krasovsky. Asymptotic distribution of zeros of polynomials satisfying difference equations. *J. Comput. Appl. Math.*, 150:57–70, 2003.
[30] P.G. Nevai and J.S. Dehesa. On asymptotic average properties of zeros of orthogonal polynomials. *SIAM J. Math. Anal.*, 10:1184–1192, 1979.
[31] B. Simon. Ratio asymptotics and weak asymptotic measures for orthogonal polynomials on the real line. *J. Aprox. Theory*, 126:198–217, 2004.

LOCAL APPROXIMATION OF INVARIANT FIBER BUNDLES: AN ALGORITHMIC APPROACH

CHRISTIAN PÖTZSCHE
School of Mathematics, Univ. of Minnesota, Minneapolis, MN 55455, USA
E-mail address: poetzsch@math.umn.edu

MARTIN RASMUSSEN
Department of Mathematics, Univ. of Augsburg, D-86135 Augsburg, Germany
E-mail address: martin.rasmussen@math.uni-augsburg.de

Abstract. This paper contains an approach to compute Taylor approximations of invariant manifolds associated with arbitrary fixed reference solutions of nonautonomous difference equations. Our framework is sufficiently general to include, e.g., stable and unstable manifolds of periodic orbits, or classical center-stable/-unstable manifolds corresponding to equilibria. In addition, our focus is to give applicable and quantitative results.

Finally, in the appendix we present a short manual to the Maple program IFB_Comp to calculate Taylor approximations of invariant manifolds.

1. Preliminaries

1.1. **Introduction.** The role of invariant manifolds as a qualitative tool in the modern theory of autonomous dynamical systems cannot be overestimated (cf., e.g., [Shu87]). However, in general, it is difficult to determine invariant manifolds explicitly. Nevertheless, in many situations, it suffices to know only their Taylor approximation up to a certain order, like, e.g., in bifurcation theory or to apply Pliss's center manifold reduction.

Although this seems classical and well-established, even recently some papers on the Taylor approximation of invariant manifolds appeared (cf. [BK98, EvP04]). Beyond such systematic approaches, concrete computations can be found at many places, like, e.g., in the monograph [Kuz95, pp. 151–165, Section 5.4]. They all have in common that one has to solve a (possibly high-dimensional) linear algebraic equation to determine the desired Taylor coefficients, and, what is more important, they apply to the setting of autonomous equations only.

In this paper we present an algorithmic approach to obtain Taylor coefficients of invariant manifolds for nonautonomous difference equations, which is based on the theoretical results developed in [PR05]. The importance of a nonautonomous theory is due to the fact that, e.g., we are able to tackle more realistic problems with time-dependent parameters, or investigate the behavior near nonconstant solutions (cf. Subsection 3.1). Differing from the formal methods developed in [PR05], the present paper is focused on applicability of results: the propositions and theorems are quantitative to a large extend and necessary transformations of difference equations are given in a constructive way, such that they can be applied to given examples without further preparations.

2000 *Mathematics Subject Classification.* Primary 37D10, 39A11; Secondary 37C60.

Research supported by the "Graduiertenkolleg: Nichtlineare Probleme in Analysis, Geometrie und Physik" (GK 283) financed by the DFG and the State of Bavaria.

The appendix contains a brief description of our `Maple` program `IFB_Comp` to calculate Taylor approximations of invariant fiber bundles for nonautonomous difference equations.

1.2. **Notation.** The field of real numbers is denoted by $\mathbb{R}$, the complex numbers by $\mathbb{C}$, the integers by $\mathbb{Z}$, and for given $\kappa \in \mathbb{Z}$ we write $\mathbb{Z}_\kappa^+ := \{k \in \mathbb{Z} : \kappa \leq k\}$, $\mathbb{Z}_\kappa^- := \{k \in \mathbb{Z} : k \leq \kappa\}$, $\mathbb{N} := \mathbb{Z}_1^+$.

For arbitrary $N \in \mathbb{N}$, we consider the N-dimensional Euclidean space $\mathbb{R}^N$ with inner product $\langle x, y\rangle := \sum_{k=1}^N x_k y_k$ and induced norm $\|x\| := \sqrt{\langle x, x\rangle}$ for vectors $x, y \in \mathbb{R}^N$ with components x_k, y_k, respectively. Elements of $\mathbb{R}^N$ are always understood as columns throughout the paper. The *orthogonal complement* $V^\perp$ of a linear subspace $V \subseteq \mathbb{R}^N$ is given by $\{y \in \mathbb{R}^N : \langle x, y\rangle = 0 \text{ for all } x \in V\}$. The r-ball with center of x is denoted as $B_r^N(x) = \{y \in \mathbb{R}^N : \|x - y\| < r\}$; we abbreviate $B_r^N := B_r^N(0)$.

We write $\mathcal{L}(\mathbb{R}^N)$ for the set of real square matrices with N rows, $\mathcal{GL}(\mathbb{R}^N)$ for the subset of regular square matrices, $\mathbb{1}_N$ is the identity matrix and 0_N the zero matrix in $\mathcal{L}(\mathbb{R}^N)$. For $T \in \mathcal{L}(\mathbb{R}^N)$, the linear subspaces $\ker T := \{x \in \mathbb{R}^N : Tx = 0\}$ and $\operatorname{im} T := \{Tx \in \mathbb{R}^N : x \in \mathbb{R}^N\}$ denote the *kernel* and *range* of T, respectively; the determinant of T is denoted by $\det T$. Finally, the *spectrum* of T is given by the set $\sigma(T) := \{\lambda \in \mathbb{C} : \det(\lambda \mathbb{1}_N - T) = 0\}$.

It is important to point out that, for a vector- or matrix-valued sequence x, we use the convenient notation $x'(k) = x(k+1)$. The *k-fiber* of a set $S \subseteq \mathbb{Z} \times \mathbb{R}^N$ is given by $S(k) := \{x \in \mathbb{R}^N : (k, x) \in S\}$.

1.3. **Linear difference equations.** With a matrix sequence $A : \mathbb{Z} \to \mathcal{L}(\mathbb{R}^N)$ we define the *transition matrix* $\Phi(k, \kappa) \in \mathcal{L}(\mathbb{R}^N)$ of the linear difference equation

$$x' = A(k)x \tag{1.1}$$

in $\mathbb{R}^N$ as the mapping

$$\Phi(k, \kappa) := \begin{cases} \mathbb{1}_N & \text{for } k = \kappa \\ A(k-1)\cdots A(\kappa) & \text{for } k > \kappa \end{cases},$$

and if $A(k)$ is invertible for $k < \kappa$, then $\Phi(k, \kappa) := A(k)^{-1} \cdots A(\kappa - 1)^{-1}$.

A projection-valued mapping $P_+ : \mathbb{Z} \to \mathcal{L}(\mathbb{R}^N)$ is called an *invariant projector* of (1.1) if

$$P_+'(k)A(k) = A(k)P_+(k) \quad \text{for } k \in \mathbb{Z} \tag{1.2}$$

holds, and an invariant projector P_+ is denoted as *regular* if

$$A(k)|_{\ker P_+(k)} : \ker P_+(k) \to \ker P_+'(k) \text{ is bijective for all } k \in \mathbb{Z}. \tag{1.3}$$

Then the restriction $\bar{\Phi}(k, \kappa) := \Phi(k, \kappa)|_{\ker P_+(\kappa)} : \ker P_+(\kappa) \to \ker P_+(k)$, $\kappa \leq k$, is a well-defined isomorphism, and we write $\bar{\Phi}(\kappa, k)$ for its inverse. Let $\mathbb{I}$ denote either $\mathbb{Z}$ or $\mathbb{Z}_\kappa^+$. Then the linear difference equation (1.1) is said to possess an *exponential dichotomy* on $\mathbb{I}$ (ED for short) with *rates* $0 < \alpha_+ < \alpha_-$ if there exists a regular invariant projector $P_+ : \mathbb{Z} \to \mathcal{L}(\mathbb{R}^N)$ such that the dichotomy estimates

$$(1.4) \qquad \sup_{l,k\in\mathbb{I}, l<k} \|\Phi(k,l)P_+(l)\| \, \alpha_+^{l-k} < \infty, \qquad \sup_{l,k\in\mathbb{I}, l\leq k} \left\|\bar{\Phi}(l,k)P_-(k)\right\| \alpha_-^{k-l} < \infty$$

are satisfied, where $P_-(k) := \mathbb{1}_N - P_+(k)$ denotes the *complementary projector*. In the following, the symbol $P_\pm$ simultaneously stands for P_+ or P_-, respectively, and we proceed accordingly with our further notation. Hence, the set

$$(1.5) \qquad \mathcal{V}^\pm := \left\{(k,x) \in \mathbb{Z}\times\mathbb{R}^N : x \in \operatorname{im} P_\pm(k)\right\}$$

is invariant w.r.t. (1.1), i.e., its fibers satisfy $\Phi(k,\kappa)\mathcal{V}^\pm(\kappa) \subseteq \mathcal{V}^\pm(k)$ for $\kappa \leq k$.

1.4. **Statement of the problem.** After these preparation we can present our primary objectives. Thereto, let $\tilde{U} \subseteq \mathbb{R}^N$ be a nonempty open convex set and $f : \tilde{U}\times\mathbb{Z} \to \mathbb{R}^N$ be a mapping. We consider the nonautonomous difference equation

$$(1.6) \qquad x' = f(x,k),$$

whose maximal forward solution satisfying the initial condition $x(k_0) = x_0$ is denoted by $\varphi(\cdot, k_0, x_0)$ for $k_0 \in \mathbb{Z}$ and $x_0 \in \tilde{U}$.

Let us assume there exists a fixed reference solution $\nu : \mathbb{Z} \to \tilde{U}$ of (1.6) with $B_r^N(\nu(k)) \subseteq \tilde{U}$ for $k \in \mathbb{Z}$ and some $r > 0$. Typical examples of such reference solutions are equilibria, periodic or homo-/heteroclinic solutions, but we do not restrict ourself to such a situation here. Rather it is our goal to describe the domain of exponential attraction for ν and to provide local approximations of it.

Thereto, we say a solution μ of (1.6) is *exponentially decaying to* ν *on* $\mathbb{Z}_\kappa^\pm$ if μ exists on $\mathbb{Z}_\kappa^\pm$ for some $\kappa \in \mathbb{Z}$ and satisfies

$$\sup_{k\in\mathbb{Z}_\kappa^\pm} \|\mu(k) - \nu(k)\| \, \alpha_\pm^k < \infty$$

for some $\alpha_+ < 1 < \alpha_-$. Our global set-up will be as follows:

Hypothesis. *Let $f : \tilde{U}\times\mathbb{Z} \to \mathbb{R}^N$ be a mapping such that the partial derivatives $D_1 f, \ldots, D_1^m f$ w.r.t. the first variable exist and are continuous for some $m \geq 2$. Moreover, we assume the following:*

(H_1) *The* variational equation

$$(1.7) \qquad x' = D_1 f(\nu(k), k)x$$

possesses an ED on $\mathbb{Z}$ with rates α_+, α_- and invariant projector P_+.

(H_2) *For each $n \in \{2, \ldots, m\}$ there exist reals $K_n \geq 0$ and points $x_n \in \tilde{U}$ such that $\|D_1^n f(x_n, k)\| \leq K_n$ and for each bounded set $\Omega \subseteq \tilde{U}$ there exists a real $K \geq 0$ with $\|D_1^m f(x,k)\| \leq K$ for all $x \in \Omega$ and $k \in \mathbb{Z}$.*

It is a consequence of these assumptions that the nonlinear difference equation (1.6) possesses two locally invariant sets $\mathcal{S}_\nu^+$ and $\mathcal{S}_\nu^-$, which are graphs of functions $s_\nu^\pm(\cdot, k)$ over the affine subspaces $\nu(k) + \mathcal{V}^\pm(k)$. To see this, apply [PR05, Theorem 3.2] to (3.1). More precisely, there exist $\rho > 0$ and mappings

$$s_\nu^\pm : \left\{(x,k) \in \mathbb{R}^N\times\mathbb{Z} : x \in (\nu(k) + \mathcal{V}^+(k)) \cap D_\rho^N(\nu(k))\right\} \to \mathbb{R}^N$$

satisfying $s_\nu^\pm(\nu(k),k) \equiv 0$ on $\mathbb{Z}$, $\lim_{x\to 0} D_1 s_\nu(x+\nu(k),k) = 0$ uniformly in $k \in \mathbb{Z}$,

$$s_\nu^\pm(\nu(k) + P_\pm(k)x, k) \in \mathcal{V}^\mp(k) \quad \text{for } k \in \mathbb{Z},\ x \in B_\rho^N,$$

such that the graphs $\mathcal{S}_\nu^\pm := \{(k, \xi + s_\nu^\pm(\xi,k)) : \xi \in (\nu(k) + \mathcal{V}^\pm(k)) \cap B_\rho^N(\nu(k))\}$ are *locally invariant fiber bundles* (IFBs for short) of (1.6). This means,

$$(k_0, x_0) \in \mathcal{S}_\nu^\pm \quad \Rightarrow \quad (k, \varphi(k; k_0, x_0)) \in \mathcal{S}_\nu^\pm \tag{1.8}$$

holds for $k \geq k_0$ as long as $\varphi(k; k_0, x_0)$ remains in the domain of definition for $s_\nu^\pm(\cdot, k)$. Moreover, we have

$$\mathcal{S}_\nu^+ \cap \mathcal{S}_\nu^- = \{(k, \nu(k)) \in \mathbb{Z}\times\mathbb{R}^N : k \in \mathbb{Z}\}.$$

In this context, $\mathcal{S}_\nu^+$ and $\mathcal{S}_\nu^-$ are denoted as *pseudo-stable* and *pseudo-unstable* fiber bundle of ν, respectively.

To illuminate this rather general framework, we close the section with a dynamic description of the sets $\mathcal{S}_\nu^+$ and $\mathcal{S}_\nu^-$ (cf. [Pöt98, p. 87, Satz 2.4.8]).

Remark 1.1.
(1) Under the assumption that the variational equation (1.7) possesses an ED with $\alpha_+ < \alpha_- = 1$ we have:

- if a solution μ of (1.6) is exponentially decaying to ν on $\mathbb{Z}_\kappa^+$, then there exists a $\kappa^* \in \mathbb{Z}_\kappa^+$ with $(k, \mu(k)) \in \mathcal{S}_\nu^+$ for all $k \in \mathbb{Z}_{\kappa^*}^+$,
- on the other hand, there exists a $\rho_1 \in (0, \rho)$ such that every solution μ with $\mu(\kappa) \in \mathcal{S}_\nu^+(\kappa) \cap B_{\rho_1}^N(\nu(\kappa))$ decays exponentially to ν on $\mathbb{Z}_\kappa^+$,
- if a solution μ exists on $\mathbb{Z}_\kappa^-$ and satisfies $\mu(k) \in B_\rho^N(\nu(k))$ for all $k \in \mathbb{Z}_\kappa^-$, then $((k, \mu(k)) \in \mathcal{S}_\nu^-$ for all $k \in \mathbb{Z}_\kappa^-$,

and $\mathcal{S}_\nu^+$ is denoted the *stable* and $\mathcal{S}_\nu^-$ the *center-unstable fiber bundle* of ν.
(2) Under the assumption that (1.7) has an ED with $1 = \alpha_+ < \alpha_-$ we have:

- if a solution μ of (1.6) is exponentially decaying to ν on $\mathbb{Z}_\kappa^-$, then there exists a $\kappa^* \in \mathbb{Z}_\kappa^-$ with $(k, \mu(k)) \in \mathcal{S}_\nu^-$ for all $k \in \mathbb{Z}_{\kappa^*}^-$,
- on the other hand, there exists a $\rho_1 \in (0, \rho)$ such that every solution μ with $\mu(\kappa) \in \mathcal{S}_\nu^-(\kappa) \cap B_{\rho_1}^N(\nu(\kappa))$ decays exponentially to ν on $\mathbb{Z}_\kappa^-$,
- if a solution μ exists on $\mathbb{Z}_\kappa^+$ and satisfies $\mu(k) \in B_\rho^N(\nu(k))$ for all $k \in \mathbb{Z}_\kappa^+$, then $((k, \mu(k)) \in \mathcal{S}_\nu^+$ for all $k \in \mathbb{Z}_\kappa^+$,

and $\mathcal{S}_\nu^+$ is denoted the *center-stable* and $\mathcal{S}_\nu^-$ the *unstable fiber bundle* of ν.

This terminology corresponds to the autonomous situation of invariant manifolds considered, e.g., in [Shu87]. It is the aim of this paper to obtain local approximations of these sets in form of Taylor expansions.

2. Sufficient Criteria for an Exponential Dichotomy

Even though an ED is a generic property in the class of linear systems with bounded coefficient sequences (cf. [AM96]), it is difficult to verify an assumption like (H_1) for a given nonautonomous equation. This section, however, contains sufficient criteria for exponential dichotomies in certain special cases.

We need some preparations from linear algebra (cf. [HS74, pp. 109–133]). Let $T \in \mathcal{L}(\mathbb{R}^N)$ and $0 < \alpha_+ < \alpha_-$. We say that T possesses an (α_+, α_-)-*spectral decomposition* if the sets

$$\sigma^+ := \{\lambda \in \sigma(T) : |\lambda| \le \alpha_+\}, \qquad \sigma^- := \{\lambda \in \sigma(T) : \alpha_- \le |\lambda|\}$$

are nonempty with $\sigma(T) = \sigma^+ \cup \sigma^-$, i.e., $\sigma(T)$ can be separated by an annulus with center 0 and radii $\alpha_+ < \alpha_-$. Having this at hand, we define

$$V_T^\pm := \bigoplus_{\substack{\lambda \in \sigma^\pm, \\ \Im\lambda = 0}} \ker (T - \lambda \mathbb{1}_N)^N \oplus \bigoplus_{\substack{\lambda \in \sigma^\pm, \\ \Im\lambda > 0}} \ker \left(T^2 - 2\Re\lambda T + |\lambda|^2 \mathbb{1}_N\right)^N$$

and $n_\pm := \dim V_T^\pm$. Let $\{x_1^\pm, \ldots, x_{n_\pm}^\pm\}$ be a basis of $V_T^\pm$. Using the regular matrix $C := (x_1^+, \ldots, x_{n_+}^+, x_1^-, \ldots, x_{n_-}^-) \in \mathcal{L}(\mathbb{R}^N)$ we introduce the projections

$$Q_T^+ := C \begin{pmatrix} \mathbb{1}_{n_+} & \\ & 0_{n_-} \end{pmatrix} C^{-1}, \qquad Q_T^- := C \begin{pmatrix} 0_{n_+} & \\ & \mathbb{1}_{n_-} \end{pmatrix} C^{-1},$$

which are complementary and fulfill $\ker Q_T^\pm = V_T^\mp$, $\operatorname{im} Q_T^\pm = V_T^\pm$.

2.1. **Autonomous equations.** In this subsection we assume that the mapping A in (1.1) does not depend on $k \in \mathbb{Z}$, i.e., we consider an autonomous linear difference equation of the form

$$x' = Ax. \tag{2.1}$$

Here, an eigenvalue λ of $A \in \mathcal{L}(\mathbb{R}^N)$ is said to be *semisimple* if its algebraic and geometric multiplicities coincide.

Proposition 2.1. *Let $0 < \alpha_+ < \alpha_-$ be reals, assume the coefficient matrix A possesses an (α_+, α_-)-spectral decomposition and the eigenvalues of A with modulus α_+ and α_- are semisimple. Then* (1.1) *possesses an ED on $\mathbb{Z}$ with rates α_+, α_- and constant invariant projector Q_A^+.*

Proof. See [Pöt98, p. 25, Satz 1.4.11] and [Kal92, p. 105, Satz 2.1.3.2]. □

2.2. **Periodic equations.** Let $\omega \in \mathbb{N}$. The difference equation (1.1) is said to be ω-*periodic* if $A(k) = A(k + \omega)$ for all $k \in \mathbb{Z}$. A 1-periodic equation (1.1) corresponds to the autonomous case (2.1).

Before stating the subsequent proposition, we note that for all $k \in \mathbb{Z}$ the matrix $M_\omega(k) := \Phi(k+\omega, k)$ has the same eigenvalues as the so-called *monodromy matrix* $M_\omega(0)$ (cf., e.g., [Zha99, p. 51, Theorem 2.8]). They are denoted as *Floquet multipliers* of (1.1). A Floquet theory for periodic difference equations can be found in [Aga92, Section 2.9, pp. 68–71].

Proposition 2.2. *Let $0 < \alpha_+ < \alpha_-$ be reals, assume the monodromy matrix $M_\omega(0)$ possesses an $(\alpha_+^\omega, \alpha_-^\omega)$-spectral decomposition and the Floquet multipliers with modulus α_+^ω and α_-^ω are semisimple. Then* (1.1) *possesses an ED on $\mathbb{Z}$ with α_+, α_- and an ω-periodic invariant projector $P_+(k) := Q_{M_\omega(k)}^+$ for all $k \in \mathbb{Z}$.*

Proof. We first prove that the projector defined by $P_+(k) := Q^+_{M_\omega(k)}$ for $k \in \mathbb{Z}$ is invariant, i.e., satisfies (1.2). Thereto, choose $k \in \mathbb{Z}$ and $x \in \mathbb{R}^N$. We decompose $x = x_1 + x_2$ with $x_1 \in \ker P_+(k)$ and $x_2 \in \operatorname{im} P_+(k)$; thus $x_1 \in V^-_{M_\omega(k)}$. Since the asserted equation is linear we assume w.l.o.g. that there exists a $\lambda \in \sigma^+$ such that $x_1 \in \ker\left((M_\omega(k)^2 - 2\Re\lambda M_\omega(k) + |\lambda|^2 \mathbb{1}_N)\right)^N$. The periodicity of A yields

$$\begin{aligned} &\left(M_\omega(k)^2 - 2\Re\lambda M_\omega(k) + |\lambda|^2 \mathbb{1}_N\right)^N x_1 = 0 \\ \Rightarrow\ & A(k+\omega)\left(M_\omega(k)^2 - 2\Re\lambda M_\omega(k) + |\lambda|^2 \mathbb{1}_N\right)^N x_1 = 0 \\ \Rightarrow\ & \left(M'_\omega(k)^2 - 2\Re\lambda M'_\omega(k) + |\lambda|^2 \mathbb{1}_N\right)^N A(k)x_1 = 0, \end{aligned}$$

hence $A(k)x_1 \in V^-_{M'_\omega(k)} = \ker P'_+(k)$ and we have $P'_+(k)A(k)x_1 = A(k)P_+(k)x_1$. Analogously, $P'_+(k)A(k)x_2 = A(k)P_+(k)x_2$ follows. This shows the invariance of P_+. Using Proposition 2.1 we have

$$\|\Phi(k\omega,\kappa\omega)P_+(\kappa\omega)\| \le \tilde K_+ \alpha_+^{k\omega-\kappa\omega}, \qquad \|\Phi(\kappa\omega,k\omega)P_-(k\omega)\| \le \tilde K_- \alpha_-^{\kappa\omega-k\omega}$$

for $k \ge \kappa$ and certain $\tilde K_\pm \ge 1$. We define

$$\hat K_+ := \max_{i=1}^{k}\max_{j=0}^{i-1} \|\Phi(i,j)P_+(j)\|\alpha_+^{j-i}, \quad \hat K_- := \max_{i=1}^{k}\max_{j=0}^{i-1} \|\Phi(j,i)P_-(i)\|\alpha_-^{i-j},$$

and it follows directly that (1.1) possess an ED on $\mathbb{Z}$ with rates α_+, α_-, bounds $\tilde K_+\hat K_+, \tilde K_-\hat K_-$ (cf. (1.4)) and invariant projector P_+. □

2.3. **Further criteria.** The next result is motivated by [Cop78, p. 70, Lemma 1].

Proposition 2.3. *Let $\kappa \in \mathbb{Z}$, $B : \mathbb{Z} \to \mathcal{L}(\mathbb{R}^N)$ and P_+ denote a regular invariant projector for* (1.1). *Moreover, assume* (1.1) *possesses an ED on $\mathbb{Z}^+_\kappa$ with α_-, α_+, P_+ and that there exists a sequence $(k_n)_{n\in\mathbb{N}}$ in $\mathbb{Z}$, $\lim_{n\to\infty} k_n = \infty$ such that*

$$\lim_{n\to\infty}\sup_{k\in J}\|A(k+k_n) - B(k)\| = 0 \quad \textit{for } J \subset \mathbb{Z} \textit{ finite.}$$

Then the linear difference equation $x' = B(k)x$ possesses an ED on $\mathbb{Z}$ with α_+, α_- and the invariant projector $Q_+(k) := \lim_{n\to\infty} P_+(k+k_n)$.

Proof. For all $n \in \mathbb{N}$, the translated equation $x' = A(k+k_n)x$ possesses the transition matrix $\Phi_n(k,\kappa) = \Phi(k+k_n, \kappa+k_n)$. Furthermore, it satisfies (1.3) with the invariant projector $P^n_+(k) := P_+(k+k_n)$, and due to the dichotomy assumptions there exist constants $K_+, K_- \ge 1$ with

$$\left\|\Phi_n(k,l)P^n_+(l)\right\| \le K_+\alpha_+^{k-l}, \qquad \left\|\bar\Phi_n(l,k)P^n_-(k)\right\| \le K_-\alpha_-^{l-k} \tag{2.2}$$

for $k \ge l \ge \kappa - k_n$. Since $\left\|P^n_+(k)\right\| \le K_+$ for all $k \ge \kappa - k_n$, passing over to a subsequence of $(k_n)_{n\in\mathbb{N}}$ yields the existence of $Q_+(k) := \lim_{n\to\infty} P^n_+(k)$ for all $k \in \mathbb{Z}$. On the other hand, $\Psi(k,l) := \lim_{n\to\infty}\Phi_n(k,l)$ is the transition matrix of $x' = B(k)x$ and taking the limit $n \to \infty$ in (2.2) leads to

$$\|\Psi(k,l)Q_+(l)\| \le K_+\alpha_+^{k-l}, \quad \left\|\bar\Psi(l,k)Q_-(k)\right\| \le K_-\alpha_-^{l-k} \quad \text{for } k \ge l > -\infty.$$

Since invariant projectors for ED on $\mathbb{Z}$ are uniquely determined (cf. [Kal94, p. 12]), one sees that Q_+ does not depend on the chosen subsequence. □

Beyond the above, there exist certain other conditions leading to an ED of equation (1.1) or (1.7). They can be subdivided into three classes using the key words: *Slowly varying coefficients* (cf. [Pöt04b, Corollary 3.6]); *Diagonal dominance* (cf. [Cop78, p. 55–56, Proposition 3] and [Pal77] for ODEs) and *Lyapunov functions* (cf. [Cop78, p. 61, Proposition 2] for ODEs).

3. Transformation of Difference Equations

In this section we describe how a nonautonomous difference equation (1.6) can be brought into a (decoupled) form, such that it is comparatively simple to calculate its IFBs, instead of working with the original system. Precisely, one has to proceed in two steps:

3.1. **Equation of perturbed motion.** Under the transformation $\mathcal{T}_k^1 : x \mapsto x - \nu(k)$ the difference equation (1.6) becomes

$$x' = D_1 f(\nu(k), k)x + f_\nu(x, k) \tag{3.1}$$

with $f_\nu(x,k) := f(x + \nu(k), k) - f(\nu(k), k) - D_1 f(\nu(k), k)x$ defined on $B_\rho^N \times \mathbb{Z}$. Note that (3.1) possesses the trivial solution.

3.2. **Lyapunov transformation.** Now it is our aim to decouple (1.7) without destroying the dynamical features of (3.1). We make use of a Lyapunov transformation (cf. [Pöt98, p. 166, Lemma A.6.1]). Thereto, let $P_+ : \mathbb{Z} \to \mathcal{L}(\mathbb{R}^N)$ be the invariant projector from (H_1) associated with the ED of (1.7). Then, due to the regularity condition (1.3), the fibers $\mathcal{V}^\pm(k)$, $k \in \mathbb{Z}$, possess constant dimensions $n_\pm$ with $n_+ + n_- = N$. For each $k \in \mathbb{Z}$, let $\{x_1(k), \dots, x_{n_+}(k)\}$ be an orthonormal basis of $\operatorname{im} P_+(k)$ and $\{y_1(k), \dots, y_{n_-}(k)\}$ be an orthonormal basis of $(\operatorname{im} P_+(k))^\perp$. Such orthonormal basis can be obtained using a Gram-Schmidt procedure (cf. [Hig96, pp. 376ff]) in lower dimensions.

Setting $C(k) := (x_1(k), \dots, x_{n_+}(k), y_1(k), \dots, y_{n_-}(k)) \in \mathcal{GL}(\mathbb{R}^N)$ yields

$$C(k)^{-1}P_+(k)C(k) = \begin{pmatrix} \mathbb{1}_{n_-} & R(k) \\ & 0_{n_+} \end{pmatrix},$$

and the mapping $\Lambda(k) := C(k)\begin{pmatrix} \mathbb{1}_{n_-} & -R(k) \\ & \mathbb{1}_{n_+} \end{pmatrix}$ is indeed a Lyapunov transformation, since we have

$$\|\Lambda(k)\| \le 2 + \|P_+(k)\|, \qquad \left\|\Lambda(k)^{-1}\right\| \le 1 + \|P_+(k)\| \quad \text{for } k \in \mathbb{Z}$$

(see [Pöt98, p. 28, Definition 1.5.1] for details); note that $P_+ : \mathbb{Z} \to \mathcal{L}(\mathbb{R}^N)$ is bounded due to (1.4). Thus, applying the transformation $\mathcal{T}_k^2 : x \mapsto \Lambda(k)x$ to (3.1) yields a nonautonomous difference equation of the form

$$\begin{aligned} x_+' &= A_+(k)x_+ + F_+(x_+, x_-, k) \\ x_-' &= A_-(k)x_- + F_-(x_+, x_-, k) \end{aligned} \tag{3.2}$$

(see [Pöt98, pp. 29–30, Lemma 1.5.4]) with $A_+ : \mathbb{Z} \to \mathcal{L}(\mathbb{R}^{n_+})$, $A_- : \mathbb{Z} \to \mathcal{GL}(\mathbb{R}^{n_-})$ given by

$$\begin{pmatrix} A_+(k) & \\ & A_-(k) \end{pmatrix} := \Lambda'(k)^{-1} D_1 f(\nu(k), k)\Lambda(k) \quad \text{for } k \in \mathbb{Z}$$

and maps $F_+ : B^{n_+}_{\rho_+} \times B^{n_-}_{\rho_-} \times \mathbb{Z} \to \mathbb{R}^{n_+}$, $F_- : B^{n_+}_{\rho_+} \times B^{n_-}_{\rho_-} \times \mathbb{Z} \to \mathbb{R}^{n_-}$ being m-times continuously differentiable w.r.t. (x_+, x_-) for some $\rho_+, \rho_- > 0$, and defined by

$$\begin{pmatrix} F_+(x_+, x_-, k) \\ F_-(x_+, x_-, k) \end{pmatrix} := \Lambda'(k) f_\nu \left(\Lambda(k)^{-1} \begin{pmatrix} x_+ \\ x_- \end{pmatrix}, k \right).$$

Then the assumptions (H_1) and (H_2) guarantee:

(i) The transition matrices Φ_+ and Φ_- of $x'_+ = A_+(k)x_+$ and $x'_- = A_-(k)x_-$, respectively, satisfy for all $k, l \in \mathbb{Z}$ the estimates

$$\|\Phi_+(k,l)\| \le K_+ \alpha_+^{k-l}, \qquad \|\Phi_-(l,k)\| \le K_- \alpha_-^{l-k} \quad \text{for } k \ge l. \tag{3.3}$$

(ii) We have $(F_+, F_-)(0,0,k) \equiv (0,0)$ on $\mathbb{Z}$ and the partial derivatives satisfy

$$\lim_{(x_+,x_-)\to(0,0)} D_{(1,2)}(F_+, F_-)(x_+, x_-, k) = 0 \quad \text{uniformly in } k \in \mathbb{Z}.$$

Remark 3.1. Let P_+ be ω-periodic. Then it is obvious from the above construction that the mapping Λ inherits the periodicity of P_+ if one chooses the basis of $\operatorname{im} P_+(k)$ and of $(\operatorname{im} P_+(k))^\perp$ accordingly.

4. Invariant Fiber Bundles

In this section we state an existence result for IFBs of the difference equation (3.2) and describe a method to compute Taylor approximations of them.

Proposition 4.1. *Assume (H_1) and (H_2) hold. Then there exist neighborhoods $U_+ \subseteq \mathbb{R}^{n_+}$, $U_- \subseteq \mathbb{R}^{n_-}$ of zero such that:*

(a) *There exists a continuous mapping $s^+ : U_+ \times \mathbb{Z} \to \mathbb{R}^{n_-}$ satisfying:*

(a_1) *Under the* gap condition

$$\alpha_+^m < \alpha_-, \tag{4.1}$$

s^+ is m-times continuously differentiable in the first argument, with $\lim_{\xi\to 0} D_1 s^+(\xi, k) = 0$ uniformly in $k \in \mathbb{Z}$,

(a_2) *the* invariance equation

$$s^+(A_+(\kappa)\xi + F_+(\kappa, \xi, s^+(\xi,\kappa)), \kappa+1) = A_-(\kappa)s^+(\xi,\kappa) + F_-(\xi, s^+(\xi,\kappa), \kappa)$$

holds for $(\xi, \kappa) \in U_+ \times \mathbb{Z}$ with $A_+(\kappa)\xi + F_+(\kappa, \xi, s^+(\xi,\kappa) \in U_+$,

(a_3) *s^+ is ω-periodic in the second argument if (3.1) is ω-periodic,*

(a_4) *its graph $S^+ := \{(\kappa, \xi, s^+(\xi,\kappa)) : \kappa \in \mathbb{Z}, \xi \in U_+\}$ is a pseudo-stable fiber bundle of (3.2) corresponding to its zero solution.*

(b) *There exists a continuous mapping $s^- : U_- \times \mathbb{Z} \to \mathbb{R}^{n_+}$ satisfying:*

(b_1) *Under the* gap condition

$$\alpha_+ < \alpha_-^m, \tag{4.2}$$

s^- is m-times continuously differentiable in the first argument, with $\lim_{\xi\to 0} D_1 s^-(\xi, k) = 0$ uniformly in $k \in \mathbb{Z}$,

(b_2) *the* invariance equation

$$s^-(A_-(\kappa)\xi + F_-(\kappa, \xi, s^-(\xi,\kappa)), \kappa+1) = A_+(\kappa)s^-(\xi,\kappa) + F_+(\xi, s^-(\xi,\kappa), \kappa)$$

holds for $(\xi, \kappa) \in U_- \times \mathbb{Z}$ with $A_-(\kappa)\xi + F_-(\kappa, \xi, s^-(\xi,\kappa)) \in U_-$,

(b_3) *s^- is ω-periodic in the second argument if* (3.1) *is ω-periodic,*

(b_4) *its graph $S^- := \{(\kappa, s^-(\xi,\kappa),\xi) : \kappa \in \mathbb{Z}, \xi \in U_-\}$ is a pseudo-unstable fiber bundle of* (3.2) *corresponding to its zero solution.*

Proof. Using a standard cut-off technique one modifies (3.2) appropriately and applies [PS04, Theorem 4.1]. The periodicity assertion follows from [Aul98, Corollary 4.2] (see also [Pöt04a, Theorem 2.4]). □

Proposition 4.2. *Assume* (H_1) *and* (H_2) *hold. Then the IFBs $\mathcal{S}^\pm$ of* (3.2) *(cf. Proposition 4.1) and $\mathcal{S}^\pm_\nu$ of* (1.6) *are related by*

$$\mathcal{S}^\pm_\nu(k) = \nu(k) + \Lambda(k)^{-1}\mathcal{S}^\pm(k) \quad \text{for } k \in \mathbb{Z}.$$

Proof. This is obvious from the transformations $\mathcal{T}^1_k, \mathcal{T}^2_k$ applied to (1.6) to obtain (3.2) given in Section 3. □

Our final goal is to obtain Taylor approximations of the IFB $\mathcal{S}^\pm_\nu$ for (1.6). It is sufficient to concentrate on the IFB $\mathcal{S}^\pm$ for (3.2), since $\mathcal{S}^\pm_\nu$ and $\mathcal{S}^\pm$ are related by Proposition 4.2.

To deduce such a result, we present a formal approach using Fréchet derivatives (cf. [Lan93, Chapter XIII]) leading to a compact convenient notation. Although partial derivatives have the advantage that our formulas could be implemented instantly on a computer, the resulting expressions turn out to be immense — in particular for higher order derivatives. Yet, some further notation is needed:

Let $k, N, M \in \mathbb{N}$. For an k-tuple of the same vector $x \in \mathbb{R}^N$ we write $x^{(k)} := (x, \dots, x)$. The linear space of symmetric k-linear mappings from $(\mathbb{R}^N)^k$ to $\mathbb{R}^M$ is denoted by $\mathcal{L}_k(\mathbb{R}^N, \mathbb{R}^M)$. With $T \in \mathcal{L}(\mathbb{R}^N)$ and $X \in \mathcal{L}_k(\mathbb{R}^N, \mathbb{R}^M)$ we abbreviate

$$Xx_1 \cdots x_k := X(x_1, \dots, x_k), \qquad X_T x_1 \cdots x_k := X(Tx_1, \dots, Tx_k).$$

Moreover, with given $j, l \in \mathbb{N}$, we write

$$P_j^<(l) := \left\{ (M_1, \dots, M_j) \;\middle|\; \begin{array}{l} M_i \subseteq \{1, \dots, l\} \text{ and } M_i \neq \emptyset \text{ for } i \in \{1, \dots, j\}, \\ M_1 \cup \ldots \cup M_j = \{1, \dots, l\}, \\ M_i \cap M_k = \emptyset \text{ for } i \neq k, i, k \in \{1, \dots, j\}, \\ \max M_i < \max M_{i+1} \text{ for } i \in \{1, \dots, j-1\} \end{array} \right\}$$

for the set of *ordered partitions* of $\{1, \dots, l\}$ with length j and $\#M$ for the cardinality of a finite set $M \subset \mathbb{N}$. For a set $M = \{m_1, \dots, m_k\} \subseteq \{1, \dots, l\}$ we write $Xx_M := Xx_{m_1} \cdots x_{m_k}$ for $k \leq l$ and vectors $x_1, \dots, x_l \in \mathbb{R}^N$.

We are interested in local approximations for the mapping $s^\pm$ from Proposition 4.1. The latter one guarantees under the gap conditions (4.1), (4.2) that $s^\pm(\cdot, k) : U_\pm \to \mathbb{R}^{n_\mp}$, $k \in \mathbb{Z}$, is m-times continuously differentiable and Taylor's theorem (cf. [Lan93, p. 350]) implies the representation

$$s^\pm(x,k) = \sum_{n=2}^{m} \frac{1}{n!} s^\pm_n(k) x^{(n)} + R^\pm(x,k)$$

with coefficient functions $s^\pm_n : \mathbb{Z} \to \mathcal{L}_n(\mathbb{R}^{n_\pm}, \mathbb{R}^{n_\mp})$ given by $s^\pm_n(k) := D^n_1 s^\pm(0,k)$ and a remainder $R^\pm$ satisfying $\lim_{x\to 0} \frac{R^\pm(x,k)}{\|x\|^m} = 0$.

- It is convenient to introduce the function $S^{\pm}: U_{\pm}\times\mathbb{Z} \to \mathbb{R}^N$,

$$S^+(x,k) := \begin{pmatrix} x \\ s^+(x,k) \end{pmatrix}, \qquad S^-(x,k) := \begin{pmatrix} s^-(x,k) \\ x \end{pmatrix}$$

  and its partial derivatives $S_n^{\pm}(k) := D_1^n S^{\pm}(0,k)$.
- We also introduce the function $g^{\pm}(x,k) := A_{\pm}(k)x + F_{\pm}(S^{\pm}(x,k),k)$ with partial derivatives

$$g_1^{\pm}(k)x_1 = A_{\pm}(k)x_1,$$

$$g_n^{\pm}(k)x_1\cdots x_n = \sum_{l=2}^{n} \sum_{(M_1,\dots,M_l)\in P_l^<(n)} D_1^l F_{\pm}(0,0,k) S^{\pm}_{\#M_1}(k)x_{M_1}\cdots S^{\pm}_{\#M_l}(k)x_{M_l}$$

  for $n \in \{2,\dots,m\}$.

Now it is a consequence of [PR05] that the sequence $s_n^{\pm}: \mathbb{Z} \to \mathcal{L}_n(\mathbb{R}^{n_\pm}, \mathbb{R}^{n_\mp})$ is the unique bounded solution of the so-called *homological equation*

$$X'_{A_{\pm}(k)} = A_{\mp}(k)X + H_n^{\pm}(k) \tag{4.3}$$

with the inhomogeneity $H_2^{\pm}(k) := D_1^2 F_{\mp}(0,0,k)$ and

$$\begin{aligned} H_n^{\pm}(k)x_1\cdots x_n :=\ & D_1^n F_{\mp}(0,0,k)x_1\cdots x_n \\ & + \sum_{l=2}^{n-1} \sum_{(M_1,\dots,M_l)\in P_l^<(n)} \big(D_1^l F_{\mp}(0,0,k) S^{\pm}_{\#M_1}(k)x_{M_1}\cdots S^{\pm}_{\#M_l}(k)x_{M_l} \\ & - s_l^{\pm}(k+1) g^{\pm}_{\#M_1}(k)x_{M_1}\cdots g^{\pm}_{\#M_l}(k)x_{M_l}\big) \quad \text{for } x_1,\dots,x_n \in \mathbb{R}^{n_\pm}. \end{aligned}$$

This yields the following

Theorem 4.3. *Assume* (H_1), (H_2) *and*

$$\sup_{k\in\mathbb{Z}} \|D_1^n f(\nu(k),k)\| < \infty \quad \textit{for } n \in \{2,\dots,m\}$$

hold. Then one has:

(a) *Under the gap condition* (4.1) *the mapping* $s^+: U_+\times\mathbb{Z} \to \mathbb{R}^{n_-}$ *from Proposition 4.1(a) possesses the derivatives*

$$D_1^n s^+(0,k) = -\sum_{j=k}^{\infty} \Phi_-(k,j+1) H_n^+(j)_{\Phi_+(j+1,k)} \quad \textit{for } n \in \{2,\dots,m\}, \tag{4.4}$$

(b) *under the gap condition* (4.2) *the mapping* $s^-: U_-\times\mathbb{Z} \to \mathbb{R}^{n_+}$ *from Proposition 4.1(b) possesses the derivatives*

$$D_1^n s^-(0,k) = \sum_{j=-\infty}^{k-1} \Phi_+(k,j+1) H_n^-(j)_{\Phi_-(j+1,k)} \quad \textit{for } n \in \{2,\dots,m\}. \tag{4.5}$$

Proof. See [PR05, Theorem 4.2]. □

Remark 4.4.
(1) To avoid a repetitive computation of the infinite series in (4.4) and (4.5), we recommend to calculate $s_n^\pm(\kappa)$ for some fixed time $\kappa \in \mathbb{Z}$ and then use the homological equation (4.3) to determine subsequent values $s_n^\pm(k)$ recursively for $k > \kappa$.
(2) In case the difference equation (3.1) is ω-periodic for some $\omega \in \mathbb{N}$, the Taylor coefficients $s_n^\pm(k)$ inherit this periodicity for $n \in \{2, \dots, m\}$. Consequently, due to the variation of constants formula applied to the homological equation (4.3) and using $s_n^\pm(k+\omega) = s_n^\pm(k)$, one gets the relations

$$s_n^+(k) = \Phi_-(k, \omega+k) s_n^+(k)_{\Phi_+(\omega+k,k)} - \sum_{i=k}^{\omega+k-1} \Phi_-(k, i+1) H_n^+(i)_{\Phi_+(i,k)},$$

$$s_n^-(k) = \Phi_+(\omega+k, k) s_n^-(k)_{\Phi_-(k,\omega+k)} + \sum_{i=k}^{\omega+k-1} \Phi_+(\omega+k, i+1) H_n^-(i)_{\Phi_-(i,\omega+k)}.$$

For $0 \le k < \omega$ they yield algebraic equations to determine $s_n^\pm(0), \dots, s_n^\pm(\omega-1)$. In addition, these formulas are generalizations of the multilinear Sylvester equations obtained in the autonomous case (see, e.g., [BK98, Theorem 2.4]).

While the infinite series (4.4), (4.5) to determine the partial derivatives in Proposition 4.1 provide an analytical solution of our problem, they seem to be of restricted practical use due to the limit process involved. However, it is possible to obtain an a priori error estimate:

Corollary 4.1 (error estimates). *Choose a real* $\gamma_n^\pm > \sup_{k\in\mathbb{Z}} \|H_n^\pm(k)\|$, *let* $\varepsilon > 0$ *be arbitrary and* $k, K \in \mathbb{Z}$. *Then, for finite approximations to the series* (4.4) *and* (4.5), *the following holds:*

(a) *In case* $K - k > \log_{\frac{\alpha_-}{\alpha_+^n}} \left(\frac{K_+^n K_- \gamma_n^+}{\varepsilon(\alpha_- - \alpha_+^n)}\right)$ *one has*

$$\left\| -\sum_{j=k}^{K} \Phi_-(k, j+1) H_n^+(j)_{\Phi_+(j,k)} - D_1^n s^+(0,k) \right\| < \varepsilon,$$

(b) *in case* $k - K > \log_{\frac{\alpha_-^n}{\alpha_+}} \left(\frac{K_+ K_-^n \gamma_n^-}{\varepsilon(\alpha_-^n - \alpha_+)}\right)$ *one has*

$$\left\| \sum_{j=K}^{k-1} \Phi_+(k, j+1) H_n^-(j)_{\Phi_-(j,k)} - D_1^n s^-(0,k) \right\| < \varepsilon,$$

with

$$K_+ := \sup_{l \le k} \|\Phi(k,l) P_+(l)\| \, \alpha_+^{l-k}, \qquad K_- := \sup_{l \le k} \left\|\bar{\Phi}(l,k) P_-(k)\right\| \alpha_-^{k-l}.$$

Proof. See [PR05, Corollary 4.1]. □

5. Examples

This section contains two examples how to apply the results above. While the first example is more on a demonstration level, the second one deals with a periodic problem. Precisely, we calculate a 4th order Taylor approximation for the stable and unstable manifolds corresponding to a hyperbolic 2-periodic orbit of the Henon map.

Example 5.1. Consider the following nonautonomous difference equation describing a Flour beetle population (cf., e.g., [CD95])

$$\begin{aligned} x_1' &= b x_3 e^{-c_1(k)x_3 - c_2(k)x_1} \\ x_2' &= (1-\mu_1)x_1 \\ x_3' &= x_2 e^{-c_3(k)x_3} + (1-\mu_2)x_3 \end{aligned} \tag{5.1}$$

with parameters $b > 0$, $\mu_1, \mu_2 \in (0,1)$ and bounded sequences $c_1, c_2, c_3 : \mathbb{Z} \to (0,\infty)$. The linearization in $(0,0,0)$ has a real eigenvalue $\rho \in (1-\mu_2, \infty)$ and a complex-conjugated pair $\lambda_{1/2}$ satisfying $|\lambda_{1/2}| < \rho$. Hence, we have a 2-dimensional pseudo-stable and a 1-dimensional pseudo-unstable fiber bundle.

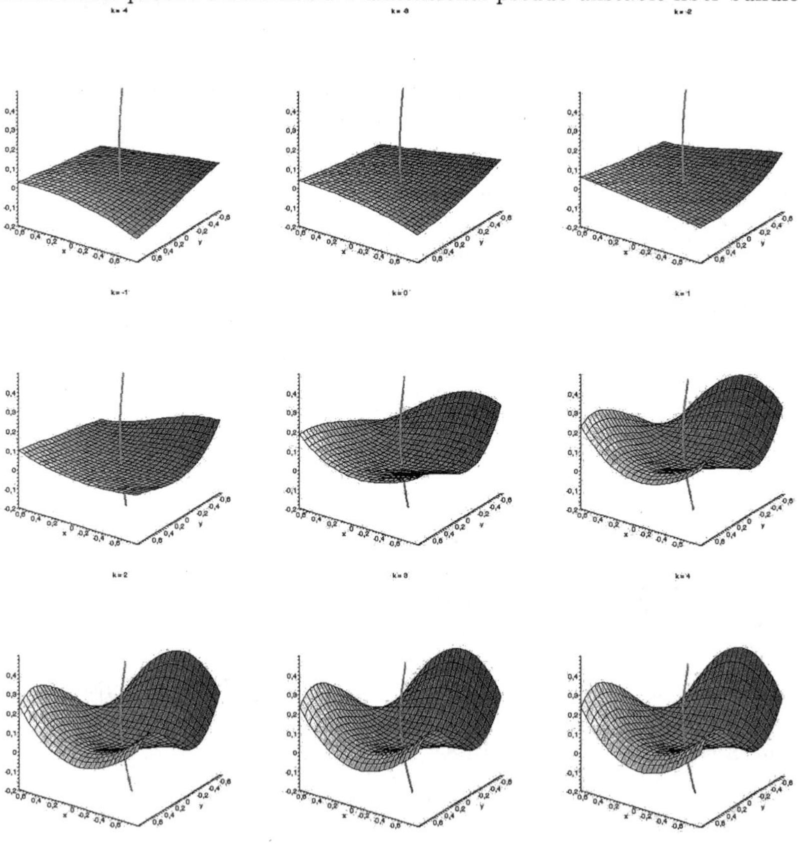

FIGURE 1. Stable and center-unstable fiber bundle corresponding to the zero solution for the flour beetle model (5.1), $k \in \{-4, \ldots, 4\}$

For our numerical calculations we fix the parameters $b := 0.65$, $\mu_1 := 0.11$, $\mu_2 := 0.58$ and set $c_1(k) := 0.92 + 0.45 \arctan(k)$, $c_2(k) := 0.9 + 0.13 \arctan(k)$, $c_3(k) := 0.18 + 0.06 \arctan(k)$. This yields the eigenvalues $-0.26 \pm 0.67i$, 1. If we apply the transformation $\mathcal{T}_k^2$ to (5.1), then the corresponding IFBs of the transformed system can be found in Figure 1.

Example 5.2. Consider the Henon map

$$\begin{aligned} x_1' &= 1 + x_2 - a x_1^2 \\ x_2' &= b x_1 \end{aligned} \tag{5.2}$$

with parameters $a := \frac{7}{5}$, $b := \frac{3}{10}$. We are going to study its behavior close to the 2-periodic solution $\nu(k) = (\frac{1}{4} + (-1)^k \frac{\sqrt{413}}{28}, \frac{3}{40} - 3(-1)^k \frac{\sqrt{413}}{280})$. The corresponding equation of perturbed motion is given by

$$x' = \begin{pmatrix} -\frac{7}{10} - (-1)^k \frac{\sqrt{413}}{10} & 1 \\ \frac{3}{10} & 0 \end{pmatrix} x + \begin{pmatrix} -\frac{7}{5} x_1^2 \\ 0 \end{pmatrix} \tag{5.3}$$

with a 2-periodic linear part. Then its monodromy matrix reads as $\Phi(2,0) = \begin{pmatrix} -\frac{167}{50} & -\frac{7}{10} + \frac{\sqrt{413}}{10} \\ -\frac{21}{100} - \frac{3\sqrt{413}}{100} & \frac{3}{10} \end{pmatrix}$ and the Floquet multipliers turn out to be $-\frac{38}{25} \pm \frac{\sqrt{5551}}{50}$. Due to Proposition 2.2 we obtain an ED on $\mathbb{Z}$ for the linear part of equation (5.3) with $\alpha_+ = \frac{1}{5}\sqrt{\frac{\sqrt{5551}}{2} - 38}$, $\alpha_- = \frac{1}{5}\sqrt{38 + \frac{\sqrt{5551}}{2}}$ and invariant projector

$$P_+(k) = \begin{pmatrix} \frac{1}{2} + \frac{\sqrt{5551}}{122} & \frac{5\sqrt{5551}\left(7-(-1)^k\sqrt{413}\right)}{11102} \\ \frac{3\sqrt{5551}(7+(-1)^k\sqrt{413})}{22204} & \frac{1}{2} - \frac{\sqrt{5551}}{122} \end{pmatrix}.$$

This leads to the Lyapunov transformation

$$\Lambda(k) = \frac{1}{\mu(k)} \begin{pmatrix} \lambda_{11}(k) & 1 \\ \lambda_{21}(k) & \lambda_{22}(k) \end{pmatrix}$$

with $\mu(k) := \frac{1}{5}\sqrt{\frac{952}{13} - \frac{33\sqrt{5551}}{52} + (-1)^k \sqrt{59}(\frac{19\sqrt{7}}{13} - \frac{7\sqrt{793}}{52})}$ and

$$\begin{aligned} \lambda_{11}(k) &= \tfrac{82176625\sqrt{5551}}{413557573106} - \tfrac{575236375}{13559264692} + (-1)^k (\tfrac{5177127375\sqrt{46787}}{5376248450378} - \tfrac{82176625\sqrt{413}}{13559264692}), \\ \lambda_{21}(k) &= \tfrac{29501408375\sqrt{7}\sqrt{793}}{21504993801512} + \tfrac{13230436625}{352540881992} + (-1)^k (\tfrac{1725709125\sqrt{46787}}{21504993801512} - \tfrac{246529875\sqrt{413}}{352540881992}) \\ \lambda_{22}(k) &= \tfrac{7}{20} - \tfrac{\sqrt{5551}}{260} + (-1)^k (\tfrac{\sqrt{413}}{20} - \tfrac{\sqrt{46787}}{260}) \end{aligned}$$

To avoid such extensive expressions we switch to a floating point notation from now on, which is sufficient for our numerical purposes. Then the transformed equation (5.3) is given by

$$\begin{aligned} y_1' = &- (0.0593 - (-1)^k 0.1828) y_1 + (0.1256 - (-1)^k 0.1612) y_1^2 \\ &- (0.3835 - (-1)^k 0.2334) y_1 y_2 + (0.0867 - (-1)^k 0.2449) y_2^2 \\ y_2' = &- (0.5950 + (-1)^k 1.8341) y_2 - (0.8491 - (-1)^k 0.5169) y_1^2 \\ &+ (0.7684 - (-1)^k 2.1688) y_1 y_2 - (1.5273 - (-1)^k 0.0515) y_2^2 \end{aligned}$$

and the invariant fiber bundles read as

$$s(x,k) = s_0(x) + (-1)^k s_1(x), \qquad r(x,k) = r_0(x) - (-1)^k r_1(x)$$

with

$$\begin{aligned}
s_0(x) &= 0.4760x^2 - 1.2467x^3 + 3.7690x^4 - 12.5193x^5 + O(x^5),\\
s_1(x) &= 0.6196x^2 - 1.3355x^3 + 3.8381x^4 - 12.5799x^5 + O(x^5),\\
r_0(y) &= 0.0031y^2 + 0.0184y^3 + 0.0106y^4 + 0.0031y^5 + O(y^5),\\
r_1(y) &= 0.0088y^2 + 0.0623y^3 + 0.0231y^4 + 0.0088y^5 + O(y^5).
\end{aligned}$$

The following figure visualizes these invariant fiber bundles.

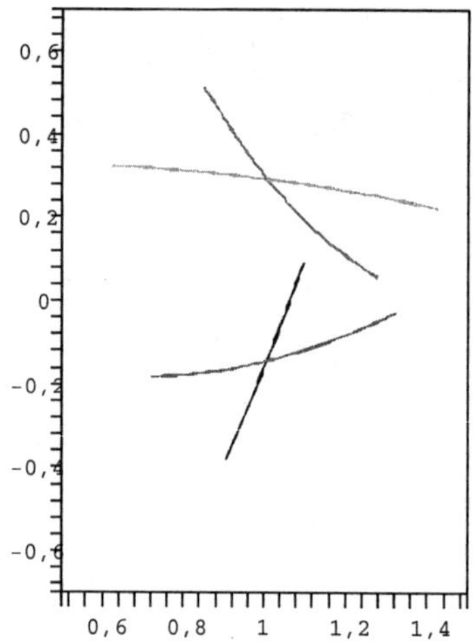

FIGURE 2. Locally stable and unstable fiber bundle corresponding to the 2-periodic orbit $\{\nu(0), \nu(1)\}$ the Henon map (5.2)

APPENDIX: A MANUAL TO IFB_COMP

To approximate the infinite sums (4.4) and (4.5) in Theorem 4.3 we have written the `Maple` program `IFB_Comp`, which can be downloaded from the URL

`http://www.math.uni-augsburg.de/ana/dyn_sys/visual_e.hmtl`

In this appendix we present a few remarks on the usage of `IFB_Comp` assuming the reader is familiar with the computer algebra system `Maple`. One essentially has to proceed in two steps:

(1) Input the system data (dimensions, linear and nonlinear part) as explained in the program.

(2) Then execute the procedures `Main` and `Output`.
- `Main = Main(`k^-`,`k^+`,`p`,`o`,`b`)` is the procedure to compute the Taylor approximation of order o for the k-fibers of the pseudo-stable or pseudo-unstable bundle for $k = k^-, \dots, k^+$. The argument p (standard value: 10) describes how many terms of the infinite sums in (4.4) and (4.5) are computed. To compute the pseudo-stable bundle choose $b := 0$, for the pseudo-unstable bundle set $b := 1$.
- `Output=Output(`b`, `k`, `x^-`, `x^+`, `p`)` is the procedure to plot the k-fiber of the corresponding bundle. As in the procedure `Main`, b stands for type of the bundle. x^- and x^+ determine the area of output which is given by $[x^-, x^+]^N$. The argument p (standard value: 1000) describes the accuracy of the output.

References

[Aga92] R.P. Agarwal, *Difference Equations and Inequalities*, Marcel Dekker, New York, 1992.

[Aul98] B. Aulbach, *The Fundamental Existence Theorem on Invariant Fiber Bundles*, J. Diff. Eqn. Appl. **3** (1998), 501–537.

[AM96] B. Aulbach and N. Van Minh, *The Concept of Spectral Dichotomy for Linear Difference Equations II*, J. Diff. Eqn. Appl. **2** (1996), 251–262.

[BK98] W.-J. Beyn and W. Kleß, *Numerical Taylor Expansion of Invariant Manifolds in Large Dynamical Systems*, Numerische Mathematik **80** (1998), 1–38.

[Cop78] W.A. Coppel, *Dichotomies in Stability Theory*, Lecture Notes in Math. 629, Springer-Verlag, New York, 1978.

[CD95] R.F. Costantino, J.M. Cushing, B. Dennis and R.A. Desharnais, *Experimentally Induced Transitions in the Dynamic Behavior of Insect Populations*, Nature **375** (1995), 227–230.

[EvP04] T. Eirola and J. von Pfaler, *Taylor Expansions for Invariant Manifolds*, Numerische Mathematik **99** (2004), 25–46.

[Hig96] N.J. Higham, *Accuracy and Stability of Numerical Algorithms*, SIAM, Philadelphia, NJ, 1996.

[HS74] M. W. Hirsch and S. Smale, *Differential Equations, Dynamical Systems, and Linear Algebra*, Academic Press, Boston, 1974.

[Kal92] J. Kalkbrenner, *Nichthyperbolische exponentielle Dichotomie* (in german), Diploma Thesis, University of Augsburg, 1992.

[Kal94] ———, *Exponentielle Dichotomie und chaotische Dynamik nichtinvertierbarer Differenzengleichungen* (in german), Ph.D. Thesis, University of Augsburg, 1994.

[Kuz95] Y.A. Kuznetsow, *Elements of Applied Bifurcation Theory*, Applied Mathematical Sciences 112, Springer-Verlag, New York, 1995.

[Lan93] S. Lang, *Real and Functional Analysis*, Graduate Texts in Mathematics 142, Springer-Verlag, New York, 1993.

[Pal77] K.J. Palmer, *A Diagonal Dominance Criterion for Exponential Dichotomy*, Bull. Austral. Math. Soc. **17(3)** (1977), 363–374.

[Pöt98] C. Pötzsche, *Nichtautonome Differenzengleichungen mit invarianten und stationären Mannigfaltigkeiten*, Diploma Thesis, University of Augsburg, 1998.

[Pöt04a] ———, *Stability of Center Fiber Bundles for Nonautonomous Difference Equations*, Fields Institute Communications **42** (2004), 295-304.

[Pöt04b] ———, *Exponential Dichotomies of Linear Dynamic Equations on Measure Chains Under Slowly Varying Coefficients*, J. Math. Anal. Appl. **289** (2004), 317–335.

[PR05] C. Pötzsche and M. Rasmussen, *Taylor Approximation of Invariant Fiber Bundles*, Nonlinear Analysis (TMA), **60(7)** (2005), 1303–1330.

[PS04] C. Pötzsche and S. Siegmund, *C^m-Smoothness of Invariant Fiber Bundles*, Topological Methods in Nonlinear Analysis, **24** (2004), 107–146.
[Shu87] M. Shub, *Global Stability of Dynamical Systems*, Springer-Verlag, New York, 1987.
[Zha99] F. Zhang, *Matrix Theory*, Springer-Verlag, New York, 1999.

Necessary and sufficient conditions for oscillation of coupled nonlinear discrete systems

Serena Matucci[a], Pavel Řehák[b] *

[a] *Department of Electronics and Telecommunications*
University of Florence, I-50139 Florence, Italy

[b] *Mathematical Institute*
Academy of Sciences of the Czech Republic
Žižkova 22, CZ61662 Brno, Czech Republic

Abstract

A new result about the oscillation of a system of two coupled second order nonlinear difference equations is presented here. This result is complementary to those of the previous paper [4] by the authors, and leads to a complete characterization of oscillation for this class of systems.

Keywords: coupled nonlinear difference system, (non)oscillatory solution, oscillatory system.

AMS Classification: 39A10, 39A11.

1 Introduction

Consider the nonlinear difference system

$$\begin{aligned} \Delta(r_k \Phi_\alpha(\Delta x_k)) &= -f(k, y_{k+1}), \\ \Delta(q_k \Phi_\beta(\Delta y_k)) &= g(k, x_{k+1}), \end{aligned} \tag{1}$$

*Supported by the Grant No. KJB1019407 of the Grant Agency of ASCR and by the Grant No. 201/04/0580 the Czech Grant Agency.

where $\{r_k\}, \{q_k\}$ are real positive sequences defined on $\mathbb{N}$, Δ is the forward difference operator defined by $\Delta x_k = x_{k+1} - x_k$, $\Phi_\lambda(u) = |u|^{\lambda-1} \operatorname{sgn} u$ with $\lambda > 1$, and $f, g : \mathbb{N} \times \mathbb{R} \to \mathbb{R}$ are continuous functions, nondecreasing with respect to the second variable, and such that $uf(k,u) > 0$, $ug(k,u) > 0$ for every $u \neq 0$ and $k \in \mathbb{N}$.

Throughout it is assumed

$$\sum_{k=1}^{\infty} \Phi_{\alpha*}\left(\frac{1}{r_k}\right) = \infty = \sum_{k=1}^{\infty} \Phi_{\beta*}\left(\frac{1}{q_k}\right), \tag{2}$$

where $\alpha*, \beta*$ denote the conjugate numbers of α, β, respectively, i.e. $\alpha* = \alpha/(\alpha-1)$, $\beta* = \beta/(\beta-1)$. Notice that (2) is satisfied if $r_k = q_k = 1$ for every $k \in \mathbb{N}$.

By a *solution* of (1) we mean a vector sequence $(x,y) = (\{x_k\}, \{y_k\})$ satisfying (1) for $k \in \mathbb{N}$. The component x $[y]$ of a solution is called *nonoscillatory* if it is eventually positive or negative. Otherwise it is called *oscillatory.* In view of the sign assumptions on f, g, it is easy to show that both sequences x, y have the same behavior with respect to oscillation, i.e. either x, y are both nonoscillatory or x, y are both oscillatory. Thus a solution (x,y) is said *oscillatory* or *nonoscillatory* according to x, y are oscillatory or nonoscillatory. Finally, system (1) is said to be *oscillatory* if all its solutions are oscillatory.

Investigation of (1) has several motivations. First of all, system (1) can be viewed as a discrete counterpart of a coupled nonlinear differential system with p-laplacian operators which appears in studying spherically symmetric solutions of certain nonlinear elliptic systems; the study of these systems has received an increasing interest in the last years, see for instance [1, 2]. System (1) also covers a wide class of fourth order nonlinear difference equations, intensively studied in the literature, see e.g. [3, 7, 8, 9, 10]. At last, our theory can be understood as an extension of the approach known from the extensive theory of second order scalar nonlinear difference equations (in a formally "self-adjoint" form). Observe that (1) consists of two second order equations. Motivated ourselves in such a way, some kind of the results (conditions) can be expected, even though the raise of the order makes usually problems much more difficult.

System of the form (1) has been considered in [5] and in [6], where some (nonoscillatory) boundary value problems are examined. In [4] we started to study oscillation of (1). The aim of this contribution is to prove the remaining case, which is not treated in [4], in order to obtain a complete characterization of oscillation of (1), from a certain point of view. The results here presented are quite original, and are not just the discretization of some existing continuous ones.

The paper is organized as follows. Since our approach is more or less based on a-priori classification of nonoscillatory solutions, first we discuss this topic along with recalling some of the important properties of our system. Then, for completeness, we present oscillatory results from the previous paper, which will be followed by proving a new (complementary) statement. In the final part, we discuss possible extensions, comparison with other results, and further related topics.

2 Notations and Preliminary Results

System (1) exhibits nice symmetry properties, which will be widely used in the subsequent considerations. Let $\tilde{f}(k,s) = -f(k,-s)$, $\tilde{g}(k,s) = -g(k,-s)$, for every $(k,s) \in \mathbb{N} \times \mathbb{R}$. Clearly, $\tilde{f}$ and $\tilde{g}$ satisfy the same assumptions as f and g, respectively.

Symmetry I. If (u,v) is a solution of (1), then $(-u,-v)$ is a solution of

$$\begin{aligned} \Delta(r_k \Phi_\alpha(\Delta x_k)) &= -\tilde{f}(k, y_{k+1}), \\ \Delta(q_k \Phi_\beta(\Delta y_k)) &= \tilde{g}(k, x_{k+1}). \end{aligned}$$

Notice that the above system has the same structure as (1).

Symmetry II. If (u,v) is a solution of (1), then $(-v,u)$ is a solution of

$$\begin{aligned} \Delta(q_k \Phi_\beta(\Delta x_k)) &= -g(k, y_{k+1}), \\ \Delta(r_k \Phi_\alpha(\Delta y_k)) &= \tilde{f}(k, x_{k+1}). \end{aligned}$$

Observe that the above transformation leads to a system where the roles of the coefficients are somehow interchanged. The resulting system has the same structure as (1).

If (x,y) is a solution of (1), we denote with $x^{[1]}$ and $y^{[1]}$ the quasidifference of x and y, respectively:

$$x_k^{[1]} = r_k \Phi_\alpha(\Delta x_k), \quad y_k^{[1]} = q_k \Phi_\beta(\Delta y_k), \qquad k \in \mathbb{N}.$$

As claimed, both components of a solution have the same behavior with respect to oscillation. Due to the sign condition on the nonlinearities f and g, also the quasidifferences of the components of any nonoscillatory solution are nonoscillatory. Very simple arguments, based on condition (2) and Symmetries I and II, leads to the following statement, which, for historical reasons, can be called a *discrete Kiguradze-type lemma.*

Lemma 1 ([4]). *System* (1) *does not have nonoscillatory solutions satisfying any of the following two conditions*

(i) $x_k y_k > 0,\ x_k x_k^{[1]} < 0$ *eventually,*

(ii) $x_k y_k < 0,\ y_k y_k^{[1]} < 0$ *eventually.*

In view of Lemma 1, any nonoscillatory solution (x, y) of (1) belongs to the one of the following four classes (for large k):

(C1) $x_k y_k > 0, x_k x_k^{[1]} > 0, y_k y_k^{[1]} < 0$; (C2) $x_k y_k > 0, x_k x_k^{[1]} > 0, y_k y_k^{[1]} > 0$;

(C3) $x_k y_k < 0, x_k x_k^{[1]} < 0, y_k y_k^{[1]} > 0$; (C4) $x_k y_k < 0, x_k x_k^{[1]} > 0, y_k y_k^{[1]} > 0$.

The following notations will be useful:

$$\begin{aligned}
\mathcal{S}^{\alpha}_{rf}(C) &= \sum_{k=1}^{\infty} \Phi_{\alpha*}\left(\frac{1}{r_k}\sum_{j=k}^{\infty}|f(j,C)|\right),\\
\mathcal{S}^{\beta}_{qg}(C) &= \sum_{k=1}^{\infty} \Phi_{\beta*}\left(\frac{1}{q_k}\sum_{j=k}^{\infty}|g(j,C)|\right),\\
\mathcal{P}^{\alpha}_{gr}(C) &= \sum_{k=1}^{\infty}\left|g\left(j, C\sum_{j=1}^{k}\Phi_{\alpha*}\left(\frac{1}{r_j}\right)\right)\right|,\\
\mathcal{P}^{\beta}_{fq}(C) &= \sum_{k=1}^{\infty}\left|f\left(j, C\sum_{j=1}^{k}\Phi_{\beta*}\left(\frac{1}{q_j}\right)\right)\right|,
\end{aligned}$$

where C is a nonzero real constant. Taking into account the monotonicity of f and g with respect to the second variable, we have

$$\sum_{k=1}^{\infty}|f(k,C)| = \infty \text{ for } C \neq 0 \implies \mathcal{S}^{\alpha}_{rf}(C) = \infty,\ \mathcal{P}^{\beta}_{fq}(C) = \infty$$

$$\sum_{k=1}^{\infty}|g(k,C)| = \infty \text{ for } C \neq 0 \implies \mathcal{S}^{\beta}_{qg}(C) = \infty,\ \mathcal{P}^{\alpha}_{gr}(C) = \infty.$$

It is not difficult to prove the next four implications. The symmetry arguments play important roles.

Lemma 2 ([4]).

(i) *If* $\mathcal{S}^{\beta}_{qg}(C) = \infty$ *for every* $C \neq 0$, *then there is no solution of class (C1).*

(ii) *If* $\mathcal{P}^{\beta}_{fq}(C) = \infty$ *for every* $C \neq 0$, *then there is no solution of class (C2).*

(iii) *If* $\mathcal{S}^{\alpha}_{rf}(C) = \infty$ *for every* $C \neq 0$, *then there is no solution of class (C3).*

(iv) *If* $\mathcal{P}^{\alpha}_{gr}(C) = \infty$ *for every* $C \neq 0$, *then there is no solution of class (C4).*

3 Main Result

We start this section with recalling the criteria (Theorems 1-4), which have already been proved in [4], in order to provide a complete survey. Moreover, this will enable to show how we proceed on investigation of oscillation, and which cases remain to be examined.

The first simple criterion follows immediately from Lemma 2.

Theorem 1. *If, for every* $C \neq 0$

$$\mathcal{S}^{\alpha}_{rf}(C) = \infty, \ \mathcal{S}^{\beta}_{qg}(C) = \infty, \ \mathcal{P}^{\beta}_{fq}(C) = \infty, \ \mathcal{P}^{\alpha}_{gr}(C) = \infty, \tag{3}$$

then (1) is oscillatory. In particular, (3) *is satisfied if, for every* $C \neq 0$, *it holds* $\sum_{k=1}^{\infty} |f(k, C)| = \sum_{k=1}^{\infty} |g(k, C)| = \infty$.

The next step is to consider the case in which at least one of the series in (3) converges. From now on, we therefore assume the condition

$$\sum_{k=1}^{\infty} |f(k, M)| < \infty, \qquad \sum_{k=1}^{\infty} |g(k, C)| = \infty \tag{4}$$

for some $M \neq 0$ and for all $C \neq 0$. Due to Symmetry II, the opposite case, i.e.,

$$\sum_{k=1}^{\infty} |f(k, C)| = \infty, \qquad \sum_{k=1}^{\infty} |g(k, M)| < \infty \tag{5}$$

does not need to be examined. Indeed, our results then can be simply reformulated by replacing f with g, α with β, and r with q. In view of (4), four cases remain to be examined:

I. $\mathcal{S}_f^\alpha(C) = \infty$, $\mathcal{P}_f(C) = \infty$ for all $C \neq 0$,

II. $\mathcal{S}_f^\alpha(C) = \infty$, $\mathcal{P}_f(M) < \infty$, for all $C \neq 0$ and at least a constant $M \neq 0$,

III. $\mathcal{S}_f^\alpha(M) < \infty$, $\mathcal{P}_f(C) = \infty$, for at least a constant $M \neq 0$ and all $C \neq 0$,

IV. $\mathcal{S}_f^\alpha(M) < \infty$, $\mathcal{P}_f(M) < \infty$, for at least a constant $M \neq 0$.

In Case I, Theorem 1 holds and system (1) is oscillatory without any further condition. Additional assumptions on nonlinearities are required to treat the remaining three cases. To this end we introduce the following concept.

Definition 1. We call (1) *strongly superlinear* [*strongly sublinear*] if $\gamma > 0$ and $\delta > 0$ exist such that $|f(k,u)|/|u|^\gamma$ and $|g(k,u)|/|u|^\delta$ are nondecreasing [nonincreasing] in $|u|$ for each fixed $k \in \mathbb{N}$, and $\gamma\delta > (\alpha-1)(\beta-1)$ [$\gamma\delta < (\alpha-1)(\beta-1)$].

The following theorem deals with Case II. In addition to strong sublinearity, we need a further conditions, which is shown to be necessary for oscillation. Similar results hold for the other cases.

Theorem 2. *Assume* $\mathcal{S}_{qg}^\beta(C) = \mathcal{P}_{gr}^\alpha(C) = \mathcal{S}_{rf}^\alpha(C) = \infty$, $\mathcal{P}_{fq}^\beta(M) < \infty$ *for all* $C \neq 0$ *and a constant* $M \neq 0$. *Let* (1) *be strongly sublinear. Then* (1) *is oscillatory if and only if*

$$\sum_{k=1}^{\infty} \left| f\left(k, \sum_{j=1}^{k} \Phi_{\beta*}\left[\sum_{i=1}^{j-1} g\left(i, \sum_{n=1}^{i} \Phi_{\alpha*}\left(\frac{C}{r_n}\right)\right)\right]\right)\right| = \infty \tag{6}$$

for every $C \neq 0$.

The necessity follows from the fact that, if (6) fails to hold, then (1) has nonoscillatory solutions, as shown in [6]. A similar remark holds for the next theorem, which deals with Case III.

Theorem 3. *Assume* $\mathcal{S}_{qg}^\beta(C) = \mathcal{P}_{gr}^\alpha(C) = \mathcal{P}_{fq}^\beta(C) = \infty$, $\mathcal{S}_{rf}^\alpha(M) < \infty$, *for all* $C \neq 0$ *and a constant* $M \neq 0$. *Let* (1) *be strongly superlinear. Then* (1) *is oscillatory if and only if*

$$\sum_{k=1}^{\infty} \left| \Phi_{\beta*}\left\{\frac{1}{q_k} \sum_{j=k}^{\infty} g\left[j, \sum_{i=j+1}^{\infty} \Phi_{\alpha*}\left(\frac{1}{r_i} \sum_{n=i}^{\infty} f(n,C)\right)\right]\right\}\right| = \infty \tag{7}$$

for every $C \neq 0$.

In order to treat Case IV, additional assumptions on nonlinearities and on the mutual asymptotic behavior of the sequences r and q are required.

Theorem 4. *Assume $\mathcal{S}^{\beta}_{qg}(C) = \mathcal{P}^{\alpha}_{gr}(C)$, $\mathcal{S}^{\alpha}_{rf}(M) < \infty$, $\mathcal{P}^{\beta}_{fq}(M) < \infty$ for all $C \neq 0$ and a constant $M \neq 0$. Let $\alpha, \beta \geq 2$, and let* (1) *be strongly sublinear with $\gamma < 1$ and $\delta \leq 1$. Further suppose that*

$$0 < \liminf_{k\to\infty} \frac{\Phi_{\beta*}(q_k)}{\Phi_{\alpha*}(r_k)} \leq \limsup_{k\to\infty} \frac{\Phi_{\beta*}(q_k)}{\Phi_{\alpha*}(r_k)} < \infty. \tag{8}$$

Then (1) *is oscillatory if and only if* (6) *holds for every $C \neq 0$.*

Comparing Theorems 2 and 4, we can observe that, if $\mathcal{S}^{\alpha}_{rf}(M) < \infty$ for a constant $M \neq 0$, then condition (6) is still necessary and sufficient for oscillation of (1) provided that some restrictions on the values of the indexes $\alpha, \beta, \gamma, \delta$ and on the coefficients r and q are assumed. An analogous result holds also for condition (7) in Theorem 3, as stated in the subsequent theorem, which is a new result. As usual, the proof consists in excluding the existence of solutions of (1) in each of the four classes (C1)-(C4). In particular, the proof that (C2) is empty is quite different from the ideas used in the proofs of the previous theorems [4], and therefore it is developed here with all the details.

Theorem 5. *Assume that $\mathcal{S}^{\beta}_{qg}(C) = \mathcal{P}^{\alpha}_{gr}(C) = \infty$, $\mathcal{S}^{\alpha}_{rf}(M) < \infty$, $\mathcal{P}^{\beta}_{fq}(M) < \infty$ for all $C \neq 0$ and for a constant $M \neq 0$. Let* (1) *be strongly superlinear with $\beta \leq 2$, $\gamma > \alpha - 1$ and $\delta \geq 1$. Further suppose that* (8) *is satisfied. Then* (1) *is oscillatory if and only if* (7) *holds for every $C \neq 0$.*

Proof. We prove only the sufficiency, since the necessity is the same as in the proof of Theorem 3 [4]. In view of Lemma 2, (1) has no solution of the type (C1) and (C4). The proof that (C3) is empty follows from the proof of [4, Theorem 3]. Hence, it is sufficient to prove that (1) has no solution of the type (C2). Suppose, by a contradiction, that such a solution (x, y) exists. In view of symmetries, we may assume $x_k > 0, y_k > 0$ for $k \geq m$, with m a sufficiently large index. Then, from (1), it results that $x^{[1]}$ is eventually decreasing to a nonnegative constant, $y^{[1]}$ is eventually increasing to a positive constant or infinity, and consequently x is eventually increasing to a positive constant or infinity, and y is eventually increasing to infinity. From (1) we therefore have

$$x^{[1]}_k \geq \sum_{j=k}^{\infty} f(j, y_{j+1}) \quad \text{and} \quad y^{[1]}_k \geq \sum_{j=m}^{k-1} g(j, x_{j+1}) \tag{9}$$

for $k \geq m$. Further, from $\Delta x_k = \Phi_{\alpha*}\left(x_k^{[1]}/r_k\right)$, taking into account that $\{x_k^{[1]}\}$ is increasing for $k \geq m$, we have

$$x_k \geq \Phi_{\alpha*}\left(x_{k-1}^{[1]}\right) R_{k-1} \tag{10}$$

for $k \geq m$, where $R_k = \sum_{j=m}^{k} \Phi_{\alpha*}(1/r_j)$. Next we want to find a suitable estimate for Δy. From $\Delta y_k = \Phi_{\beta*}\left(y_k^{[1]}/q_k\right)$, taking into account the second inequality in (9) and the condition $\beta* \geq 2$ (since $\beta \leq 2$), we obtain

$$\Delta y_k \geq \Phi_{\beta*}\left(\frac{1}{q_k}\sum_{j=m}^{k-1} g(j, x_{j+1})\right) \geq C_1 \Phi_{\beta*}\left(\frac{1}{q_k}\right)\sum_{j=m}^{k-1} g(j, x_{j+1}),$$

where C_1 is a suitable positive constant. Since g is strongly superlinear and $\delta \geq 1$, the right-hand side can be rewritten as follows

$$\begin{aligned} C_1 \Phi_{\beta*}\left(\frac{1}{q_k}\right)\sum_{j=m}^{k-1} \frac{g(j, x_{j+1})}{x_{j+1}^{\delta}} x_{j+1}^{\delta} &\geq C_1 \Phi_{\beta*}\left(\frac{1}{q_k}\right)\sum_{j=m}^{k-1} \frac{g(j, x_m)}{x_m^{\delta}} x_{j+1}^{\delta} \\ &\geq C_2 \Phi_{\beta*}\left(\frac{1}{q_k}\right)\sum_{j=m}^{k-1} g(j, x_m) x_{j+1}, \end{aligned}$$

where $C_2 > 0$ is a suitable constant, which exists as x being eventually positive increasing. The subsequent inequalities hold for $k \geq m$, taking into account, respectively, the inequality (10), the fact that $x^{[1]}$ is eventually decreasing, the first inequality in (9), the superlinearity of f, and the fact that y is eventually increasing:

$$\begin{aligned} \Delta y_k &\geq C_2 \Phi_{\beta*}\left(\frac{1}{q_k}\right)\sum_{j=m}^{k-1} g(j, x_m)\Phi_{\alpha*}\left(x_j^{[1]}\right) R_j \\ &\geq C_2 \Phi_{\beta*}\left(\frac{1}{q_k}\right)\Phi_{\alpha*}\left(x_k^{[1]}\right)\sum_{j=m}^{k-1} g(j, x_m) R_j \\ &\geq C_2 \Phi_{\beta*}\left(\frac{1}{q_k}\right)\Phi_{\alpha*}\left(\sum_{n=k}^{\infty} \frac{f(n, y_{n+1})}{y_{n+1}^{\gamma}} y_{n+1}^{\gamma}\right)\sum_{j=m}^{k-1} g(j, x_m) R_j \\ &\geq C_2 \Phi_{\beta*}\left(\frac{1}{q_k}\right)\Phi_{\alpha*}\left(\sum_{n=k}^{\infty} \frac{f(n, y_m)}{y_m^{\gamma}} y_{n+1}^{\gamma}\right)\sum_{j=m}^{k-1} g(j, x_m) R_j \end{aligned}$$

$$\geq C_3 y_{k+1}^{\gamma/(\alpha-1)} \Phi_{\beta*}\left(\frac{1}{q_k}\right) \Phi_{\alpha*}\left(\sum_{n=k}^{\infty} f(n, y_m)\right) \sum_{j=m}^{k-1} g(j, x_m) R_j .$$

Here $C_3 = C_2 y_m^{\gamma/(1-\alpha)}$. Dividing the above inequality by y_{k+1}^{ω}, where $\omega = \gamma/(\alpha-1)$, and summing it from m to ∞, we get

$$\sum_{k=m}^{\infty} C_3 \Phi_{\beta*}\left(\frac{1}{q_k}\right) \Phi_{\alpha*}\left(\sum_{n=k}^{\infty} f(n, y_m)\right) \sum_{j=m}^{k-1} g(j, x_m) R_j \leq \sum_{k=m}^{\infty} \frac{\Delta y_k}{y_{k+1}^{\omega}}. \tag{11}$$

Notice that

$$\sum_{k=m}^{\infty} \Phi_{\beta*}\left(\frac{1}{q_k}\right) \Phi_{\alpha*}\left(\sum_{n=k}^{\infty} f(n, y_m)\right) \sum_{j=m}^{k-1} g(j, x_m) R_j < \infty \tag{12}$$

since

$$\sum_{k=m}^{\infty} \frac{\Delta y_k}{y_{k+1}^{\omega}} \leq \int_{y_m}^{\infty} \frac{dt}{t^{\omega}} < \infty,$$

being $\omega > 1$ and y increasing. Interchanging the order of sums in (12) we get

$$\sum_{j=m}^{\infty} g(j, x_m) R_j \sum_{k=j+1}^{\infty} \Phi_{\beta*}\left(\frac{1}{q_k}\right) \Phi_{\alpha*}\left(\sum_{n=k}^{\infty} f(n, y_m)\right) < \infty.$$

Since (8) holds, this yields

$$\sum_{j=m}^{\infty} g(j, x_m) R_j \mathcal{A}_{j+1} < \infty, \tag{13}$$

where

$$\mathcal{A}_j = \sum_{k=j}^{\infty} \Phi_{\alpha*}\left(\frac{1}{r_k} \sum_{n=k}^{\infty} f(n, y_m)\right).$$

Without loss of generality, we may assume that m is so large that $\mathcal{A}_m \leq \min\{x_m, 1\}$. Since $\delta \geq 1$ and $\{\mathcal{A}_j\}$ is decreasing, from (13) we get

$$\sum_{j=m}^{\infty} g(j, x_m) \sum_{i=m}^{j} \Phi_{\alpha*}\left(\frac{1}{r_i}\right) \mathcal{A}_{j+1}^{\delta} < \infty.$$

Interchange the order of sums again to obtain

$$\sum_{i=m}^{\infty} \Phi_{\alpha*}\left(\frac{1}{r_i}\right) \sum_{j=i}^{\infty} \frac{g(j, x_m)}{x_m^{\delta}} (x_m \mathcal{A}_{j+1})^{\delta} < \infty,$$

and from here the superlinearity of g yields

$$\sum_{i=m}^{\infty} \Phi_{\alpha*}\left(\frac{1}{r_i}\right) \sum_{j=i}^{\infty} g(j, \mathcal{A}_{j+1}) < \infty.$$

Choosing m so large that $\sum_{j=m}^{\infty} g(j, \mathcal{A}_{j+1}) \leq 1$, since $\beta* \geq 2$, from the last inequality we have

$$\sum_{i=m}^{\infty} \Phi_{\alpha*}\left(\frac{1}{r_i}\right) \Phi_{\beta*}\left(\sum_{j=i}^{\infty} g(j, \mathcal{A}_{j+1})\right) < \infty.$$

Finally, in view of (8), we obtain

$$\sum_{i=m}^{\infty} \Phi_{\beta*}\left(\frac{1}{q_i} \sum_{j=i}^{\infty} g\left[j, \sum_{k=j+1}^{\infty} \Phi_{\alpha*}\left(\frac{1}{r_k} \sum_{n=k}^{\infty} f(n, y_m)\right)\right]\right) < \infty.$$

which contradicts (7). □

4 Extensions and Remarks

As an extension of (1), consider the system with deviating arguments

$$\begin{aligned} \Delta(r_k \Phi_\alpha(\Delta x_k)) &= -f(k, y_{k+\varrho}), \\ \Delta(q_k \Phi_\beta(\Delta y_k)) &= g(k, x_{k+\sigma}), \end{aligned} \tag{14}$$

where $\varrho, \sigma \in \mathbb{Z}$. All the results stated for (1) can be easily rewritten for (14). The only changes in the conditions are some shift of the indices in the sums, while everything else, like the structure of the solution space (classes (C1)-(C4)), divergence or convergence of the series in relevant conditions, and requirements on strong super/sub-linearity, remains the same.

Another possible generalization of (1) can be done when dropping the assumption on f and g to be nondecreasing with respect to the continuous variable, and supposing that continuous functions F_i, G_i, $i = 1, 2$, exist such that $F_1(k,u) \le f(k,u) \le F_2(k,u)$, $G_1(k,u) \le g(k,u) \le G_2(k,u)$ for every $(k,u) \in \mathbb{N}\times\mathbb{R}$, with F_i, G_i, $i = 1,2$, nondecreasing with respect to the second variable. Note that, in this case, we have to require strong super/sub-linearity of the functions F_i, G_i, $i = 1, 2$, and that necessary conditions for oscillation do not coincide with the sufficient ones any more.

An important special case of (1) occurs when

$$\alpha, \beta = 2,\ r_k, q_k \equiv 1,\ \sigma = \varrho = 0,\ g(k,u) = u/\psi_k,\ \{\psi_k\} : \mathbb{N} \to \mathbb{R}^+. \tag{15}$$

In this case, (1) reduces to the fourth order equation

$$\Delta^2(\psi_k \Delta^2 y_k) + f(k, y_k) = 0. \tag{16}$$

Equations of this form, but also with different shifts of the index at the second y, have been considered by various authors in the last years, see e.g. [9], [10], and the references therein. In this special case we have

$$\sum_{k=m}^{\infty} \left| \Phi_{\beta*}\left(\frac{1}{q_k}\sum_{j=k}^{\infty} g(j,C)\right)\right| = |C| \sum_{k=m}^{\infty} \frac{k-m+1}{\psi_k},$$

$$\sum_{k=m}^{\infty} \left| g\left(k, \sum_{j=m}^{k-1} \Phi_{\alpha*}\left(\frac{C}{r_j}\right)\right)\right| = |C| \sum_{k=m}^{\infty} \frac{k-m}{\psi_k},$$

and therefore either both series converges or both diverges. In [10], equation (16) was studied under the condition that both the above series are divergent. Now, if we rewrite Theorems 4 and 5 for system (14) with the special choice (15), we obtain two main theorems of [10]. In [9], the case where both series are convergent is considered. Now, similarly as above, if we rewrite Theorems 4 and 5 for system (14) with the special choice (15) and using Symmetry II, then we obtain two main theorems of [9]. We want to remark that viewing fourth order equations as coupled systems enables better understanding the structure of solution space. In particular, the cited main theorems in [9], [10] are obtained in our setting with half of effort using symmetry argument. Note that in [9] and [10], it is impossible to consider the cases corresponding to Theorems 2 and 3.

As a final remark, we underline that the question, whether the oscillation of (1) is possible if at least one of the series in (3) converges, is still not completely answered. We have completely discussed the case (4) and, thanks to symmetry, also the case (5). Hence, it remains to examine the case where both the series $\sum_{k=1}^{\infty} |f(k,M)|$, $\sum_{k=1}^{\infty} |g(k,M)|$ converge for some $M \neq 0$. In this case, three subcases have to be distinguished:

(a) $\mathcal{P}_{gr}^{\alpha}(M) < \infty$, $\mathcal{P}_{fq}^{\beta}(M) < \infty$ for some $M \neq 0$. In this case, system (1) has a nonoscillatory solution, as shown in [6].

(b) $\mathcal{S}_{qg}^{\beta}(M) < \infty$, $\mathcal{S}_{rf}^{\alpha}(M) < \infty$ for some $M \neq 0$. In this case, system (1) has a nonoscillatory solution, as shown in [6].

From (a) and (b) we see that the convergence of at least three series in (3) implies that oscillation of (1) is impossible. Hence, there remain two subcases which however can be unified into the following one, thanks to the symmetry of (1).

(c) $\mathcal{P}_{fq}^{\beta}(M) < \infty$, $\mathcal{S}_{qg}^{\beta}(M) < \infty$ [or, symmetrically, $\mathcal{P}_{gr}^{\alpha}(M) < \infty$, $\mathcal{S}_{rf}^{\alpha}(M) < \infty$] for some $M \neq 0$. This is the most difficult case, which remains unsolved. Note only that, in some typical special cases like (15), subcase (c) cannot happen (in the sense that the remaining two series are divergent).

References

[1] Clémen, P., Fleckinger, J., Mitidieri, E., de Thélin, F., *Existence of positive solutions for a nonvariational quasilinear elliptic system*, J. Diff. Equat. **166** (2000), 455–477.

[2] Fleckinger, J., Pardo, R. and de Thélin, F., *Four parameter bifurcation for a p-Laplacian system,* Electron. J. Differ. Equ. **2001**, 6, (2001), 1–15.

[3] Hooker, J. W., Patula, W. T., *Growth and oscillation properties of solutions of a fourth order linear difference equation,* J. Aust. Math. Soc., Ser. B 26 (1985), 310–328.

[4] Marini, M., Matucci, S., Řehák, P., *Oscillation of coupled nonlinear discrete systems*, J. Math. Anal. Appl. **295** (2004), 459–472.

[5] Marini, M., Matucci, S., Řehák, P., *Strongly decaying solutions of nonlinear forced discrete systems*, in "Proceedings of ICDEA 2001, Augsburg" B. Aulbach, S. Elyadi, G. Ladas Eds., to appear.

[6] Marini, M., Matucci, S., Řehák, P., *Boundary value problems for non-linear higher order difference systems*, submitted.

[7] Popenda, J., Schmeidel, E., *On the solutions of fourth order difference equations*, Rocky Mount. J. Math. **25** (1995), 1485–1499.

[8] Taylor jr, W.E. *Fourth order difference equations: oscillation and nonoscillation*, Rocky Mount. J. Math. **23** (1993), 781–795.

[9] Thandapani, E., Arockiasamy, I.M., *Fourth-Order nonlinear oscillations of difference equations*, Comp. Math. Appl. **24** (2001), 357–368.

[10] Yan, J., Liu, B., *Oscillatory and asymptotic behaviour of fourth order nonlinear difference equations*, Acta Mathematica Sinica **13** (1997), 105–115.

Non-standard Finite Difference Methods for Dissipative Singular Perturbation Problems*

Jean M.-S. Lubuma and Kailash C. Patidar †,
Department of Mathematics and Applied Mathematics, University of Pretoria,
Pretoria 0002, South Africa

(Received 15 November 2004; In final form 10 June, 2005)

Differential equations in which a very small parameter is multiplied to the highest derivative occur in many fields of science and engineering. Such differential equations form a class of "singular perturbation problems". Classical methods fail in the numerical treatment of these problems. In particular, the standard finite difference method is not reliable. Following Mickens modelling rules [9], we design non-standard finite difference schemes. The schemes thus obtained replicate the dissipativity properties of the solution of the differential equations. The schemes are analyzed for convergence. Several numerical examples are given to support the predicted theory. More importantly, these numerical examples demonstrate uniform convergence of the non-standard schemes.

AMS Subject Classification (2000): 65L10, 65L12, 65L70, 65L99.

Keywords : Singular perturbations, dissipativity, non-standard methods, boundary value problems, ordinary differential equations.

1 Introduction

We consider the singularly perturbed two point boundary value problems of the form

$$\begin{gathered} -\varepsilon y'' + a(x)y' + b(x)y = f(x) \text{ on } (0,1) \\ y(0) = \alpha_0,\ y(1) = \alpha_1, \end{gathered} \tag{1.1}$$

---

*The research contained in this paper was supported by the South African National Research Foundation.

†Corresponding author, E-mail: kailash.patidar@up.ac.za, Fax.: +27-12-4203893.

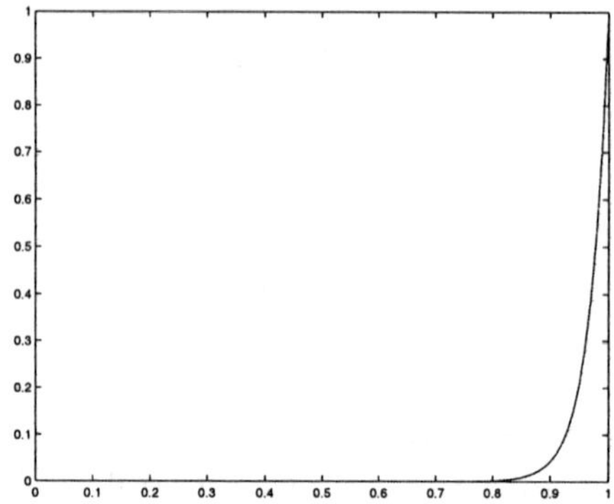

Figure 1. Solution of Problem (1.2) for $\varepsilon = 10^{-3}$.

where α_0, α_1 are given constants and ε is a small positive parameter. Further, $f(x)$, $a(x)$ and $b(x)$ are sufficiently smooth functions satisfying the conditions

$$a(x) \geq a > 0 \quad \text{and} \quad b(x) \geq b > 0.$$

Problems in which a small parameter is multiplied to a highest derivative arise in various fields of science and engineering, for instance fluid mechanics, fluid dynamics, elasticity, quantum mechanics, chemical reactor theory, hydrodynamics, etc. The main concern with such problems is the rapid growth or decay of the solution in one or more narrow "layer region(s)". The specific problem under consideration in this paper is called dissipative because the rapidly varying component of the solution decays exponentially (dissipates) away from a localized breakdown or discontinuity point in the layer region(s) as $\varepsilon \to 0$. A typical illustration of this phenomenon is the problem

$$-\varepsilon y'' + y = 0, \; y(0) = 0, \; y(1) = 1 \tag{1.2}$$

whose solution $y(x) = \sinh(x/\sqrt{\varepsilon})/\sinh(1/\sqrt{\varepsilon})$ is plotted in Figure 1.

Since the pioneer works of R.E. Mickens in the mid-1980's the non-standard approach has shown great potential in the design of reliable schemes that preserve significant properties of the solutions of differential models in science and engineering (see, e.g. [1,4,9,10]). To the best knowledge of the authors, the idea of the non-standard methods has not been implemented for the singular perturbation problems (SPPs) so far. Hence in this paper, we develop such non-standard methods for SPPs. A definition of Mickens non-standard finite difference scheme is formalized as follows in [1]:

Definition 1.1 A difference equation to determine approximate solutions y_m to the solution $y(x)$ of the problem (1.1) is called a non-standard finite difference method if at least one of the following conditions is met:

(i) The classical denominator h or h^2 of the discrete first or second order derivative is replaced by a nonnegative function ϕ such that

$$\phi(z) = z + O(z^2) \text{ or } \phi(z) = z^2 + O(z^3) \text{ as } 0 < z \to 0. \tag{1.3}$$

(ii) Nonlinear terms that occur in the differential equation are approximated in a nonlocal way.

The power of the non-standard finite difference methods is measured via qualitative stability as depicted below [1].

Definition 1.2 A difference equation in y_m is called qualitatively stable with respect to some property P of the differential equation (1.1) or of its solution whenever the discrete equation or its solution replicates the property P for any value of Δx.

Mickens [9] set five rules for the constructions of the discrete models that have the capability to replicate the properties of the exact solution. Definition 1.1 retains only two of these rules. The first rule is essential in that the qualitative behavior of the exact solution is reflected by the denominator function $\phi_{\Delta x}$. However, we do not need the second rule in this paper as we are dealing with linear problems only. The other rules are expressed in terms of Definition 1.2. For example, the schemes under consideration in this paper are stable with respect to the order of the differential equation.

The paper is organized as follows. In Section 2, we derive exact schemes for particular cases of problem (1.1). Section 3 is devoted to the construction and the analysis of the non-standard difference schemes for (1.1) and related problems. Several numerical examples demonstrating the applicability of the methods and their uniform convergence are given in Section 4. Discussion on the results as well as further research plans are indicated in Section 5.

2 Exact Schemes

Let n denote a positive integer and let the interval $[0, 1]$ be divided into n equal parts through the nodes $x_m = mh$, $m = 0(1)n$ where $h = 1/n$ is the "mesh size".

Consider the constant coefficient homogeneous problem corresponding to (1.1):

$$-\varepsilon y'' + ay' + by = 0 \text{ on } (0, 1). \tag{2.1}$$

Eq. (2.1) has two linearly independent solutions, namely, $\exp(\lambda_1 x)$ and $\exp(\lambda_2 x)$ with $\lambda_{1,2} = (-a \pm \sqrt{a^2 + 4b\varepsilon})/(-2\varepsilon)$. We denote the approximate solution to $y(x)$ at the grid points $x_{j's}$ by $\nu_{j's}$. Therefore, the theory of difference equations ([8,9]), shows that the second order linear difference equation

$$\begin{vmatrix} \nu_{m-1} & \exp(\lambda_1 x_{m-1}) & \exp(\lambda_2 x_{m-1}) \\ \nu_m & \exp(\lambda_1 x_m) & \exp(\lambda_2 x_m) \\ \nu_{m+1} & \exp(\lambda_1 x_{m+1}) & \exp(\lambda_2 x_{m+1}) \end{vmatrix} = 0$$

or equivalently,

$$-\exp\left(-\frac{ah}{2\varepsilon}\right)\nu_{m+1} + 2\cosh\left(\frac{h\sqrt{a^2+4b\varepsilon}}{2\varepsilon}\right)\nu_m - \exp\left(\frac{ah}{2\varepsilon}\right)\nu_{m-1} = 0 \quad (2.2)$$

is the exact difference scheme of Eq. (2.1) in the sense that the difference equation (2.2) has the same general solution $\nu_m = c_1\exp(\lambda_1 x_m) + c_2\exp(\lambda_2 x_m)$ as the differential equation (2.1) (cf. [9]).

Two particular cases of (1.1) will motivate the analysis in the rest of this paper. These are

$$-\varepsilon y'' + a(x)y' = f(x),\ y(0) = \alpha_0,\ y(1) = \alpha_1; \quad (2.3)$$

i.e., $b(x) \equiv 0$ and

$$-\varepsilon y'' + b(x)y = f(x),\ y(0) = \alpha_0,\ y(1) = \alpha_1; \quad (2.4)$$

i.e., $a(x) \equiv 0$.

After algebraic manipulations, we deduce from (2.2), that the exact schemes for the constant coefficient problems, corresponding to (2.3) and (2.4) are

$$-\varepsilon\frac{\nu_{m+1} - 2\nu_m + \nu_{m-1}}{\frac{h\varepsilon}{a}\left(\exp\left(\frac{ah}{\varepsilon}\right) - 1\right)} + a\frac{\nu_m - \nu_{m-1}}{h} = f \quad (2.5)$$

and

$$-\varepsilon\frac{\nu_{m+1} - 2\nu_m + \nu_{m-1}}{\frac{4}{\rho^2}\sinh^2\left(\frac{\rho h}{2}\right)} + b\nu_m = f,\ \rho = \sqrt{\frac{b}{\varepsilon}}. \quad (2.6)$$

3 The Non-standard Finite Difference Schemes

Consider Eq. (2.3) which, at a fixed node x_m, reads

$$-\varepsilon y''(x_m) + a_m y'(x_m) = f_m \tag{3.1}$$

where $a_m = a(x_m)$ and $f_m = f(x_m)$.

Motivated by Eq. (2.5), we may approximate Eq. (3.1) by the non-standard scheme

$$-\varepsilon \frac{\nu_{m+1} - 2\nu_m + \nu_{m-1}}{\psi_m^2} + a_m \frac{\nu_m - \nu_{m-1}}{h} = f_m \tag{3.2}$$

where

$$\psi_m^2 \equiv \psi_m^2(h, \varepsilon) := \frac{h\varepsilon}{a_m}\left(\exp\left(\frac{a_m h}{\varepsilon}\right) - 1\right). \tag{3.3}$$

We observe that

$$\psi_m^2(h, \varepsilon) = h^2 + O\left(\frac{h^3}{\varepsilon}\right) \tag{3.4}$$

and thus this function satisfies the condition (1.3) in Definition 1.1.

Subsequently, we will rather consider the following variant of the scheme (3.2):

$$-\varepsilon \frac{\nu_{m+1} - 2\nu_m + \nu_{m-1}}{\widetilde{\psi}_m^2} + \widetilde{a}_m \frac{\nu_m - \nu_{m-1}}{h} = f_m, \tag{3.5}$$

where $\widetilde{a}_m = (a_m + a_{m+1})/2$ and $\widetilde{\psi}_m^2$ is defined by replacing a_m by $\widetilde{a}_m$ in (3.3).

For problem (2.4), we have constructed the following non-standard scheme in [7]:

$$-\varepsilon \frac{\nu_{m+1} - 2\nu_m + \nu_{m-1}}{\widetilde{\phi}_m^2} + \widetilde{b}_m \nu_m = f_m \tag{3.6}$$

where

$$\widetilde{\phi}_m \equiv \widetilde{\phi}_m(h, \varepsilon) := \frac{2}{\widetilde{\rho}_m}\sinh\left(\frac{\widetilde{\rho}_m h}{2}\right) = h + O\left(\frac{h^3}{\varepsilon}\right) \tag{3.7}$$

with $\widetilde{b}_m = (b_{m-1} + b_m + b_{m+1})/3$ and $\widetilde{\rho}_m = \sqrt{\widetilde{b}_m/\varepsilon}$.

It is to be noted from the form of the exact scheme (2.2) that the extraction of the denominator function for the second order discrete derivative in this scheme is not straightforward. Therefore while considering the general non-homogeneous problem (1.1), one can think of choosing either of the denominator functions used in (3.5) and (3.6). However, owing to the fact that the behavior of the solution of the differential equation (1.1) is similar to that of (2.3), we prefer to select the denominator function used in (3.5). This leads then to the following non-standard scheme for (1.1):

$$-\varepsilon\frac{\nu_{m+1} - 2\nu_m + \nu_{m-1}}{\widetilde{\psi}_m^2} + \widetilde{a}_m\frac{\nu_m - \nu_{m-1}}{h} + \widetilde{b}_m y_m = f_m. \tag{3.8}$$

In what follows, the vectors $y = [y_1, y_2, \cdots, y_{n-1}]^T$ and $\nu = [\nu_1, \nu_2, \cdots, \nu_{n-1}]^T$ will be used where, we recall that, y_j and ν_j denote the exact and approximate solutions at the node x_j. Further, we denote by M various positive constants, which are independent of h and ε.

We study the two non-standard schemes (3.5) and (3.8). The non-standard scheme (3.6) is extensively analyzed in [7].

We denote the discrete operators in the non-standard schemes (3.5) and (3.8) by L_{21}^h and L_{210}^h, so that $L_{21}^h\nu = f$ and $L_{210}^h\nu = f$. The corresponding continuous operators will be denoted by L_{21} and L_{210}, respectively.

The operator L_{21} satisfies the following continuous maximum principle:

Lemma 3.1 Assume that $y(0) \geq 0$ and $y(1) \geq 0$. Then $L_{21}y(x) \geq 0$ for all $x \in (0,1)$ implies that $y(x) \geq 0$ for all $x \in [0,1]$.

Proof. See [12].

It is not hard to show that the non-standard scheme (3.5) satisfies the following discrete maximum principle:

Lemma 3.2 Let $\{\nu_j\}$ be a set of values at the grid points x_j satisfying $\nu_0 \geq 0$, $\nu_n \geq 0$ and $L_{21}^h\nu_j \geq 0$, $j = 1(1)n-1$, then $\nu_j \geq 0\ \forall\ j = 0(1)n$.

Moreover, the scheme (3.5) satisfies the following uniform stability estimate:

Lemma 3.3 If Z_i is any mesh function such that $Z_0 = Z_n = 0$, then

$$|Z_i| \leq \frac{1}{a}\max_{1\leq j\leq n-1}\left|L_{21}^h Z_j\right| \quad \text{for} \quad 0 \leq i \leq n.$$

Proof. Let $\widetilde{M} := \frac{1}{a}\max_{1\leq j\leq n-1}\left|L_{21}^h Z_j\right|$ and introduce two mesh functions η_i^+ and η_i^- defined by $\eta_i^\pm = \widetilde{M}x_i \pm Z_i$. Clearly, $\eta_0^\pm = 0$ and $\eta_n^\pm \geq 0$. Now for

$1 \leq i \leq n-1$, $L_{21}^h \eta_i^{\pm} = L_{21}^h(\widetilde{M} x_i) \pm L_{21}^h Z_i = \widetilde{M} a_i \pm L_{21}^h Z_i$. Then the discrete maximum principle in Lemma 3.2 implies that $\eta_i^{\pm} \geq 0$ for all $0 \leq i \leq n$. This gives us $|Z_i| \leq \widetilde{M}$.

At the grid point x_j, we consider a mesh function $y_j - \nu_j$ (where as mentioned earlier y_j is the exact solution at x_j). We observe that the local truncation error is

$$L_{21}^h(y_j - \nu_j) = -\varepsilon\left(y_j'' - \frac{y_{j+1} - 2y_j + y_{j-1}}{\widetilde{\psi}_j^2}\right) + \left(a_j y_j' - \widetilde{a}_j \frac{y_j - y_{j-1}}{h}\right).$$

Thus

$$\left|L_{21}^h(y_j - \nu_j)\right| \leq M\left|hy'(\xi_1) + hy''(\xi_2) + h^3 y^{(iv)}(\xi_3)\right| \tag{3.9}$$

where $x_{j-1} < \xi_i < x_{j+1}$ for $i = 1, 2, 3$.

Applying then Lemma 3.3, to this mesh function $y_j - \nu_j$, we get

$$|y_j - \nu_j| \leq M\left|hy'(\xi_1) + hy''(\xi_2) + h^3 y^{(iv)}(\xi_3)\right|. \tag{3.10}$$

Since $f(x)$ in (2.3) is sufficiently smooth, the solution $y(x)$ of this problem satisfies [11]:

$$\left|y^{(k)}(x)\right| \leq M\left(1 + \varepsilon^{-k}\exp\left(-a(1-x)/\varepsilon\right)\right) \tag{3.11}$$
$$\forall\ x \in [0,1] \text{ and } k = 0(1)4.$$

Using this estimate into (3.10), we get the following result:

Theorem 3.1 Let $a(x)$ and $f(x)$ are sufficiently smooth functions in the problem (2.3) so that $y(x) \in C^4[0,1]$. Then the numerical solution ν obtained via the non-standard finite difference method (3.5) satisfies the estimate

$$\max_{0 \leq j \leq n} |y_j - \nu_j| \leq Mh\left[1 + \sup_{x\in[0,1]} \frac{\exp\left(-\frac{a(1-x)}{\varepsilon}\right)}{\varepsilon^4}\right]. \tag{3.12}$$

Note that the continuous and discrete operators L_{210} and L_{210}^h also satisfies the maximum principles stated in the lemmas 3.1 and 3.2. Further, since $\widetilde{a}_i + \widetilde{b}_i x_i$ is always greater than a for all $1 \leq i \leq n-1$, the discrete operator L_{210}^h satisfies the uniform stability estimate provided in Lemma 3.3. The estimate given in (3.11) also holds for the solution $y(x)$ of (1.1). Taking into account of all these information, we have the following result:

Theorem 3.2 Let $a(x)$, $b(x)$ and $f(x)$ be sufficiently smooth functions in the problem (1.1) so that $y(x) \in C^4[0,1]$. Then the numerical solution ν obtained via the non-standard finite difference method (3.8) satisfies the estimate given in Theorem 3.1.

With reference to problem (2.4), we quote the analogue of Theorem 3.1 obtained in [7] by a completely different argument.

Theorem 3.3 Assume that the functions $b(x)$ and $f(x)$ in (2.4) are smooth enough such that the solution $y(x)$ is in $C^4[0,1]$. Then we have the error estimate

$$\max_j |y(x_j) - \nu_j| \leq Mh^2 \left[1 + \sup_{x\in[0,1]} \frac{\exp\left(-x\sqrt{b/\varepsilon}\right) + \exp\left(-(1-x)\sqrt{b/\varepsilon}\right)}{\varepsilon^2} \right].$$

Furthermore, the following qualitative stability result was proved in [7]:

Theorem 3.4 The non-standard finite difference scheme (3.6) is qualitatively stable with respect to the property of dissipativity. In other words, the special solutions $d(x,\tau)|_{\tau=x_j}$ and $e(x,\tau)|_{\tau=x_j}$ of

$$-\varepsilon d''(x,\tau) + b(\tau)d(x,\tau) = 0 \quad \text{and} \quad -\varepsilon e''(x,\tau) + b(\tau)e(x,\tau) = 0$$

are also solutions of

$$-\varepsilon \frac{\nu_{j+1,j} - 2\nu_{j,j} + \nu_{j-1,j}}{\widetilde{\phi}_j^2} + \widetilde{b}_j \nu_{j,j} = 0.$$

Here $\tau \in [0,1]$, ‘ $''$ ’ denotes the differentiation with respect to x,

$$d_\tau(x) \equiv d(x,\tau) = \exp\left(-x\sqrt{b(\tau)/\varepsilon}\right),$$

$$e_\tau(x) \equiv e(x,\tau) = \exp\left(-(1-x)\sqrt{b(\tau)/\varepsilon}\right)$$

and

$$\nu_{j,k} = d_\tau(x_j)_{|_{\tau=x_k}} \quad \text{or} \quad \nu_{j,k} = e_\tau(x_j)_{|_{\tau=x_k}}.$$

4 Test Examples and Numerical Results

In this section we present some numerical results corresponding to constant or variable coefficients in problem (1.1). In each example, the exact solution is provided.

Example 4.1 [13]. $a(x) \equiv 0,\ b(x) = 1/\varepsilon,\ f(x) = (x - 1 - x\exp(-1/\sqrt{\varepsilon}))/\varepsilon$,

$$y(x) = x - 1 - x\exp(-1/\sqrt{\varepsilon}\,) + \exp(-x/\sqrt{\varepsilon}\,).$$

Example 4.2 [14]. $a(x) \equiv 0,\ b(x) = 1 + x(1-x)$,

$$\begin{aligned} f(x) = 1 + x(1-x) &+ \left[2\sqrt{\varepsilon} - x^2(1-x)\right]\exp[-(1-x)/\sqrt{\varepsilon}] \\ &+ \left[2\sqrt{\varepsilon} - x(1-x)^2\right]\exp[-x/\sqrt{\varepsilon}], \end{aligned}$$

$$y(x) = 1 + (x-1)\exp[-x/\sqrt{\varepsilon}\,] - x\ \exp[-(1-x)/\sqrt{\varepsilon}\,].$$

Example 4.3 [6]. $a(x) = 1,\ b(x) \equiv 0,\ f(x) = 0$,

$$y(x) = [1 - \exp\{(x-1)/\varepsilon\}]/[1 - \exp(-1/\varepsilon)].$$

Example 4.4 [6]. $a(x) = 1,\ b(x) \equiv 0,\ f(x) = \exp(x)$,

$$y(x) = \frac{1}{1-\varepsilon}\left[\exp(x) - \frac{1 - \exp\{1 - (1/\varepsilon)\} + \{\exp(1) - 1\}\exp\{(x-1)/\varepsilon\}}{1 - \exp(-1/\varepsilon)}\right].$$

Example 4.5 [2]. $a(x) = 1,\ b(x) = 1 + \varepsilon,\ f(x) = 0$,

$$y(x) = \exp(x) + 2^{-1/\varepsilon}(x+1)^{(1+1/\varepsilon)}.$$

Example 4.6 [5]. $a(x) = 1/(x+1),\ b(x) = 1/(x+2),\ f(x) = (-\varepsilon + a(x) + b(x))\exp(x) + b(x)2^{-1/\varepsilon}(x+1)^{(1+1/\varepsilon)}$;

$$y(x) = \exp(x) + 2^{-1/\varepsilon}(x+1)^{(1+1/\varepsilon)}.$$

Maximum errors at all the mesh points are evaluated using the formula:

$$E_{n,\varepsilon} := \max_{0 \le j \le n} |y(x_j) - \nu_j|,$$

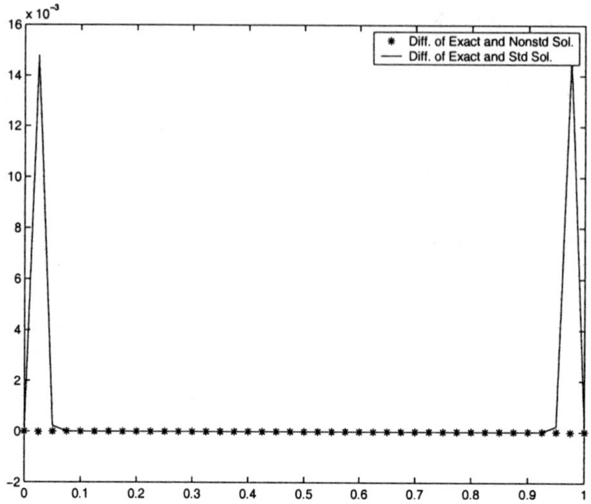

Figure 2. Errors of Std. & Non-std. Solutions for Ex. 4.2 for $n = 40$, $\varepsilon = 10^{-5}$.

for different values of n and ε, where $\nu_j \equiv \nu_j^n$ is the approximate solution obtained via the respective nonstandard methods.

The numerical rates of convergence are computed using the formula [3]:

$$r_k \equiv r_{k,\varepsilon} := \log_2 \left(E_{n_k,\varepsilon} / E_{2n_k,\varepsilon} \right), \quad k = 1, 2, \cdots .$$

Further, we compute

$$E_n = \max_{0<\varepsilon\leq 1} E_{n,\varepsilon}$$

whereas the numerical rate of uniform convergence is given by

$$R_n := \log_2 \left(E_n / E_{2n} \right).$$

5 Discussion

We have described some non-standard finite difference methods for solving dissipative singular perturbation problems. The main feature of the methods

Table 1. Results for Example 4.1 (Max. Errors)

ε	$n=8$	$n=16$	$n=32$	$n=64$	$n=128$	$n=256$
10^{-1}	0.11E-15	0.44E-15	0.22E-15	0.56E-15	0.18E-14	0.14E-14
10^{-2}	0.11E-15	0.11E-15	0.11E-15	0.11E-15	0.22E-15	0.18E-14
10^{-3}	0.00E+00	0.00E+00	0.11E-15	0.11E-15	0.22E-15	0.11E-15
10^{-4}	0.00E+00	0.00E+00	0.00E+00	0.00E+00	0.00E+00	0.00E+00
10^{-10}	0.00E+00	0.00E+00	0.00E+00	0.00E+00	0.11E-15	0.11E-15

Table 2. Results for Example 4.2 (Max. Errors)

ε	$n=8$	$n=16$	$n=32$	$n=64$	$n=128$	$n=256$
10^{-1}	0.27E-02	0.65E-03	0.16E-03	0.41E-04	0.10E-04	0.25E-05
10^{-2}	0.82E-02	0.20E-02	0.50E-03	0.13E-03	0.31E-04	0.79E-05
10^{-3}	0.97E-02	0.24E-02	0.59E-03	0.15E-03	0.37E-04	0.91E-05
10^{-4}	0.95E-02	0.26E-02	0.84E-03	0.19E-03	0.72E-04	0.21E-04
10^{-6}	0.95E-02	0.25E-02	0.63E-03	0.16E-03	0.47E-04	0.47E-04
10^{-8}	0.95E-02	0.25E-02	0.63E-03	0.16E-03	0.40E-04	0.10E-04
10^{-10}	0.95E-02	0.25E-02	0.63E-03	0.16E-03	0.40E-04	0.10E-04
E_n	0.95E-02	0.25E-02	0.63E-03	0.16E-03	0.40E-04	0.10E-04

Table 3. Results for Example 4.3 (Max. Errors)

ε	$n=8$	$n=16$	$n=32$	$n=64$	$n=128$	$n=256$
10^{-1}	0.78E-15	0.67E-15	0.78E-15	0.17E-14	0.33E-14	0.12E-13
10^{-2}	0.11E-15	0.67E-15	0.78E-15	0.33E-15	0.12E-13	0.53E-13
10^{-3}	0.00E+00	0.00E+00	0.11E-15	0.33E-15	0.22E-15	0.44E-15
10^{-4}	0.00E+00	0.00E+00	0.00E+00	0.00E+00	0.00E+00	0.00E+00
10^{-10}	0.00E+00	0.00E+00	0.00E+00	0.00E+00	0.00E+00	0.00E+00

Table 4. Results for Example 4.4 (Max. Errors)

ε	$n=8$	$n=16$	$n=32$	$n=64$	$n=128$	$n=256$
10^{-1}	0.12E-01	0.32E-02	0.81E-03	0.20E-03	0.51E-04	0.13E-04
10^{-2}	0.75E-01	0.33E-01	0.11E-01	0.31E-02	0.80E-03	0.20E-03
10^{-4}	0.89E-01	0.49E-01	0.26E-01	0.13E-01	0.65E-02	0.32E-02
10^{-6}	0.89E-01	0.49E-01	0.26E-01	0.13E-01	0.66E-02	0.33E-02
10^{-8}	0.89E-01	0.49E-01	0.26E-01	0.13E-01	0.66E-02	0.33E-02
10^{-10}	0.89E-01	0.49E-01	0.26E-01	0.13E-01	0.66E-02	0.33E-02
E_n	0.89E-01	0.49E-01	0.26E-01	0.13E-01	0.66E-02	0.33E-02

Table 5. Results for Example 4.5 (Max. Errors: using (2.2))

ε	$n=16$	$n=32$	$n=64$	$n=128$	$n=256$	$n=512$
2^{-1}	0.89E-15	0.31E-13	0.87E-13	0.24E-12	0.73E-11	0.18E-10
2^{-2}	0.50E-14	0.16E-13	0.46E-13	0.59E-12	0.58E-12	0.22E-11
2^{-4}	0.33E-15	0.26E-13	0.35E-13	0.21E-12	0.97E-12	0.34E-12
2^{-6}	0.67E-15	0.56E-15	0.12E-13	0.43E-13	0.46E-12	0.21E-12
2^{-8}	0.94E-15	0.12E-14	0.95E-14	0.23E-13	0.53E-13	0.84E-13
2^{-10}	0.89E-15	0.13E-14	0.49E-14	0.87E-14	0.97E-14	0.33E-14
10^{-3}	0.56E-15	0.15E-14	0.54E-14	0.13E-13	0.10E-13	0.27E-14

Table 6. Results for Example 4.5 (Max. Errors: using (3.8))

ε	$n = 16$	$n = 32$	$n = 64$	$n = 128$	$n = 256$	$n = 512$
2^{-1}	0.59E-04	0.15E-04	0.37E-05	0.93E-06	0.23E-06	0.58E-07
2^{-4}	0.46E-03	0.12E-03	0.29E-04	0.72E-05	0.18E-05	0.45E-06
2^{-8}	0.43E-02	0.15E-02	0.45E-03	0.12E-03	0.30E-04	0.75E-05
10^{-3}	0.53E-02	0.25E-02	0.11E-02	0.38E-03	0.11E-03	0.29E-04
10^{-4}	0.56E-02	0.28E-02	0.14E-02	0.68E-03	0.32E-03	0.14E-03
10^{-6}	0.57E-02	0.29E-02	0.14E-02	0.72E-03	0.36E-03	0.18E-03
10^{-8}	0.57E-02	0.29E-02	0.14E-02	0.72E-03	0.36E-03	0.18E-03
10^{-10}	0.57E-02	0.29E-02	0.14E-02	0.72E-03	0.36E-03	0.18E-03
E_n	0.57E-02	0.29E-02	0.14E-02	0.72E-03	0.36E-03	0.18E-03

Table 7. Results for Example 4.6 (Max. Errors: using (3.8))

ε	$n = 16$	$n = 32$	$n = 64$	$n = 128$	$n = 256$	$n = 512$
2^{-1}	0.97E-02	0.50E-02	0.25E-02	0.13E-02	0.64E-03	0.32E-03
2^{-2}	0.17E-01	0.84E-02	0.43E-02	0.21E-02	0.11E-02	0.54E-03
2^{-4}	0.27E-01	0.13E-01	0.67E-02	0.33E-02	0.17E-02	0.83E-03
2^{-6}	0.38E-01	0.17E-01	0.80E-02	0.39E-02	0.19E-02	0.95E-03
2^{-8}	0.54E-01	0.25E-01	0.10E-01	0.46E-02	0.21E-02	0.10E-02
2^{-10}	0.60E-01	0.30E-01	0.15E-01	0.64E-02	0.27E-02	0.12E-02

Table 8. Results for Example 4.2 (Rates of Convergence): $n_k = 8 \times 2^{k-1}$, $k = 1(1)5$

ε	r_1	r_2	r_3	r_4	r_5
10^{-1}	0.20E+01	0.20E+01	0.20E+01	0.20E+01	0.20E+01
10^{-2}	0.20E+01	0.20E+01	0.20E+01	0.20E+01	0.20E+01
10^{-4}	0.19E+01	0.15E+01	0.22E+01	0.14E+01	0.17E+01
10^{-8}	0.19E+01	0.20E+01	0.20E+01	0.20E+01	0.20E+01
10^{-11}	0.19E+01	0.20E+01	0.20E+01	0.20E+01	0.20E+01
R_n	0.19E+01	0.20E+01	0.20E+01	0.20E+01	0.20E+01

Table 9. Results for Example 4.4 (Rates of Convergence): $n_k = 8 \times 2^{k-1}$, $k = 1(1)5$

ε	r_1	r_2	r_3	r_4	r_5
10^{-1}	0.20E+01	0.20E+01	0.20E+01	0.20E+01	0.20E+01
10^{-2}	0.10E+01	0.15E+01	0.18E+01	0.19E+01	0.20E+01
10^{-4}	0.87E+00	0.94E+00	0.97E+00	0.98E+00	0.99E+00
10^{-6}	0.87E+00	0.94E+00	0.97E+00	0.98E+00	0.99E+00
10^{-8}	0.87E+00	0.94E+00	0.97E+00	0.98E+00	0.99E+00
10^{-10}	0.87E+00	0.94E+00	0.97E+00	0.98E+00	0.99E+00
R_n	0.87E+00	0.94E+00	0.97E+00	0.98E+00	0.99E+00

Table 10. Results for Example 4.6 (Rates of Convergence: using (3.8)): $n_k = 16 \times 2^{k-1}$, $k = 1(1)5$

ε	r_1	r_2	r_3	r_4	r_5
2^{-1}	0.95E+00	0.97E+00	0.99E+00	0.99E+00	0.10E+01
2^{-2}	0.95E+00	0.97E+00	0.99E+00	0.99E+00	0.10E+01
2^{-4}	0.97E+00	0.98E+00	0.99E+00	0.10E+01	0.10E+01
2^{-6}	0.11E+01	0.11E+01	0.10E+01	0.10E+01	0.10E+01
2^{-8}	0.13E+01	0.12E+01	0.12E+01	0.11E+01	0.10E+01
2^{-10}	0.93E+00	0.11E+01	0.13E+01	0.13E+01	0.12E+01
10^{-3}	0.92E+00	0.98E+00	0.11E+01	0.13E+01	0.13E+01

is that the dissipative nature of the problems is incorporated in the schemes via Mickens' rule of renormalization of the denominator of the discrete derivative. The methods have been analyzed for convergence. Six examples have been solved to demonstrate the applicability of the proposed methods.

For Examples 4.2, 4.4, and 4.6 we have computed rates of convergence (see Tables 8, 9 and 10), which show the uniform second (first) order convergence for the problems without (with) first derivative term.

Further, we would like to mention that in the construction of ϕ in Section 3, we assumed that $a(x)$ and $b(x)$ are constant. So when we really have the constant $a(x)$ and $b(x)$ then we will have exact schemes and no question of stability occur. One also expects excellent numerical solutions in such cases. To see this one can refer to the numerical results presented in the Tables 1 and 3. For the most general problem, we have the exact scheme for the homogeneous problem only and the results with this scheme for the respective problem are presented in Table 5.

To further corroborate the use of the non-standard methods, in Figure 2, the difference between the exact solution and the approximate solution obtained via the non-standard finite difference method (3.6) with $\widetilde{b}_m = b_m$ and corresponding standard finite difference method is plotted for Example 4.2. One can see clearly that in the boundary layer regions, the standard method performs badly. The same can be seen for the other examples also.

In view of the above numerical experiments, the authors believe that the proposed method is uniformly first (second) order convergent in the case when the governing singularly perturbed differential equation possess (do not possess) the first derivative term. In other words, the authors feel that the estimates in Theorem 3.1 and Theorem 3.3 (which provides the convergence for a fixed ε) are not sharp since they do not carefully take into account the singular behavior at the end points. The authors are investigating this issue along with the extension of the nice outcomes of the proposed method to other type of singularly perturbed problems. These include dispersive problems, turning point problems, nonlinear problems, etc.

References

[1] R. Anguelov and J.M.-S. Lubuma, Contributions to the mathematics of the nonstandard finite difference method and applications, *Numer. Methods Partial Differential Eq.* **17** (2001), 518-543.

[2] P.P.N. de Groen and P.W. Hemker, Error bounds for exponentially fitted galerkin methods applied to stiff two-point boundary value problems, *Numerical Analysis of Singular Perturbation Problems*, P.W. Hemker and J.J.H. Miller (eds.), Academic Press, New York, 1979, 217- 249.

[3] E.P. Doolan, J.J.H. Miller and W.H.A. Schilders, *Uniform Numerical Methods for Problems with Initial and Boundary Layers*, Boole Press, Dublin, 1980.

[4] H. Al-Kahby, F. Dannan and S. Elaydi, Non-standard discretization methods for some biological models, In: R.E. Mickens (ed.), *Applications of nonstandard finite difference schemes*, World Scientific, Singapore, 2000, pp. 155-180.

[5] C.G. Eugene, Uniformly high-order difference schemes for a singularly perturbed two-point boundary value problem, *Math. Comp.* **48(178)** (1987), 551-564.

[6] J. Lorenz, Combination of initial and boundary value methods for a class of singular perturbation problems, *Numerical Analysis of Singular Perturbation Problems*, P.W. Hemker and J.J.H. Miller (eds.), Academic Press, New York, 1979, 295-315.

[7] J.M.-S. Lubuma and K.C. Patidar, Non-standard finite difference method for self-adjoint singular perturbation problems, In: T. Simos (ed.), *VSP Lecture Series on Computer and Computational Sciences* **1** (2004), 328-331.

[8] R.E. Mickens, *Difference Equations: Theory and Applications*, Van Nostrand Reinhold, New York, 1990.

[9] R.E. Mickens, *Nonstandard Finite Difference Models of Differential Equations*, World Scientific, Singapore, 1994.

[10] R.E. Mickens (editor), *Applications of Nonstandard Finite Difference Schemes*, World Scientific, Singapore, 2000.

[11] J.J.H. Miller, E. O'Riordan and G.I. Shishkin, *Fitted Numerical Methods for Singular Perturbation Problems*, World Scientific, Singapore, 1996.

[12] M.H. Protter and H.F. Weinberger, *Maximum Principles in Differential Equations*, Prentice-Hall, New Jersey, 1967.

[13] A.H. Schatz and L.B. Wahlbin, On the finite element method for singularly perturbed reaction diffusion problems in two and one dimension, *Math. Comp.* **40** (1983), 47-89.

[14] V. Vukoslavčević and K. Surla, Finite element method for solving self adjoint singularly perturbed boundary value problems, *Mathematica Montisnigri* **VII** (1996), 89-86.

ON A CLASS OF GENERALIZED AUTOREGRESSIVE PROCESSES

KAMAL C. CHANDA
TEXAS TECH UNIVERSITY
LUBBOCK, TX 79409-1042

Abstract

We have introduced a class of generalized autoregressive processes for which each of the autoregressive parameters is the sum of a constant and a function of a set of other random variables. It is shown that under some mild regularity conditions on these random variables the constant parts of the autoregressive parameters can be estimated from the given data set in a manner similar to those for the classical autoregressive processes. Although the sampling properties of these estimators are different from those for the classical types, we have shown that these estimators are consistent and asymptotically normally distributed.

1 INTRODUCTION

The identification and estimation of parameters in an autoregressive process of order p ($AR(p)$) from a set of data from such a process are well discussed in statistical literature (see for example Priestley (1981)). But sometimes the autoregressive parameters are contaminated with random errors thereby creating some problems of identification and estimation for the set of parameters defining the stochastic processes. The present article deals with various special situations relating to such processes and suggests statistical methods to estimate the uncontaminated parts of these autoregressive "parameters." It is shown in subsequent sections that these estimators when suitably estimated on the basis of some mild regularity conditions prove to be consistent and asymptotically normally distributed. Only the summary details are discussed - the complete details appear in Chanda (2004).

2 AUTOREGRESSIVE MODELS (CLASSIC)

Let $\{\varepsilon_t\}$ be an i.i.d. sequence of random variables with $E\varepsilon_1 = 0, E\varepsilon_1^2 = \sigma^2 (0 < \sigma < \infty)$ and let us define the process $\{X_t\}$ by

$$X_t = \varepsilon_t + \sum_{i=1}^{p} \beta_i X_{t-i}, \tag{2.1}$$

where $\beta_i (1 \leq i \leq p)$ are constants satisfying the property that all roots of the polynomial $\xi^p - \sum_{i=1}^{p} \beta_i \xi^{p-i}$ are inside the unit circle. Then it can be shown that if we set $\mathbf{X}_t = (X_t, \ldots, X_{t-p+1})^T$, $\mathbf{B} = \begin{pmatrix} \beta_1 & \cdots & \beta_{p-1} & \beta_p \\ 1 & \cdots & 0 & 0 \\ 0 & \cdots & 1 & 0 \end{pmatrix}$ and $\boldsymbol{\eta} = (1, 0, \ldots, 0)^T$ then

$$\mathbf{X}_t = \boldsymbol{\eta}^T \varepsilon_t + \mathbf{B}\, \mathbf{X}_{t-1} \tag{2.2}$$

which leads to the following almost sure representation (a.s.) of X_t in the form

$$X_t = \varepsilon_t + \sum_{r=1}^{\infty} W_{rt}, \tag{2.3}$$

where $W_{rt} = \boldsymbol{\eta}^T \mathbf{B}^r \boldsymbol{\eta}\, \varepsilon_{t-r}$.

It is well known that the regularity condition imposed above on the autoregressive parameters $\beta_i (1 \leq i \leq p)$ implies that the eigenvalues of $\mathbf{B}$ are all inside the unit circle. This will guarantee the a.s. representation (2.3) above. Note that if we write $\boldsymbol{\Gamma}_p = (\gamma_{i-j}), \gamma_v = EX_1 X_{1+v} (-\infty < v < \infty)$ and $\boldsymbol{\gamma}_p = (\gamma_1, \ldots, \gamma_p)^T$ then we have that $\boldsymbol{\Gamma}_p \boldsymbol{\beta} = \boldsymbol{\gamma}_p$ where $\boldsymbol{\beta} = (\beta_i, \ldots, \beta_p)^T$. This implies that if we estimate $\hat{\boldsymbol{\beta}} = (\hat{\beta}_1, \ldots, \hat{\beta}_p)$ as the solution of the equation $\hat{\boldsymbol{\Gamma}}_p \hat{\boldsymbol{\beta}} = \hat{\boldsymbol{\gamma}}$ where $\hat{\gamma}_v$ is the serial covariance of $\{X_t\}$ of lag v then the results in the following Theorem holds

Theorem 2.1.

$$\hat{\boldsymbol{\beta}} \xrightarrow{P} \boldsymbol{\beta} \textit{ as } n \to \infty, \tag{2.4}$$

and

$$\mathcal{L}(\sqrt{n}(\hat{\boldsymbol{\beta}} - \boldsymbol{\beta})) \to N(\mathbf{0},\, \sigma^2 \boldsymbol{\Gamma}_p^{-1}) \textit{ as } n \to \infty. \tag{2.5}$$

The proof of this Theorem is given in Priestley (1981).

3 GENERALIZED AUTOREGRESSIVE MODELS

We now introduce a class of autoregressive schemes $\{X_t\}$ which have the same representation as (2.1) except that the β_i's are now contaminated with errors which, themselves, are functions of random variables. The main problem is now to find out under what conditions on the uncontaminated parts of the autoregressive parameters in the model and regularity conditions (suitably formulated) on these random errors we would be able to establish an a.s. representation for $\{X_t\}$ similar to (2.3) and devise methods for estimating these uncontaminated parameters such that these estimators are consistent and asymptotically normally distributed.

Two different models will be considered in what follows.

3.1 GENERALIZED AUTOREGRESSIVE MODELS (TYPE A)

Let X_t be determined by the following relation.

$$X_t = \varepsilon_t + \sum_{i=1}^{p} (\beta_i + f_{it}) X_{t-i}, \tag{3.1}$$

where $\{\varepsilon_t\}$ is as defined above, $f_{it} = f_i(\varepsilon_{t-1}, \ldots, \varepsilon_{t-q})$ and q is a known integer ≥ 1. Then we can show that under some mild regularity conditions on $\{\varepsilon_t\}$ and $f_{it}(1 \leq i \leq p)$ to be stated later we have the following a.s. representation for X_t, viz.,

$$X_t = \varepsilon_t + \sum_{r=1}^{\infty} W_{rt} \tag{3.2}$$

where

$$W_{rt} = \boldsymbol{\eta}^T \Pi_{u=1}^{r} (\mathbf{B} + \mathbf{F}_{t-u+1}) \boldsymbol{\eta} \varepsilon_{t-r} (r \geq 1),$$

$\boldsymbol{\eta}, \mathbf{B}$ are defined as above and

$$\mathbf{F}_t = \begin{pmatrix} f_{1t} & \cdots & f_{pt} \\ 0 & \cdots & 0 \\ \vdots & & \\ 0 & \cdots & 0 \end{pmatrix}.$$

Assume that the eigenvalues of $\mathbf{B}$ are all inside the unit circle and are all distinct. Then there exists a nonsignular matrix $\mathbf{R}$ such that $\mathbf{Q} := \mathbf{R}^{-1}\mathbf{B}\mathbf{R}$ is a diagonal matrix with the diagonals representing the eigenvalues of $\mathbf{B}$. Then on simplification we can write

$$|W_{rt}| \leq M\ \Pi_{u=1}^{r} S_{t-u+1}|\varepsilon_{t-r}|, \tag{3.3}$$

where $S_t = \rho_0 + ||\mathbf{R}^{-1}\mathbf{F}_t\mathbf{R}||_\infty, \rho_0 = \max|$ eigenvalues of $\mathbf{B}|$ and we define for any $p \times p$ matrix $\mathbf{C} = (c_{ij}),\ ||\mathbf{C}||_\infty = \max_{1 \leq i \leq p} \sum_{j=1}^{p} |c_{ij}|$. Then subject to existence we can show that

$$E|W_{rt}|^k \leq M\left[\rho_0 + (E||\mathbf{R}^{-1}\mathbf{F}_t\mathbf{R}||_\infty^{2kq})^{1/2kq}\right]^{kr} \leq M\lambda^{kr} \tag{3.4}$$

where $\lambda = \left[\rho_0 + (E||\mathbf{R}^{-1}\mathbf{F}_t\mathbf{R}||_\infty^{2kq})^{1/2kq}\right]$.

We now consider a few special cases.

(a) Let $f_{it} = \sum_{j=1}^{q} c_j^{(i)} \varepsilon_{t-j}$. Then $S_t = \rho_0 + \sum_{j=1}^{q} \pi_j |\varepsilon_{t-j}|, \pi_1, \ldots, \pi_q$ are appropriately defined nonnegative constants with

$$\lambda = \rho_0 + \nu_{2kq}^{1/2kq} \sum_{s=1}^{q} \pi_s,\ (\nu_s = E|\varepsilon_1|^s). \tag{3.5}$$

(b) Let $f_{it} = \sum_{\substack{1 \leq s_1 + \cdots + s_q \leq m \\ s_i \geq 0}} c_{s_1,\ldots,sq}^{(i)} |\varepsilon_{t-1}|^{s_1} \ldots |\varepsilon_{t-q}|^{s_q}$.

Then $S_t = \rho_0 + \sum_{\substack{1 \leq s_1 + \cdots + s_q \leq M \\ s_i \geq 0}} \theta_{s_1,\ldots,s_q} |\varepsilon_{t-1}|^{s_1} \ldots |\varepsilon_{t-q}|^{s_q}$ where $\theta_{s_1,\ldots,s_q}$ are appropriately defined nonnegative constants with

$$\lambda = \rho_0 + \sum_{v=1}^{m} d_v \tau^v \tag{3.6}$$

where $\tau = \nu_{2kvq}^{1/2kq}$ and $d_v = \sum_{\substack{s_1 + \cdots + s_q = v \\ s_i \geq 0}} \theta_{s_1}, \ldots, s_q$.

We now consider the problem of estimating the parameters $\beta_i (1 \leq i \leq p)$.

We assume that conditions (3.3) and (3.4) hold and that

$$E(\sum_{i=1}^{p} f_{it}X_{t-i}X_{t-j}) = 0 \tag{3.7}$$

for $j \geq m$ for some positive integer m. There are many special situations where condition (3.6) holds. For details relating to these conditions see Chanda (2004). We now state the following

Theorem 3.1. *Assume that the matrix* $\mathbf{\Gamma} = (\gamma_{i-j}; 1 \leq i \leq p, m \leq j \leq m + p - 1)$ *is nonsingular. Then if* $\boldsymbol{\beta} = (\beta_1, \ldots, \beta_p)^T$ *and* $\boldsymbol{\gamma} = (\gamma_m, \ldots, \gamma_{m+p-1})^T$

$$\boldsymbol{\beta} = \mathbf{\Gamma}^{-1}\boldsymbol{\gamma} \tag{3.8}$$

Also, if based on the dataset $\{X_t, 1 \leq i \leq n\}$ *we estimate* β *by* $\hat{\boldsymbol{\beta}} = (\hat{\beta}_1, \ldots, \hat{\beta}_p)$ *where*

$$\hat{\boldsymbol{\beta}} = \hat{\mathbf{\Gamma}}^{-1}\hat{\boldsymbol{\gamma}} \tag{3.9}$$

$\hat{\mathbf{\Gamma}} = (\hat{\gamma}_{i-j}; 1 \leq i \leq p, m \leq j \leq m + p - 1)$ *and* $\hat{\gamma}_j = \sum_{t=1}^{n-|j|}(X_t - \overline{X})(X_{t+|j|} - \overline{X})/n, \overline{X} = \sum_{t=1}^{n} X_t/n$, *then*

$$\hat{\boldsymbol{\beta}} \xrightarrow{P} \boldsymbol{\beta} \text{ as } n \to \infty, \tag{3.10}$$

and

$$\mathcal{L}(\sqrt{n}(\hat{\boldsymbol{\beta}} - \boldsymbol{\beta})) \to N(\mathbf{0}, \mathbf{\Lambda}) \text{ as } n \to \infty$$

where $\mathbf{\Lambda}$ *is a finite matrix, and* $\mathbf{0} = (0, \ldots, 0)^T$.

For proof of this Theorem see Chanda (2004).

3.2 GENERALIZED AUTOREGRESSIVE MODELS (TYPE B)

Let X_t be determined by the relation

$$X_t = \varepsilon_t + \sum_{i=1}^{p}(\beta_i + f_{it})X_{t-i}, \tag{3.11}$$

where now $f_{it} = Y_{it}, \mathbf{Y}_t = (Y_{1t}, \ldots, Y_{pt})^T, \{\mathbf{Y}_t\}$ is an i.i.d. sequence of vectors with $E\mathbf{Y}_t = 0, E\mathbf{Y}_t\mathbf{Y}_t^T = \mathbf{\Gamma}$ and $\{\mathbf{Y}_t\}$ is independent of $\{\varepsilon_t\}$ (see Feigin and

Tweedie (1985) and Nicholls and Quinn (1982)). Again it is easy to see that subject to existence

$$X_t = \varepsilon_t + \sum_{r=1}^{\infty} W_{rt}, \tag{3.12}$$

where now

$$W_{rt} = \boldsymbol{\eta}^T \Pi_{u=1}^{r} (\mathbf{B} + \mathbf{H}_{t-u+1}) \boldsymbol{\eta} \varepsilon_{t-r},$$

$\boldsymbol{\eta}$ and $\mathbf{B}$ are as defined above and

$$\mathbf{H}_t = \begin{pmatrix} Y_{1t} & \dots & Y_{pt} \\ 0 & \dots & 0 \\ \vdots & & \vdots \\ 0 & \dots & 0 \end{pmatrix}.$$

Then subject to existence

$$E|W_{rt}|^k \leq M\lambda^{kr}, \tag{3.13}$$

where

$$\lambda = \rho_0 + \sum_{\ell=1}^{p} w_\ell E^{1/k} |Y_{\ell t}|^k$$

and w_ℓ's are appropriately defined nonnegative constants. More information is available about this model. In fact we can show that the constants $\beta_i (1 \leq i \leq p)$ will satisfy the same relations as for the classic autoregressive models. In other words

$$\boldsymbol{\Gamma}_p \boldsymbol{\beta} = \boldsymbol{\gamma}_p \tag{3.14}$$

where $\boldsymbol{\beta}, \boldsymbol{\Gamma}_p$ and $\boldsymbol{\gamma}_p$ are defined as above. This implies that we may use the same estimating equations as above, viz.,

$$\hat{\boldsymbol{\Gamma}}_p \hat{\boldsymbol{\beta}} = \hat{\boldsymbol{\gamma}}_p. \tag{3.15}$$

But, of course, the sampling properties of the estimator $\hat{\boldsymbol{\beta}}$ will be different from what we know about the classic autoregressive situation. The asymptotic properties can, however, be derived by using the special methodology devised by Chanda (1993). We shall only mention the property of $\hat{\beta}$ when $p = 1$. The sampling properties of the estimators of σ^2 and EY_1^2 can be derived similarly. The following Theorem provides information relating to the sampling properties of $\hat{\beta}$.

Theorem 3.2.
$$\hat{\beta} \xrightarrow{P} \beta \text{ as } n \to \infty, \tag{3.16}$$

and
$$\mathcal{L}(\sqrt{n}(\hat{\beta} - \beta)) \to \mathcal{N}(0, \delta^2) \text{ as } n \to \infty, \tag{3.17}$$

where δ^2 is a finite constant assumed to be > 0.

In order to compute δ^2 we need to make use of the facts that

$$\hat{\beta} = \hat{\gamma}_1/\hat{\gamma}_0, \tag{3.18}$$

$$\hat{\gamma}_v \xrightarrow{P} \gamma_v \text{ as } n \to \infty$$

and that if we write $\mathbf{Z} = (\hat{\gamma}_0, \hat{\gamma}_v)^T$ and $\boldsymbol{\zeta} = (\gamma_0, \gamma_v)$ then

$$\mathcal{L}(\sqrt{n}(\mathbf{z} - \boldsymbol{\zeta})) \to \mathcal{N}(\mathbf{0}, \boldsymbol{\Lambda}), \text{ as } n \to \infty, \tag{3.19}$$

where $\boldsymbol{\Lambda} = (\lambda_{ij}), \lambda_{11} = \sum_{s=-\infty}^{\infty} (EX_1^2 X_{1+s}^2 - \gamma_0^2), \lambda_{12} = \lambda_{21} = \sum_{s=-\infty}^{\infty} (EX_1^2 X_{1+s} X_{1+s+v} - \gamma_0 \gamma_v)$ and $\lambda_{22} = \sum_{s=-\infty}^{\infty} (EX_1 X_{1+v} X_{1+s} X_{1+s+v} - \gamma_v^2)$.

Routine but tedious computation will provide us expressions for λ_{ij} in terms of the parameters σ^2, EY_1^2 and β. By using appropriate moment methods we can estimate the last three parameters and therefore estimate λ_{ij} and eventually δ^2. The methodology suggested in Chanda (1993) will then guarantee the consistency and asymptotic normality of $\hat{\delta}^2$. This result for the situation $p = 1$ can easily be extended to the general case. The estimator $\hat{\beta}$ will, therefore, have the properties of consistency and asymptotic normality.

References

[1] Chanda, Kamal C. (1993). Asymptotic properties of serial covariances for nonlinear stationary processes. *J. Multivariate Anal.* **47** 163-171.

[2] Chanda, Kamal C. (2004). Some comments on a class of generalized autoregressive processes. Submitted for publication.

[3] Feigan, Paul D. and Tweedie, Richard L. (1985). Random coefficient autoregressive processes: a Markov chain analysis of stationarity and finiteness of moments. *J. Time Ser. Anal.* **6** 1-14.

[4] Nicholls, D.F. and Quinn, B.G. (1982). *Random Coefficient Autoregressive Models: An Introduction.* Springer-Verlag Lecture Notes in Statistics. **11**, New York.

[5] Priestley, M.B. (1981). *Spectral Analysis and Time Series.* Academic Press Inc. (London) Ltd.

On
$$x_{n+1} = \frac{\beta_n x_n + \gamma_n x_{n-1}}{A_n + B_n x_n}$$
with Period-two Coefficients

C.H. Gibbons and C. B. Overdeep
Mathematical Sciences Department
Salve Regina University
Newport, RI 02841
Department of Mathematics
Western Oregon University
Monmouth, OR 97361

Abstract

We extend the known results of solutions of the autonomous counterpart of the difference equation in the title to the situation where any of the parameters are a period-two sequence with non-negative values and the initial conditions are positive.

Keywords: boundedness, global attractivity, difference equation, periodic coefficients, period-two solutions
AMS 2000 Mathematics Subject Classification: 39A10, 39A11

1 Introduction and Preliminaries

Our aim in this paper is to investigate the trichotomy character and the global asymptotic stability of solutions of the nonautonomous difference equation

$$x_{n+1} = \frac{\beta_n x_n + \gamma_n x_{n-1}}{A_n + B_n x_n}, \quad n = 0, 1, \ldots \tag{1}$$

where the initial conditions x_{-1} and x_0 are positive and the parameters β_n, γ_n, A_n, and B_n are nonnegative period-two sequences with γ_n not identically zero.

We will focus on the regions in the parameter space that extend the behavior of the autonomous case; the remaining regions in the parameter space are currently under investigation.

The autonomous version of Eq.(1) given by

$$x_{n+1} = \frac{\beta x_n + \gamma x_{n-1}}{A + B x_n}, \quad n = 0, 1, \ldots \tag{2}$$

was investigated in [KLP] when the parameters β, γ, A, and B are positive and the initial conditions x_{-1} and x_0 are positive. A change of variables reduces the number of parameters to two and so we will deal with the rational equation:

$$x_{n+1} = \frac{q x_n + r x_{n-1}}{1 + x_n}, \quad n = 0, 1, \ldots. \tag{3}$$

Eq.(3) has zero as an equilibrium solution and when $q+r > 1$ Eq.(3) also has the positive equilibrium solution:

$$\bar{x} = q + r - 1.$$

The main result which is known about the zero equilibrium of Eq.(3) is described by Theorem A while the main results known about the positive equilibrium of Eq.(3) are contained in Theorem B and Theorem C.

Theorem A

(a) Assume

$$q + r \leq 1.$$

Then the zero equilibrium of Eq.(3) is globally asymptotically stable.

(b) Assume

$$q + r > 1.$$

Then the zero equilibrium of Eq.(3) is unstable. More precisely, it is a saddle point when

$$1 - q < r < 1 + q$$

and a repeller when

$$r > 1 + q.$$

Theorem B *Suppose that r=1 and let* $x_{-1} + x_0 > 0$. *Then the positive equilibrium* $\bar{y} = q$ *of Eq.(3) is globally asymptotically stable.*

Theorem C

(a) Assume

$$r = 1 + q.$$

Then every solution of Eq.(3) converges to a period-two solution.

(b) Assume

$$1 < r < 1 + q.$$

Then every positive solution of Eq.(3) converges to the positive equilibrium of Eq.(3).

(c) Assume

$$r > 1 + q.$$

Then Eq.(3) has unbounded solutions.

Our goal in this paper is to try to extend the above Theorems A, B, and C when one or more of the parameters are replaced with a period-two sequence.

By change of variables, any combination of period-two parameters applied to Eq.(1) can be reduced to

$$x_{n+1} = \frac{q_n x_n + r_n x_{n-1}}{1 + x_n}, \quad n = 0, 1, \ldots \tag{4}$$

where the initial conditions $x_{-1} + x_0 > 0$ and the parameters q_n, r_n are nonnegative period-two sequences with r_n not identically zero.

Equation (4) can be decoupled into a difference equation representing the odd terms $\{x_{2n+1}\}$ and the even terms $\{x_{2n}\}$ of the solution. For the odd terms of the solution,

$$\begin{aligned} y_{n+1} &= [r_0 y_n^3 + (q_0 q_1 + r_0 - q_0 r_0) y_n^2 + (-q_0^2 q_1 - q_0 r_1 - q_0 r_0) y_n \\ &+ q_0 r_0 r_1 y_{n-1}] \\ &/ \ [(q_1+1) y_n^2 + (-q_0 q_1 - r_1 - q_0 + 1) y_n + r_0 r_1 y_{n-1} - q_0] \\ & \quad n = 0, 1, \ldots . \end{aligned} \tag{5}$$

For Eq. (5), the equilibria are the nonnegative solutions of

$$\begin{aligned} (q_1 + 1 - r_0)\bar{y}^3 &+ (-2q_0 q_1 + q_0 r_0 - q_0 + r_0 r_1 - r_0 - r_1 + 1)\bar{y}^2 \\ &+ (-q_0 + q_0^2 q_1 + q_0 r_1 + q_0 r_0 - q_0 r_0 r_1)\bar{y} = 0. \end{aligned} \tag{6}$$

The possible equilibrium values for Eq.(6) are

$$\bar{y} = 0,$$

$$\bar{y} = q_0,$$

and

$$\bar{y} = \frac{q_0 q_1 - (r_0 - 1)(r_1 - 1)}{q_1 + 1 - r_0}, \qquad \text{when } q_1 + 1 \neq r_0.$$

Similarly, for the even terms of the solution,

$$\begin{aligned} z_{n+1} &= [r_1 z_n^3 + (q_0 q_1 + r_1 - q_1 r_1) z_n^2 + (-q_0 q_1^2 - r_0 q_1 - q_1 r_1) z_n \\ &+ q_1 r_0 r_1 z_{n-1}] \\ &/ \ [(q_0+1) z_n^2 + (-q_0 q_1 - r_0 - q_1 + 1) z_n + r_0 r_1 z_{n-1} - q_1] \\ & \quad n = 0, 1, \ldots . \end{aligned} \tag{7}$$

For Eq. (7), the equilibria are the nonnegative solutions of

$$\begin{aligned} (q_0 + 1 - r_1)\bar{z}^3 &+ (-2q_0 q_1 + q_1 r_1 - q_1 + r_0 r_1 - r_0 - r_1 + 1)\bar{z}^2 \\ &+ (q_0 q_1^2 - q_1 + q_1 r_0 + q_1 r_1 - q_1 r_0 r_1)\bar{z} = 0. \end{aligned} \tag{8}$$

The possible equilibrium values for Eq.(8) are

$$\bar{z} = 0,$$

$$\bar{z} = q_1,$$

and

$$\bar{z} = \frac{q_0 q_1 - (r_0 - 1)(r_1 - 1)}{q_0 + 1 - r_1}, \qquad \text{when } q_0 + 1 \neq r_1.$$

2 Background

The following identities, which we state without proof, will be of importance.

$$
\begin{aligned}
x_{2n+1} - x_{2n-1} &= \frac{x_{2n}(q_0 - x_{2n-1}) + x_{2n-1}(r_0 - 1)}{1 + x_{2n}} && (9)\\
x_{2n+1} - q_0 &= \frac{r_0(x_{2n-1} - \frac{q_0}{r_0})}{1 + x_{2n}} && (10)\\
x_{2n+2} - x_{2n} &= \frac{x_{2n+1}(q_1 - x_{2n}) + x_{2n}(r_1 - 1)}{1 + x_{2n+1}} && (11)\\
x_{2n+2} - q_1 &= \frac{r_1(x_{2n} - \frac{q_1}{r_1})}{1 + x_{2n+1}} . && (12)
\end{aligned}
$$

We may consider Eq.(4) as a system of difference equations as follows:

$$
\begin{aligned}
u_n = x_{2n-1} \quad &, \quad v_n = x_{2n} \\
u_{n+1} = x_{2n+1} &= \frac{q_0 x_{2n} + r_0 x_{2n-1}}{1 + x_{2n}} = \frac{q_0 v_n + r_0 u_n}{1 + v_n} \\
v_{n+1} = x_{2n+2} &= \frac{q_1 x_{2n+1} + r_1 x_{2n}}{1 + x_{2n+1}} = \frac{q_1 u_{n+1} + r_1 v_n}{1 + u_{n+1}} \\
&= \frac{q_1 \left(\frac{q_0 v_n + r_0 u_n}{1 + v_n}\right) + r_1 v_n}{1 + \frac{q_0 v_n + r_0 u_n}{1 + v_n}} \\
&= \frac{r_1 v_n^2 + (q_0 q_1 + r_1) v_n + q_1 r_0 u_n}{(q_0 + 1) v_n + r_0 u_n + 1} .
\end{aligned}
$$

Then

$$
T\begin{pmatrix} u \\ v \end{pmatrix} = \begin{pmatrix} f(u,v) \\ g(u,v) \end{pmatrix} \tag{13}
$$

where

$$
f(u,v) = \frac{q_0 v + r_0 u}{1 + v}
$$

$$
g(u,v) = \frac{r_1 v^2 + (q_0 q_1 + r_1) v + q_1 r_0 u}{(q_0 + 1) v + r_0 u + 1}
$$

$$\frac{\partial f}{\partial u} = \frac{r_0}{1+v} \tag{14}$$

$$\begin{aligned} \frac{\partial f}{\partial v} &= -\frac{(r_0 u - q_0)}{(1+v)^2} \\ &= -\frac{r_0(u - \frac{q_0}{r_0})}{(1+v)^2} \end{aligned} \tag{15}$$

$$\begin{aligned} \frac{\partial g}{\partial u} &= -\frac{r_0(1+v)(-q_1 + r_1 v)}{((q_0+1)v + r_0 u + 1)^2} \\ &= -\frac{r_0 r_1 (1+v)(v - \frac{q_1}{r_1})}{((q_0+1)v + r_0 u + 1)^2} \end{aligned} \tag{16}$$

$$\begin{aligned} \frac{\partial g}{\partial v} &= [r_1(q_0+1)v^2 + 2r_1 v + 2r_0 r_1 uv + r_0(-q_1 + r_1)u \\ &+ q_0 q_1 + r_1] \\ &/ ((q_0+1)v + r_0 u + 1)^2 \,. \end{aligned} \tag{17}$$

3 The Case $r_0, r_1 < 1$

When $r_0, r_1 < 1$, the odd and even terms of the solution are bounded from above. This allows us to apply Linearized Stability results. The following lemma formalizes the result.

Lemma 1 *When $r_0 < 1$, the odd terms of the solution eventually enter the interval $(0, q_0/r_0]$ and remain in this interval. Similarly, when $r_1 < 1$, the even terms of the solution eventually enter the interval $(0, q_1/r_1]$ and remain in this interval.*

Proof. Suppose $r_0 < 1$ for all cases. If $x_{-1} \in (0, q_0/r_0]$, then identity (10) guarantees that the next odd term of the solution is also in the interval as $q_0 < q_0/r_0$. If $x_{-1} \in (q_0/r_0, \infty)$, then identity (9) shows that the odd terms of the solution will decrease monotonically as long as x_{2k+1} remains in this interval. A similar argument can be made using identities (12) and (11) for the even terms of the solution with respect to the interval $(0, q_1/r_1]$. □

Evaluating the Jacobian of T at $(0,0)$ using (14)-(17), (see (13)) and applying Linearized Stability (see [KL] Theorem 1.1.1(c)), the following inequality guarantees local asymptotic stability:

$$q_0 q_1 < (1 - r_0)(1 - r_1).$$

Theorem 1 *The zero solution of Eq.(4) is locally asymptotically stable when*

$$r_0, r_1 < 1$$

and

$$q_0 q_1 < (1 - r_0)(1 - r_1).$$

4 The Case $r_0 = r_1 = 1$

When $r_0 = r_1 = 1$, the odd and even terms of the solution are contained in invariant intervals. As before, this allows us to apply Linearized Stability results. The following lemma formalizes the result.

Lemma 2 *When $r_0 = r_1 = 1$, all of the odd terms of the solution are contained in either l $[q_0, x_{-1}]$ or $[x_{-1}, q_0)$ while all of the even terms of the solution are contained in either $[q_1, x_0]$ or $[x_0, q_1]$*

Proof. For all cases, suppose $r_0 = r_1 = 1$. If $x_{-1} > q_0$, then identities (9) and (10) show that the odd terms of the solution decrease and are bounded below by q_0. If $x_{-1} < q_0$, then the odd terms of the solution increase and are bounded above by q_0. Therefore, if $x_{-1} > q_0$, then the odd terms of the solution are in the interval $[q_0, x_{-1}]$ and if $x_{-1} < q_0$, then the odd terms of the solution are in the interval $[x_{-1}, q_0]$. A similar argument can be made using identities (11) and (12) for the even terms of the solution with intervals $[q_1, x_0]$ and $[x_0, q_1]$. □

Evaluating the Jacobian T using (14)-(17)(see (13)) at the solution (q_0, q_1) results in the conditions

$$\frac{1}{1 + q_0} < 1 \quad \text{and} \quad \frac{1}{1 + q_1} < 1$$

being required for stability. We have the following theorem.

Theorem 2 *When $r_0 = r_1 = 1$ and $q_0, q_1 > 0$, then the period-two solution $q_0, q_1, q_0, q_1, \ldots$ of Eq.(4) is locally asymptotically stable.*

5 The Case $r_0, r_1 > 1$

In this section we consider three separate cases.

Theorem 3 *Infinitely many prime period two solutions of Eq.(4) exist when*

$$r_0 = q_1 + 1 \quad \textit{and} \quad r_1 = q_0 + 1.$$

Proof. We prove the case for the odd terms of the solution, the case for the even terms is similar and is omitted.

Let

$$r_0 = q_1 + 1 \quad \text{and} \quad r_1 = q_0 + 1.$$

Clearly, $r_0 > 1$ and $r_1 > 1$ from which it follows that $q_0 > \frac{q_0}{r_0}$ and $q_1 > \frac{q_1}{r_1}$. We will show that the odd and even terms of the solution must be greater than q_0 and q_1, respectively. For the sake of contradiction, suppose all of the odd terms are bounded above by $\frac{q_0}{r_0}$. It follows from Eq.(9) that the odd terms must be strictly increasing. As the odd terms are bounded above, then the odd terms converge to a finite limit $L_o < q_0$. Taking the limit of (9) leads to a contradiction:

$$\begin{aligned}\lim_{n\to\infty}(x_{2n+1} - x_{2n-1}) = 0 &= \lim_{n\to\infty} \frac{x_{2n}(q_0 - x_{2n-1}) + x_{2n-1}(r_0 - 1)}{1 + x_{2n}} \\ &= \lim_{n\to\infty}\left(\frac{x_{2n}}{1 + x_{2n}}(q_0 - L_o) + \frac{L_o(r_0 - 1)}{1 + x_{2n}}\right) \neq 0.\end{aligned}$$

Thus we may assume without loss of generality that the odd and even terms of the solution are greater than q_0 and q_1, respectively.

$$r_0 x_{2n-1} - q_0 > 0 \quad \text{and} \quad r_1 x_{2n} - q_1 > 0$$

and so we can take the odd and even terms of Eq.(4)

$$\begin{aligned} x_{2n+1} &= q_0 + \frac{r_0 x_{2n-1} - q_0}{1 + x_{2n}} \\ x_{2n+2} &= q_1 + \frac{r_1 x_{2n} - q_1}{1 + x_{2n+1}} \end{aligned}$$

and apply the change of variables

$$x_{2n-1} = q_0 + y_{2n-1} \quad \text{and} \quad x_{2n} = q_1 + y_{2n}.$$

This results in

$$y_{2n+1} = \frac{q_0 q_1 + r_0 y_{2n-1}}{(1+q_1) + y_{2n}} \tag{18}$$

$$y_{2n+2} = \frac{q_0 q_1 + r_1 y_{2n}}{(1+q_0) + y_{2n+1}}. \tag{19}$$

Next, let

$$y_{2n-1} = (1+q_0)z_{2n-1} \quad \text{and} \quad y_{2n} = (1+q_1)z_{2n}.$$

Eq.(18) becomes

$$\begin{aligned}
(1+q_0)z_{2n+1} &= \frac{q_0 q_1 + r_0(1+q_0)z_{2n-1}}{1+q_1+(1+q_1)z_{2n}} \\
z_{2n+1} &= \frac{\frac{q_0 q_1}{1+q_0} + r_0 z_{2n-1}}{(1+q_1)(1+z_{2n})} \\
&= \frac{\frac{q_0 q_1}{(1+q_0)(1+q_1)} + \frac{r_0}{1+q_1} z_{2n-1}}{1+z_{2n}} \\
&= \frac{K + z_{2n-1}}{1+z_{2n}}.
\end{aligned} \tag{20}$$

Similarly, Eq.(19) becomes

$$\frac{\frac{q_0 q_1}{(1+q_0)(1+q_1)} + \frac{r_1}{1+q_0} z_{2n}}{1+z_{2n+1}} \tag{21}$$

which reduces to

$$z_{2n+2} = \frac{K + z_{2n}}{1+z_{2n+1}}$$

and so the change of variables has reduced the periodic equation to its autonomous form when $r_0 = q_1 + 1$ and $r_1 = q_0 + 1$. We apply the result established in [GKL] to complete the proof. □

Theorem 4 *Every solution of Eq.(4) converges to the period-two solution*

$$\frac{q_0 q_1 - (r_0-1)(r_1-1)}{q_1 + 1 - r_0}, \frac{q_0 q_1 - (r_0-1)(r_1-1)}{q_0 + 1 - r_1}, \ldots$$

when

$$1 < r_0 < q_1 + 1 \quad \textit{and} \quad 1 < r_1 < q_0 + 1.$$

Proof. Using a similar argument to the previous section, we can show that the odd terms of the solution are greater than q_0 and the even terms of the solution are greater than q_1 when $1 < r_0 < q_1 + 1$ and $1 < r_1 < q_0 + 1$. Through the same change of variables, the odd terms reduce to Eq.(20) and the even terms to Eq.(21). We then apply the result in [GO] and the proof is complete. □

Theorem 5 *Eq.(4) posscsses unbounded solutions when*

$$r_0 > q_1 + 1 \quad \textit{and} \quad r_1 > q_0 + 1.$$

Proof. Using a similar argument to the previous section, we can show that the odd terms of the solution are greater than q_0 and the even terms of the solution are greater than q_1 when $r_0 > q_1 + 1$ and $r_1 > q_0 + 1$. Through the same change of variables, the odd terms reduce to Eq.(20) and the even terms to Eq.(21). We then apply the result in [GO] and the proof is complete.
□

Acknowledgements
The authors would like to thank the referee for helpful comments and suggestions.

References

[GKL] C. H. Gibbons, M. R. S. Kulenović, and G. Ladas, On the Recursive Sequence $y_{n+1} = \frac{\alpha+\beta y_{n-1}}{\gamma+y_n}$, *Math. Sci. Res. Hot-Line* 4(2)(2000), 1-11 .

[GO] C. H. Gibbons and C. B. Overdeep, On the Trichotomy Character of $x_{n+1} = \frac{\alpha_n+\gamma_n x_{n-1}}{A_n+B_n x_n}$ with Period-two Coefficients, *to appear.*

[KL] M. R. S. Kulenović and G. Ladas, *Dynamics of Second Order Rational Difference Equations, Open Problems and Conjectures*, Chapman&Hall/CRC Press, 2001.

[KLP] M. R. S. Kulenović, G. Ladas, and N. R. Prokup, On the Recursive Sequence $x_{n+1} = \frac{\alpha x_n+\beta x_{n-1}}{1+x_n}$, *J. Differ. Equation Appl.* 6(5)(2000), 563-576 .

PERIODICALLY FORCED NONLINEAR DIFFERENCE EQUATIONS WITH DELAY

ABDUL-AZIZ YAKUBU
DEPARTMENT OF MATHEMATICS
HOWARD UNIVERSITY
WASHINGTON, D. C. 20059
(AYAKUBU@HOWARD.EDU)

Abstract: Periodically forced dynamical systems are of great importance in modeling biological processes in periodically varying environments. In this paper, we survey the fundamental results of Elaydi and Yakubu, Elaydi and Sacker, Cushing and Henson, Franke and Selgrade, Franke and Yakubu on periodically forced (nonautonomous) difference equations without delay. Extensions of these results to periodically forced nonlinear difference equations with delay are posed as open problems.

Key words: autonomous difference equations, delay, periodic difference equations.

1. INTRODUCTION

Nonautonomous nonlinear discrete-time models that incorporate complex population dynamics with delay have rarely been studied. However, extensive work using continuous time nonlinear differential equations with or without delay have been carried out by large number of researchers (see for example [2, 4, 10, 19-22, 30, 34, 38-40, 54], to name a few). In this paper, we survey some of the fundamental results on periodically forced (nonautonomous) nonlinear difference equations without delay, and posed open problems on periodic nonlinear difference equations with delay.

In Section 2, we introduce a very general autonomous nonlinear difference equation model without delay. The classical Beverton-Holt and Ricker models are examples of the general model. Elaydi and Yakubu proved that non-trivial cyclic orbits of autonomous systems are not globally asymptotically stable [15]. Elaydi and Sacker's extension of this result to non-autonomous difference equations is discussed in Section 3. Elaydi and Sacker [11-14] proved that for a $k-$ *periodic* dynamical system, if a periodic orbit of period r is globally asymptotically stable (no multiple attractors) then r must be a divisor of k. An extension of this result to discrete-time nonautonomous equations with delay is an open question.

Cushing and Henson's results on monotone nonautonomous nonlinear difference equation models without delay are discussed in Section 4. Cushing and Henson derived conditions for the existence of globally attracting cycles and conditions under which the attracting cycle is attenuant [5]. Elaydi and Sacker [12], Kocic [33] and Kon [35, 36] have extended these results from $2-$ *periodic* equations to $p-$ *periodic* equations with $p > 2$. What are the corresponding conditions for monotone discrete-time nonautonomous equations with delay?

The autonomous Beverton-Holt and Ricker models are capable of supporting only single attractors (no multiple attractors) while the corresponding nonautonomous models are known to have multiple attractors for some choice of parameter regimes [2, 3, 6-32, 34, 43-54]. Henson [27], Franke and Selgrade [16] as well as Franke and Yakubu [17, 18] have developed mathematical framework for studying attractors of nonautonomous discrete-time systems without delay. In particular, Franke and Selgrade proved that an attractor for a nonautonomous discrete-time system is the union of attractors of the corresponding autonomous system. This result is stated in Section 5. An extension of the attractor structure theorem of Franke and Selgrade to discrete-time nonautonomous equations with delay is an open question. We state these open problems in Section 5.

2. Global Stability Of Periodic Orbits: Autonomous Difference Equations

A continuous map on a metric space X,

$$F : X \to X$$

generates an *autonomous* difference equation

$$x(n+1) = F(x(n)),\ n \in \mathcal{Z}^+ \quad (1)$$

where for each $x(0) \in X$, $x(n) = F^n(x(0))$, $F^0(x(0)) = x(0)$, $F^n = F \circ F \circ ... \circ F$, and $\mathcal{Z}^+$ is the set of nonnegative integers.

The single species single patch autonomous ecological parametric models, the Beverton-Holt model,

$$x(n+1) = \frac{\mu K x(n)}{K + (\mu - 1)x(n)}, \qquad (2)$$

and the Ricker model,

$$x(n+1) = x(n)\exp(r(1 - \frac{x(n)}{K})), \quad (3)$$

are examples of System (1) with $X = [0, \infty)$; where μ, K and r are positive constants; and $x(n)$ is the population size at generation n [2, 5-20, 23-32, 33, 38, 43-47, 49-54]. The coefficients r and μ are the *inherent growth rates* of the species, and the carrying capacity K is a characteristic of the habitat or environment.

In the Beverton-Holt model, if $\mu \leq 1$ then every initial population size converges to the equilibrium point $x_\infty = 0$ and the population goes extinct. However, if $\mu > 1$ then every positive initial population size converges to the equilibrium point $x_\infty = K$ and the population persists [5]. That is, if $\mu \leq 1$ then $x_\infty = 0$ is globally stable in $[0, \infty)$, otherwise $x_\infty = 0$ is unstable and $x_\infty = K$ is globally stable in $(0, \infty)$.

The set of fixed points of Ricker's model is $\{0, K\}$. The fixed point $x_\infty = 0$ is unstable while $x_\infty = K$ is globally stable in $(0, \infty)$ whenever $0 < r \leq 2$ [9, 13, 20, 23-27, 34, 38, 41, 43-47, 49-54]. As the value of the parameter r increases past 2, the positive fixed point undergoes period-doubling bifurcation. Further increase in r values lead to a whole infinite sequence of such period-doubling bifurcations, and the Ricker model supports stable cycles. When $r > 3.102$ the Ricker model supports chaotic dynamics [9, 13, 20, 23-27, 34, 38, 41, 43-47, 49-54].

Hastings [25, 26], Gyllenburg *et al.* [20], Doebeli [6, 7], Yakubu [52], Yakubu and Castillo-Chavez [53] have studied discrete-time, autonomous, single-species metapopulation models that implicitly assume that dispersal is either bidirectional or unidirectional. A two-patch version of these metapopulation models under unidirectional dispersal is given by the following system of coupled nonlinear *autonomous* difference equations:

$$\left.\begin{array}{rcl} x_1(n+1) & = & (1-d)h_1(x_1(n)), \\ x_2(n+1) & = & dh_1(x_1(n)) + h_2(x_2(n)), \end{array}\right\} (4)$$

where for each Patch $i \in \{1, 2\}$, $x_i(n)$ is the population size at generation n, $h_i(x_i(n)) = x_i(n)g_i(x_i(n))$, the map $g_i : [0, \infty) \to [0, \infty)$ is the per capita growth rate and the constant fraction $d \in (0, 1)$ is the proportion of the population that disperses from Patch 1 to 2 (unidirectional dispersal). System (4) is an example of System (1) with $X = [0, \infty) \times [0, \infty)$.

When

$$\lim_{n \to \infty} (1-d)h_1(x_1(n)) = x_{1d} > 0$$

for all $x_1(n) > 0$ and

$$\lim_{n\to\infty} dh_1(x_{1d}) + h_2(x_2(n)) = x_{2d} > 0$$

for all $x_2(n) > 0$ then

$$(x_{1d}, x_{2d})$$

is a globally attracting fixed point of System (4) [52].

The Beverton-Holt and Ricker models are not capable of supporting multiple attractors [2, 5-20, 23-32, 33, 38, 43-47, 49-54]. We use the following example to illustrate multiple (coexisting) attractors in System (4).

Example 1. *In System (4), let*

$$h_1(x_1(n)) = h_2(x_1(n)) = x_1(n)\exp(r(1 - \frac{x_1(n)}{K}))$$

where $r = 2.1$, $K = 1$ *and* $d = 0.01$.

In Example 1, the pre-dispersal local populations are governed by identical Ricker model. That is, without dispersal, each local population is on the same single 2-cycle attractor. However, with dispersal, the unidirectional model supports two coexisting 2-cycle attractors (multiple attractors) at (1.3466, 0.6511) → (0.6438, 1.3612) and (1.3466, 1.3926) → (0.6438, 0.6171); see FIG. 1.

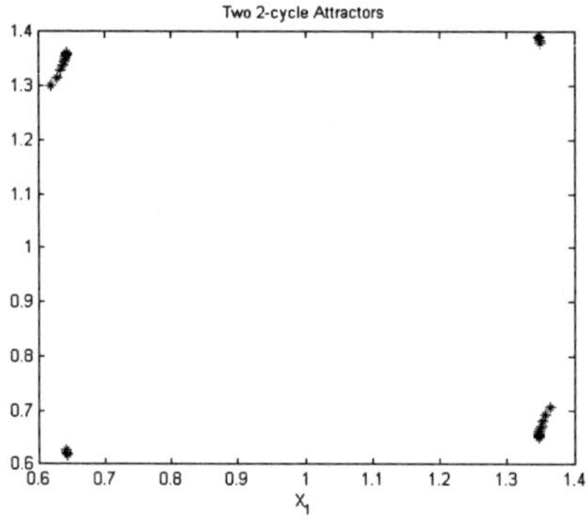

FIG. 1: Two 2-cycle attractors at (1.3466, 0.6511) → (0.6438, 1.3612) and (1.3466, 1.3926) → (0.6438, 0.6171).

Example 1 shows that dispersal is capable of generating multiple attractors in discrete-time autonomous models where the pre-dispersal local populations are on single locally asymptotically stable cyclic attractors. What is the maximum number of coexisting attractors that Example 1 can support? What is the structure of the coexisting attractors, and what is the nature of their basins of attraction [1]? Are cyclic orbits globally asymptotically stable [11-15]?

Using the Dynamics software of Nusse and Yorke [47], Yakubu and Castillo-Chavez [52, 53] observed that as the complexity of the pre-dispersal local dynamics increases the long-term dynamics of a metapopulation becomes less predictable to the point that it becomes difficult to determine, with any degree of certainty, the fate of such metapopulation. The following fundamental result of Elaydi and Yakubu proves that non-trivial cyclic orbits of autonomous systems are not globally asymptotically stable [15].

Theorem 1 [15]: *Let*

$$F : X \to X$$

be a continuous map on a connected metric space X. If a $k-$ cycle c_k is globally asymptotically stable, then c_k must be a fixed point.

In the next section, we discuss an extension of Theorem 1 to non-autonomous difference equations.

3. Non-autonomous Difference Equations

A continuous map,

$$F : \mathcal{Z}^+ \times X \to \mathcal{Z}^+ \times X$$

generates a *non-autonomous* difference equation

$$x(n+1) = F(n, x(n)) \quad (5)$$

where $n \in \mathcal{Z}^+$.

Mathematical and theoretical ecologists have developed non-autonomous models to answer questions on the effects of either periodic environments or periodic growth rates on populations [2, 5-20, 23-32, 33, 38, 43-47, 49-54]. To make Equation (5) $p-$*periodic,* we assume that there exists a smallest positive integer p satisfying

$$F(n+p, x(n)) = F(n, x(n)). \quad (6)$$

For each $i \in \{0, 1, ..., p-1\}$, define

$$F_i : X \to X$$

by $F_i(x) = F(i, x)$. For $i \geq p$, let

$$F_i(x) = F_{i \bmod (p)}(x).$$

Definition 1: *A $p-$periodic dynamical system is a sequence of maps $\{F_0, F_1, ..., F_{p-1}\}$ from X to X such that $F_i = F_{i \bmod (p)}$ for all $i \in \mathcal{Z}^+$, where p is the smallest such integer.*

By definition 1, a $p-periodic$ dynamical system is a *finite* sequence of p maps. To define the *orbit of a point* $x_0 \in X$, let $x_i = F_{i-1}(x_{i-1})$ for each $i \geq 1$.

Definition 2: *The orbit of a point* $x_0 \in X$ *under the* $p-periodic$ *dynamical system* $\{F_0, F_1, ..., F_{p-1}\}$ *is* $\{x_0, F_0(x_0), F_1(x_1), ..., F_i(x_i), ...\}$.

Definition 3: *An orbit* $\{x_0, x_1, ..., x_{k-1}, ...\}$ *is a* $k-cycle$ of the $p-periodic$ *dynamical system* $\{F_0, F_1, ..., F_{p-1}\}$ *if* $x_i = x_{i \bmod(k)}$ *for all* $i \in \mathcal{Z}^+$, *where* k *is the smallest such positive integer.*

Consequently, the orbit of a fixed point under the $p-periodic$ dynamical system $\{F_0, F_1, ..., F_{p-1}\}$ is $\{x_0, x_0, ..., x_0, ...\}$. Fixed point dynamics are rare in periodically forced discrete-time systems [17, 18].

Definition 4: *An ordered set of points*

$$C = \{c_0, c_1, ..., c_{r-1}\}, \quad c_i \in X$$

is a geometric $r-cycle$ of the $p-periodic$ *dynamical system* $\{F_0, F_1, ..., F_{p-1}\}$ *if*

$$F_{(i+nr) \bmod p}(c_i) = c_{i+1 \bmod(r)}$$

for all $n \in \mathcal{Z}$ *and* $i \in \{0, 1, ..., r-1\}$, *where* $r > 0$ *is an integer and* $r \leq p$.

Next, we state fundamental results of Elaydi and Sacker on global asymptotic stability of periodic orbits in non-autonomous equations [11-14].

Theorem 2 [11-14]: *Assume that each* $F_i \in \{F_0, F_1, ..., F_{p-1}\}$ *is a continuous map on a connected metric space,* X. *Let* $c_r = \{\overline{x}_1, \overline{x}_2, ..., \overline{x}_r\}$ *be a geometric* $r-cycle$ *of the* $p-periodic$ *Equation (5). If* c_r *is globally asymptotically stable then* $r \mid p$, *that is* r *divides* p.

The following result shows how to construct periodic dynamical systems that support globally asymptotically stable cycles.

Theorem 3 [11-14]: *Let* $r \mid p$, *where* r *and* p *are any two positive integers. Then there exists a* $p-periodic$ *dynamical system that has a globally asymptotically stable geometric* $r-cycle$ *with minimum period* r.

For an application of Theorem 3, consider the periodically forced Beverton-Holt model

$$x(n+1) = \frac{\mu K(n)x(n)}{K(n) + (\mu - 1)x(n)}, \quad (7)$$

where

$$K(n+p) = K(n) > 0.$$

Equation (7) is an example of System (5). To illustrate parameter regimes for the existence of a globally asymptotically stable cycle in Model (7), the $p-$*periodic* Beverton-Holt model, let M_n satisfy the following linear difference equation.

$$M_{n+1} = K_{n+1}M_n + \mu^{n+1}K_nK_{n-1}...K_0, \quad M_0 = 1. \qquad (8)$$

Then,

$$M_{p-1} = \Pi_{j=0}^{p-2}K_{j+1} + \sum_{m=0}^{p-2}(\Pi_{i=m+1}^{p-2}K_{i+1})\mu^{m+1} \cdot K_mK_{m-1}...K_0.$$

In [11-14], Elaydi and Sacker showed that when Model (7) has an $r-$*cycle* then

$$M_{p-1} = \left(\frac{\mu^p - 1}{\mu^r - 1}\right)K_p...K_rM_{r-1}. \qquad (9)$$

Theorem 4 [11-14]: *Suppose $\mu > 1$, $K_i > 0$, $0 \leq i \leq p-1$ and $r \mid p$. Then Model (7) has a globally asymptotically stable $r-$cycle if and only if condition (9) holds.*

Equation (7) has a globally asymptotically stable cycle of period p. In [14], Elaydi and Sacker proved that condition (9) holds if and only if $r \mid p$. That is, in Equation (7) any geometric cycle must have the same minimal period p as the equation whenever $\mu > 1$ [14].

When both intrinsic growth rate μ and the carrying capacity K are periodic with minimal common period $p \geq 2$, then the Beverton-Holt equation becomes

$$x(n+1) = \frac{\mu(n)K(n)x(n)}{K(n) + (\mu(n)-1)x(n)}, \qquad (10)$$

where

$$K(n+p) = K(n) > 0,$$

and

$$\mu(n+p) = \mu(n) > 0.$$

Using Equation (10), Elaydi and Sacker showed that it is possible to have a geometric cycle with minimal period $r < p$ whenever both μ and K are non-constant and periodic. This is impossible in Equation (7) with constant μ [14].

4. Monotone Non-autonomous Difference Equations

Are populations adversely affected by a periodic environment (relative to a constant environment of the same average carrying capacity)? Results based on the logistic differential equation imply that a periodic carrying capacity K is deleterious. That is, the average of the resulting population oscillations is less than the average of K. Cushing and Henson [5] used a general class of $2-$*periodic* monotone population models to obtain similar results for non-autonomous difference equations. Their model is based on Equation (7) with $p = 2$, and it has the general form

$$x(n+1) = x(n)f(\frac{x(n)}{1+\alpha(-1)^n}), \qquad (11)$$

where $\alpha \in [0,1)$ and the function

$$h(x) \equiv xf(x)$$

satisfies the following monotone conditions:

$$\left.\begin{array}{c} h \in C^0([0,\infty),\ [0,\infty)) \cap C^2((0,\infty),\ [0,\infty)) \\ h'(x) > 0,\ h''(x) < 0 \text{ for all } x \in (0,\infty) \\ h(0) = 0 \text{ and } \lim_{x\to\infty} h(x) < \infty. \end{array}\right\} \quad (12)$$

If $\alpha = 0$, Equation (11) reduces to an autonomous equation. In this case, as in the Beverton-Holt model, if $h'(0) \leq 1$ then every initial population size converges to the equilibrium point $x_\infty = 0$ and the population goes extinct. However, if $h'(0) > 1$ then every positive initial population size converges to the positive equilibrium point $x_{\infty e}$ and the population persists [5]. That is, if $h'(0) \leq 1$ then $x_\infty = 0$ is globally stable in $[0,\infty)$, otherwise $x_\infty = 0$ is unstable and $x_{\infty e}$ is globally stable in $(0,\infty)$.

To state the fundamental result of Cushing and Henson, we need the following notation, assumptions and definition.

$$\left.\begin{array}{c} f \in C^0([0,\infty),\ [0,\infty)) \cap C^2((0,\infty),\ [0,\infty)) \\ f''(x) > 0 \text{ for all } x \in (0,\infty)\ . \end{array}\right\} \quad (13)$$

$$h(1) = 1\ . \quad (14)$$

When $\alpha > 0$, Equation (11) is a $2-$*periodic* non-autonomous system. If $h'(0) > 1$ then the system has a unique positive *strict* 2-cycle (c_0, c_1) that is globally attracting for every positive initial population size $(c_0 \neq c_1)$ [5].

The 2-cycle (c_0, c_1) is said to be *attenuant* if and only if

$$\frac{1}{2}(c_0 + c_1) < 1\ . \quad (15)$$

Next, we state the result of Cushing and Henson on the deleterious effects of periodic environments.

Theorem 5 [5]: *Assume Condition (13) and Equation (14) hold. For each $\alpha \in (0,1)$ the positive, globally attracting 2-cycle of the periodically forced Equation (11) is attenuant.*

Theorem 5 of Cushing and Henson has been extended to the $p-$*periodic* Equation (7) with $p > 2$ in 4 papers using 3 different proofs [12, 33, 35, 36].

5. Multiple Attractors In Nonautonomous Difference Equations

Henson [27], Franke and Selgrade [16] as well as Franke and Yakubu [17, 18] have developed mathematical framework for studying attractors of periodically forced discrete-time systems. Nonautonomous systems are capable of supporting multiple attractors where the corresponding autonomous systems support single attractors. For example, the classic autonomous Beverton-Holt and Ricker models sustain only single attractors (no multiple attractors) while the corresponding nonautonomous models are known to have multiple attractors for some choice of parameter regimes [17, 18].

Franke and Yakubu illustrated multiple attractors via a tangent bifurcation in the $2-$*periodic* Ricker model

$$x(n+1) = x(n)\exp(r + \alpha(-1)^n - x(n)). \qquad (16)$$

Franke and Yakubu showed that, when α is fixed at 0.01 while r is increased from 1.8 to 2.3, a tangent bifurcation occurs and Model (16) has two stable $2-$*cycles* where the corresponding Ricker's model has only *one* attractor (see FIG. 2 from [18]).

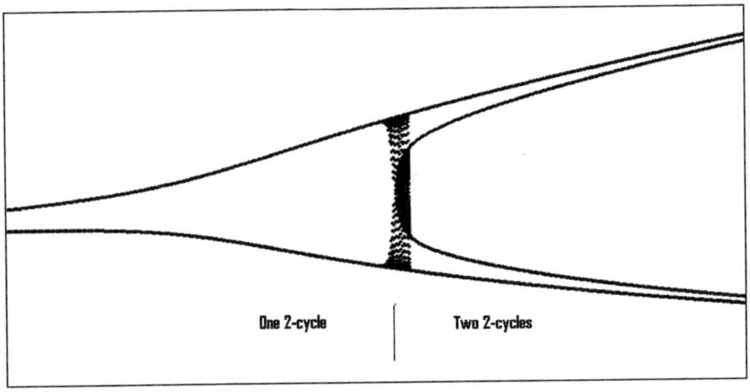

FIG. 2: Two 2-cycle attractors in Equation (16) (Multiple attractors), where on the horizontal axis $1.8 \leq r \leq 2.3$ and on the vertical axis $0 \leq y \leq 4$.

In [16], Franke and Selgrade proved that an attractor for a periodically forced discrete-time system is the union of attractors of the corresponding autonomous system. We summarize this in the following result:

Theorem 6 [16]: *Let Λ be an attractor for the $p-$periodic dynamical system $\{F_0, F_1, ..., F_{p-1}\}$. Then*

$$\Lambda = \cup_{i=0}^{i=p-1}\Lambda_i,$$

where Λ_i is an attractor for the map

$$F_{i+p-1} \circ ... \circ F_{i+1} \circ F_i : X \to X,$$

for each $i \in \{0, 1, ..., p-1\}$.

To illustrate Theorem 6 in a specific example, Franke and Selgrade considered the 2-dimensional predator-prey nonautonomous system

$$\left.\begin{array}{rcl} x_1(n+1) & = & x_1(n)(2 - x_1(n) - 0.5x_2(n)), \\ x_2(n+1) & = & x_2(n)(0.8(1+\alpha(-1)^n) + 1.3x_2(n)). \end{array}\right\} (17)$$

When $\alpha = 0$, System (17) reduces to an autonomous system with an attracting invariant loop that emerged from a Neimark-Sacker (Hopf) bifurcation. However, as α increases from 0, the attracting invariant loop splits into two attracting loops. FIG. 3 illustrates two attracting loops in System (17) where $\alpha = 0.1$.

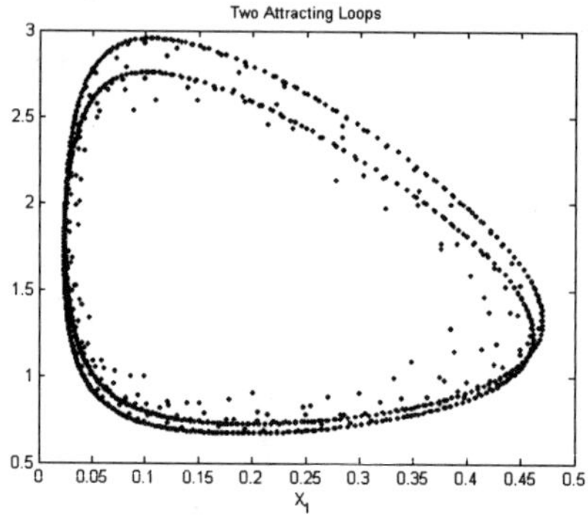

FIG. 3: Two attracting loops in System (17) with $\alpha = 0.1$ (nonautonmous) where there is only one loop when $\alpha = 0$ (autonomous).

6. Open Questions: Periodically Forced Nonlinear Systems With Delay

In this section, we use nonautonomous nonlinear discrete-time models *with delay* to generate some open problems. Motivated by biological applications, Cushing and Henson used System (11) to study the long-term dynamical behaviors of a class of periodically forced difference equations without delay [5]. Now, we use System (11) to introduce a class of periodically forced (*nonautonomous)* models with delay. With a reproductive delay of two generations, System (11) becomes,

$$x(n+1) = x(n-1)f(\frac{x(n-1)}{1+\alpha(-1)^n}). \quad (18)$$

To study System (18), we consider the following more general periodically forced discrete-time system with delay.

$$x(n+1) = F(n, x(n), x(n-1)) \quad (19)$$

where $F(n+p, x(n), x(n-1)) = F(n, x(n), x(n-1))$ for all $n \in \mathcal{Z}^+$. Notice that Model (18) is an example of (19).

Problem 1: *In System (18), let*

$$f\left(\frac{x(n-1)}{1+\alpha(-1)^n}\right) = \frac{\alpha+\beta(-1)^n}{1+x(n-1)}$$

where α and β are positive constants and $\alpha - \beta > 0$. Prove that the resulting system is capable of supporting a globally attracting 2-cycle. Is the 2-cycle attenuant?

Problem 2: *Extend the result of Cushing and Henson [5] on the globally attracting 2-cycle of the periodically forced Equation (11) without delay (Theorem 5) to the periodically forced Equation (18) with delay.*

Problem 3: *Extend the result of Elaydi and Sacker [11-14] on the global asymptotic stability of periodic orbits in nonautonomous equations without delay (Theorem 2) to nonautonomous equations with delay (Equation (19)).*

Problem 4: *Extend the result of Franke and Selgrade [16] on the structure of attractors of nonautonomous equations without delay (Theorem 6) to that of nonautonomous equations with delay (Equation (19)).*

Problem 5: *What is role of "delay" in generating multiple attractors in Equation (19) (see FIG. 2)?*

In some discrete-time epidemic models, the rate of arrival of new susceptibles per generation is determined by the population of adults r generations previously, where $r = 1, 2, 3, ...$ To model these biological processes, we consider the following higher order $p-$ *periodic* nonautonomous difference equation:

$$x(n+1) = F(n, x(n), x(n-1),\ x(n-2), x(n-3), ..., x(n-r)) \quad (20)$$

where

$$F : \mathcal{Z}^+ \times \Re_+^{r+1} \to \mathcal{Z}^+ \times \Re_+^{r+1}$$

satisfies

$$F(n+p, x(n), x(n-1),\ x(n-2), x(n-3), ..., x(n-r)) =$$

$$F(n, x(n), x(n-1),\ x(n-2), x(n-3), ..., x(n-r))$$

for all $n \in Z^+$.

The literature on linear versions of Equation (20) is extensive [10, 31, 32-38, 42, 48, 54]. Others have studied nonlinear versions of (20) (for example, see [34]). Here we pose the following question on extensions of Theorem 2 to include Equation (20).

Problem 6: *Extend the result of Elaydi and Sacker [11-14] on the global asymptotic stability of periodic orbits in nonautonomous equations without delay (Theorem 2) to the higher order Equation (20).*

7. Conclusion

This survey paper focuses on fundamental results of Elaydi and Yakubu, Elaydi and Sacker, Cushing and Henson, Franke and Selgrade, Franke and Yakubu on periodically forced (nonautonomous) difference equations without delay. Elaydi and Yakubu proved that non-trivial periodic orbits of autonomous periodic systems are not globally asymptotically stable. Elaydi and Sacker extended this result to nonautonomous systems. To be more specific, for a $k-$*periodic* dynamical system, Elaydi and Sacker proved that if a periodic orbit of period r is globally asymptotically stable then r must be a divisor of k. Motivated by applications from population biology, Cushing and Henson studied a very general class of periodically forced monotone systems, while Franke and Selgrade provided a mathematical framework for studying attractors of discrete nonautonomous systems. In a more recent paper, Franke and Yakubu demonstrated multiple attractors via cusp bifurcations in periodic dynamical systems.

The extensions of these results to nonautonomous nonlinear discrete-time systems with delay are interesting open questions.

ACKNOWLEDGEMENT: The author thanks Saber Elaydi for helpful comments and suggestions on the manuscript.

References

1. J. C. Alexander, J. A. Yorke, Z. You and I. Kan, *Riddled Basins,* Intern. J. Bifurc. Chaos, 2(4),795-813(1992).

2. M. Begon, J. L. Harper and C. R. Townsend, Ecology: individuals, populations and communities, *Blackwell Science Ltd* (1996).

3. R. F. Costantino, J. M. Cushing, B. Dennis and R. A. Desharnais, Resonant popopulation cycles in temporarily fluctuating habitats, *Bull. Math. Biol.* 60, 247-273 (1998).

4. J. M. Cushing, Periodic time-dependent predator-prey systems, *SIAM J. Appl. Math.* 32, 82-95 (1977).

5. J. M. Cushing and S. M. Henson, Global dynamics of some periodically forced, monotone difference equations, *J. Diff. Equations Appl. 7, 859-872 (2001).*

6. M. Doebeli, Dispersal and dynamics, *Theor. Popl. Biol.* 47:82-106 (1995).

7. M. Doebeli and G. D. Ruxton, Evolution of dispersal rates in metapopulation model: Branching and cyclic dynamics in phenotype space, *Evolution* 5(6), 1730-1741 (1997).

8. D. J. Earn, S. A. Levin and P. Rohani, Coherence and conservation, *Science*, Vol. 290, Nov. 17 (2000).

9. S. N. Elaydi, Discrete Chaos. *Chapman & Hall/CRC,* Boca Raton, FL, 2000.

10. S. N. Elaydi, Periodicity and stability of linear Volterra difference equations, *J. Math. Anal. Appl.,* 181, 483-492 (1994).

11. S. N. Elaydi and R. J. Sacker, Global Stability of Periodic Orbits of Nonautonomous Difference Equations and Population Biology, *J. Differential. Eq. (In Press).*

12. S. N. Elaydi and R. J. Sacker, Global Stability of Periodic Orbits of Nonautonomous Difference Equations and Population Biology, *Proceedings of ICDEA8, Brno (2003).*

13. S. N. Elaydi and R. J. Sacker, Global Stability of Periodic Orbits of Nonautonomous Difference Equations In Population Biology and Cushing-Henson Conjectures, *J. Diff. Equations Appl. (In Press).*

14. S. N. Elaydi and R. J. Sacker, Nonautonomous Beverton-Holt Equations and the Cushing-Henson Conjectures, *J. Diff. Equations Appl (In Press).*

15. S. N. Elaydi and A.-A. Yakubu, Global Stability of Cycles: Lotka-Volterra Competition Model With Stocking, *J. Diff. Equations Appl. 8(6), 537-549 (2002).*

16. J. E. Franke and J. F. Selgrade, Attractor for Periodic Dynamical Systems, *J. Math. Anal. Appl.* 286, 64-79(2003).

17. J. E. Franke and A.-A. Yakubu, Periodic Dynamical Systems in Unidirectional Metapopulation models, *J. Diff. Equations Appl. (In Press).*

18. J. E. Franke and A.-A. Yakubu, Multiple Attractors Via Cusp Bifurcation In Periodically Varying Environments, *J. Diff. Equations Appl. (In Press).*

19. J. Gűémez and M. A. Matías, Control of chaos in unidimensional maps, *Physics Letters A* 181, 29-32 (1993).

20. M. Gyllenberg, G. Söderbacka and S. Ericsson, Does migration stabilize local population dynamics? Analysis of a discrete metapopulation model, *Math. Biosc.* 118, 25-49 (1993).

21. I. Hanski, Single-species metapopulation dynamics-concepts, models and observations, *Biol. J. Linn. Soc.* 42, 17-38 (1991).

22. I. A. Hanski and M. E. Gilpin, Metapopulation Biology: ecology, genetics, and evolution, *Academic Press Ltd. San Diego, California* (1997).

23. M. P. Hassell, The dynamics of competition and predation, *Studies in Biol.* 72, The Camelot Press Ltd. (1976).

24. M. P. Hassell, J. H. Lawton and R. M. May, Patterns of dynamical behavior in single species populations, *J. Anim. Ecol.* 45, 471-486 (1976).

25. A. Hastings, Complex interactions between dispersal and dynamics: Lessons from coupled logistic equations, *Ecology* 75, 1362-1372 (1993).

26. A. Hastings, Dynamics of a single species in a spatially varying environment: The stabilizing role of high dispersal rates, *J. Math. Bio.* 16, 49-55 (1982).

27. S. M. Henson, Multiple attractors and resonance in periodically forced population models, *Physics D* 140, 33-49 (2000).

28. S. M. Henson, R. F. Costantino, J. M. Cushing, B. Dennis and R. A. Desharnais, Multiple attractors, saddles, and population dynamics in periodic habitats, *Bull. Math. Biol.* 61, 1121-1149 (1999).

29. S. M. Henson, R. F. Costantino, R. A. Desharnais, J. M. Cushing and B. Dennis, Basins of attraction: population dynamics with two stable 4-cycles (Preprint).

30. S. M. Henson and J. M. Cushing, Hierarchical models of intraspecific competition: scramble versus contest, *J. Math. Biol.* 34, 755–772 (1996).

31. S. M. Henson and J. M. Cushing, The effect of periodic habitat fluctuations on a nonlinear insect population model, *J. Math. Biol.* 36, 201-226 (1997).

32. D. Jillson, Insect populations respond to fluctuating environments, Nature 288, 699-700 (1980).

33. V. L. Kocic, A note on nonautonomous Beverton-Holt model, *J. Diff. Equations Appl.* (In press).

34. V. L. Kocic and G. Ladas, Global behavior of nonlinear difference equations of higher order with applications, *Mathematics and its Applications,* 256, Kluwer Academic Publishers Group, Dordrecht, 1993.

35. R. Kon, A note on attenuant cycles of population models with periodic carrying capacity, *J. Diff. Equations Appl.* (In press).

36. R. Kon, Attenuant cycles of population models with periodic carrying capacity, *J. Diff. Equations Appl.* (In press).

37. U. Krause and M. Pituk, Boundedness and stability for higher order difference equations, *J. Diff. Equations Appl.* 10, 343-356 (2004).

38. Y. A. Kuznetsov, Elements of Applied Bifurcation Theory, *Springer-Verlag* (1995).

39. S. A. Levin: Dispersion and population interactions, *Amer. Naturalist.* 108, 207-228 (1974).

40. S. A. Levin and R. May, A note on delay difference equations, Theor. Pop. Biol. 9, 178-187 (1976).

41. J. Li, Periodic solutions of population models in a periodically fluctuating environment, *Math. Biosc.* 110, 17-25 (1992).

42. E. Liz and J. B. Ferreiro, A note on the global stability of generalized difference equations, *Appl. Math. Lett* 15, 655-659 (2002).

43. R. M. May and G. F. Oster, Bifurcations and dynamic complexity in simple ecological models, *Amer. Naturalist* 110, 573-579 (1976).

44. R. M. May, Simple mathematical models with very complicated dynamics, *Nature* 261, 459-469 (1977).

45. R. M. May, Stability and complexity in model ecosystems, *Princeton University Press* (1974).

46. A. J. Nicholson, Compensatory reactions of populations to stresses, and their evolutionary significance, *Aust. J. Zool.* **2**, 1-65 (1954).

47. H. E. Nusse and J. A. Yorke, Dynamics: Numerical Explorations, *Springer-Verlag*, New York (1997).

48. R. Ogita, H. Matsunaga and T. Hara, Asymptotic stability condition for a class of linear delay difference equations of higher order, *J. Math. Anal. Appl.* 248, 83-96 (2000).

49. W. E. Ricker, Stock and recruitment, *Journal of Fisheries Research Board of Canada* II(5), 559-623 (1954).

50. J. F. Selgrade and H. D. Roberds, On the structure of attractors for discrete, periodically forced systems with applications to population models, *Physica D* 158, 69-82 (2001).

51. A.-A. Yakubu, Multiple attractors in juvenile-adult single species models, *J. Diff. Equations Appl.* 9(12), 1083-1093 (2003).

52. A.-A. Yakubu, Discrete-time metapopulation dynamics and unidirectional dispersal, *Journal of Difference Equations and Appl., 9(7), 633-653 (2003)*.

53. A.-A. Yakubu and C. Castillo-Chavez, Interplay between local dynamics and dispersal in discrete-time metapopulation models, *Journal of Theoretical Biology*, 218, 273-288 (2002).

54. P. Yodzis, Introduction to Theoretical Ecology, *Harper & Row*, Publishers, New York (1989).

Regularity of Difference Equations

Jarmo Hietarinta
Department of Physics, University of Turku
Turku, Finland
e-mail: Jarmo.Hietarinta@utu.fi

1 The general setting

There are many different points of view one can take on dynamics and equations, for example:

- Given data, find the equation describing it (data driven search for equations).
- Start with assumptions of mathematical nature and derive equations with certain properties (mathematics driven search of equations).
- Choose a method of analysis, apply to various equations (method driven study of equations).
- Given a new equation, study its properties using all available methods (equation driven mathematical study).

In this contribution we review certain methods for *identifying whether or not a given equation has regular solutions*, i.e., whether or not the equation is *integrable.* This is done without actually solving the equations. In the above classification this approach can be seen as "mathematics driven search of equations", or "method driven study of equations". Due to this last characterization we would like to propose this as a new tool in the toolbox for anybody studying difference equations.

Our emphasis is on regular or integrable equations. Integrable systems are interesting, because they are, at the same time, *rare and ubiquitous*:

- Integrable equations have many special mathematical properties, that is why they are *rare.* These properties lead to interesting connections between different branches of mathematics.

- Integrable equations appear as scaling limits (e.g., for long waves) of many non-integrable equations, therefore they are *ubiquitous*[1]. Indeed, although complete integrability is structurally unstable, many properties associated with integrability persist in nearby non-integrable systems.
- Integrable equations are *useful*, because, due to the above, they can serve as starting points for perturbative expansions (e.g., KAM theory of nearly integrable systems).

We will next give some examples of integrable equations.

Soliton equations

Integrability is perhaps best known trough solitons[2]. The prototypical soliton equation is the Korteweg–de Vries equation:

$$\frac{\partial^3 u}{\partial x^3} + u\frac{\partial u}{\partial x} + \frac{\partial u}{\partial t} = 0, \tag{1}$$

which describes, e.g., the moderately large and long surface waves in shallow channels. It has a dispersive part (first and last terms in (1)) which alone would lead to decaying waves, and a shock part (second and third term), and their delicate balance allows for conserved travelling waves, called *solitons*, whose scattering is *elastic*.

Another important example is provided by the nonlinear Schrödinger equation

$$i\frac{\partial \psi}{\partial t} = \frac{\partial^2 \psi}{\partial x^2} + 2|\psi|^2\psi,$$

which describes, e.g., the motion of light pulses in optical fibers.

Soliton equations have many important properties, for example 1) they have multisoliton solutions, expressible in closed form; 2) they have an infinite number of conservation laws; 3) their initial value problem can be solved for arbitrarily long times (the Inverse Scattering Transform Method). The underlying mathematical theory is the Sato theory[3], from which one can derive hierarchies of integrable soliton equations.

Integrable mechanical systems

If we next go from PDE's to ODE's the concept of integrability changes to some extent, due to the finite dimensionality of the phase space. In the setting of Hamiltonian mechanics we have *Liouville integrability*, defined as the existence of a sufficient number of analytic and independent conserved quantities "in involution" (= mutually commuting under the Poisson bracket).

The Keplerian two-body system

$$H = \tfrac{1}{2m_1}\vec{p}_1{}^2 + \tfrac{1}{2m_2}\vec{p}_2{}^2 - \frac{m_1 m_2 G}{|\vec{x}_1 - \vec{x}_2|} \tag{2}$$

is a typical example, in fact it is "superintegrable" (= having still more conserved quantities than is required for the usual integrability), its conserved quantities are the total momentum, angular momentum, and the Runge-Lentz vector.

The importance of integrable systems as starting points for perturbative expansions in realistic systems is illustrated by the role of the Keplerian two-body system (2) in the study of planetary systems: For computing true planetary motion one can use perturbation expansion around the integrable system consisting of several noninteracting sun-planet two-body systems.

Integrability and acceleration algorithms

Let us next look at examples of integrable systems in the realm of discrete equations. It turns out that several numerical *acceleration algorithms* (for partial sums) are in fact integrable lattice equations[4].

As an example consider the Shanks-Wynn ϵ-algorithm: A set of initial sequences $\epsilon_0^{(m)} = 0$, $\epsilon_1^{(m)} = S_m$ is assumed, and from this one generates new sequences $\epsilon_n^{(m)}$ (that approach the limit S_∞ faster) by

$$(\epsilon_{n+1}^{(m)} - \epsilon_{n-1}^{(m+1)})(\epsilon_n^{(m+1)} - \epsilon_n^{(m)}) = 1.$$

But this is nothing else than the integrable discrete potential KdV equation. Similarly, Bauer's η-algorithm (in terms of $X_k^{(m)} := [\eta_k^{(m)}]^{(-1)^{k+1}}$)

$$X_{n+1}^{(m)} - X_{n-1}^{(m+1)} = \frac{1}{X_n^{(m+1)}} - \frac{1}{X_n^{(m)}}$$

is the integrable discrete KdV equation.

Integrable discretization

For numerical computations one has to convert continuum equations into difference equations. In this conversion one hopes to preserve the important properties of the initial equation.

Let us consider the Verhulst or logistic equation

$$\frac{du}{dt} = \alpha u(1 - \beta u), \tag{3}$$

with solution

$$u(t) = \frac{u_0}{\beta u_0 + (1 - \beta u_0)e^{-\alpha t}}. \tag{4}$$

How should one discretize (3) in order to get behavior similar to (4)?

One might first try the naive discretization of the derivative on the LHS:

$$u(t + \Delta t) - u(t) = \Delta t\, \alpha\, u(t)(1 - \beta u(t)). \tag{5}$$

If we change variables according to $u(t) = u(n\Delta t) = \frac{1+\alpha\Delta t}{\alpha\beta\Delta t} x_n$ we can write (5) as

$$x_{n+1} = a x_n (1 - x_n),$$

which is the logistic map, with usually quite different behavior when compared to (4).

The discretization that preserves the original behavior, including solutions like (4), is given by

$$u(t + \Delta t) - u(t) = \Delta t\, \alpha\, u(t + \Delta t)(1 - \beta u(t)), \tag{6}$$

which contains $t + \Delta t$ also on the RHS. The original equation was linearizable with the transformation $u = 1/(w + \beta)$, and this second discretization preserves this property. Indeed, one finds quickly the solution to (6)

$$u(t) \equiv u(n\Delta t) = \frac{u_0}{\beta u_0 + (1 - \beta u_0)(1 - \alpha\Delta t)^n},$$

which samples the continuum solution (4), and therefore (6) is the best possible discretization of the original equation.

Examples of integrable ordinary difference equations (OΔE)

Many integrable OΔE's have been found. The most actively studied collection is perhaps the 1D-version of the Quispel-Roberts-Thompson (QRT) family[5]:

$$x_{n+1} = \frac{f_1(x_n) - x_{n-1} f_2(x_n)}{f_2(x_n) - x_{n-1} f_3(x_n)},$$

where f_i are certain 4th order polynomials. In the special case where $f_3 = 0$ one gets

$$x_{n+1} + x_{n-1} = R(x_n), \text{ with } R \text{ rational.}$$

One example is the MacMillan map

$$x_{n+1} + x_{n-1} = \frac{a + bx_n}{1 - x_n^2},$$

and we will introduce others later.

Another sub-class is obtained if we choose $f_2 = 0$, one example is given by the discrete Painlevé equation is $\mathrm{dP_{III}}$:[6]

$$x_{n+1}x_{n-1} = \frac{cd(x_n - a\lambda^n)(x_n - b\lambda^n)}{(x_n - c)(x_n - d)}.$$

2 How to identify integrable equations?

The main question we want to address here is: How can we identify equations with regular behavior, without actually solving the equation? In fact we need some sort of algorithmic method for this purpose.

For ODE's two methods have often been used:

- Local analysis (for complex time) to check whether solutions have movable singularities (Painlevé method).
- Asymptotic analysis for growth of the solution (Nevanlinna theory).

Both methods can also be applied to difference equations, but here we discuss only the first one (for the second one see [7]).

In the late 1800's Fuchs, Kowalevskaya, and Painlevé proposed that one should classify those ODE's whose solutions do not have movable singularities. Here *movable* means that the position of the singularity is arbitrary (it is one of the initial values), and *singularity* is defined as a branch point (multivaluedness of solution). (For a general overview, see [8].)

The above idea was refined to the *Painlevé method*: To identify ODE's with regular behavior, verify that all Laurent series solutions have only integer powers and that they contain enough arbitrary constants. That is: (i) First find all possible *leading behaviors* by balancing various terms in the equation. All leading terms must be of the form $(z - z_0)^n$, $n \in \mathbb{Z}$, (z is the independent variable, z_0 an initial value). (ii) Next verify that each leading term can separately be completed into a power series solution, with the required number of free parameters (there must be enough integer *resonances*). [If some parameters are missing they are probably associated with nonpolynomial (and therefore singular) parts, not obtainable by a Laurent series expansion.]

This idea was used by Painlevé and Gambier in 1890-1910[9], they identified all second order ODE's whose movable singularities are just poles.

These equations were classified into about 50 types, most of which were linearizable or solvable by elliptic functions. But there were six new equations, now called *the 6 Painlevé equations*, that define new transcendental functions.

Passing the Painlevé test means that all solutions of the differential equation are regular, but it is believed that such equations also have other properties of integrability, e.g., a Lax pair. The method has been widely used in recent years, and extended to PDE's as well[8].

What about a Painlevé test for difference equations? If we want to push the analogy with the continuum case, perhaps we should again study what happens at a singularity (i.e., where the solution blows up). This idea was developed further by Grammaticos, Ramani, and Papageorgiou[10], who proposed *The Singularity Confinement Criterion* as such a method:

If the dynamics leads to a singularity, then after a few steps one should be able to get out of it, and this should take place without essential loss of information.

This principle has been used actively and has led, e.g., to the finding of discrete analogies of Painlevé equations (see, e.g.,[6]).

3 The singularity confinement criterion in practice

We will now discuss, trough examples, how this criterion can be applied to some OΔE's.

3.1 Singularity confinement for d-PI

As an example consider the discrete Painlevé I equation (d-PI)

$$x_{n+1} + x_n + x_{n-1} = \frac{a_n}{x_n} + b, \; a_n = \alpha + \beta n \tag{7}$$

[One justification of calling this a discrete Painlevé equations is that its continuum limit is the first Painlevé equation: If one sets $\epsilon n = z$, $x_n = f(z)$, $x_{n\pm1} = f(z \pm \epsilon)$, $f(z) = \frac{b}{6}(1 - 2\epsilon^2 y(z))$, and $\alpha + \beta n = -\frac{b^2}{36}(3 + 2\epsilon^4 z)$ then in the limit $\epsilon \to 0$ one obtains P_I, $y'' = 6y^2 + z$, at the leading order ϵ^4.]

Consider first the autonomous case

$$x_{n+1} = -x_n - x_{n-1} + \frac{a}{x_n} + b.$$

Clearly this equation is singular at $x = 0$. Assume that we reach the singularity at $x_0 = 0$ with a finite $x_{-1} = \mathbf{u} \neq 0$. The sequence then continues as follows:

$$\begin{aligned}
x_1 &= -0 - \mathbf{u} + a/0 + b = \infty, \\
x_2 &= -\infty - 0 + a/\infty + b = -\infty, \\
x_3 &= +\infty - \infty + a/\infty + b = \ ?
\end{aligned}$$

To resolve "$\infty - \infty$" we assume $x_0 = \epsilon$ (small) and redo the calculations:

$$\begin{aligned}
x_{-1} &= \mathbf{u}, \\
x_0 &= \epsilon, \\
x_1 &= \tfrac{a}{\epsilon} + b - \mathbf{u} - \epsilon \\
x_2 &= -\left[\tfrac{a}{\epsilon} + b - \mathbf{u} - \epsilon\right] - \epsilon + \frac{a}{\left[\tfrac{a}{\epsilon} + b - \mathbf{u} - \epsilon\right]} + b \\
&= -\tfrac{a}{\epsilon} + \mathbf{u} + \epsilon + (\mathbf{u} \quad b)/a\,\epsilon^2 + O(\epsilon^3), \\
x_3 &= -\left[-\tfrac{a}{\epsilon} + \mathbf{u} + \epsilon + (\mathbf{u} - b)/a\,\epsilon^2 + O(\epsilon^3)\right] - \left[\tfrac{a}{\epsilon} + b - \mathbf{u} - \epsilon\right] \\
&\quad + \frac{a}{\left[-\tfrac{a}{\epsilon} + \mathbf{u} + O(\epsilon)\right]} + b \\
&= -\epsilon + (b - 2\mathbf{u})/a\,\epsilon^2 + O(\epsilon^3), \\
x_4 &= -[O(\epsilon)] - \left[-\tfrac{a}{\epsilon} + \mathbf{u} + O(\epsilon)\right] + \frac{a}{\left[-\epsilon + (b - 2\mathbf{u})/a\,\epsilon^2 + O(\epsilon^3)\right]} + b \\
&= \mathbf{u} + O(\epsilon).
\end{aligned}$$

The essential result from this computation is that after exiting from the singularity the initial information $\mathbf{u}$ is recovered in x_4.

In this case the *singularity pattern* is $\ldots, 0, \infty, -\infty, 0, \ldots$.

3.2 Nonconfined singularity

The singularity is not always confined. A worst case example is given by

$$x_{n+1} - 2x_n + x_{n-1} = \frac{a}{x_n} + b. \tag{8}$$

After starting with $x_{-1} = \mathbf{u}$, $x_0 = \epsilon$ the subsequent terms are

$$x_k = k\tfrac{a}{\epsilon} + \ldots,$$

and the singularity is not confined, ever.

3.3 Successful use of singularity confinement

It has turned out that singularity confinement is only a necessary condition for integrability. Nevertheless, it has been successfully used as a guide for *de-autonomizing* (= adding explicit n-dependence to) discrete equations. The idea is to use some given integrable autonomous equation as a starting point, but allowing n-dependence in some coefficients; the new equation is, however, required to have the same singularity pattern. From this requirement one gets equations for the n-dependent coefficients.

Let us apply this to the previous example (7) but with possibly n-dependent a_n. The singularity pattern computation now proceeds as follows:

$$\begin{aligned}
x_{-1} &= \mathbf{u},\\
x_0 &= \epsilon,\\
x_1 &= \tfrac{a_0}{\epsilon} + b - \mathbf{u} - \epsilon,\\
x_2 &= -\tfrac{a_0}{\epsilon} + \mathbf{u} + \tfrac{a_1}{a_0}\epsilon + \tfrac{a_1}{a_0}(\mathbf{u} - b)/a_0\,\epsilon^2 + O(\epsilon^3),\\
x_3 &= -\tfrac{a_2+a_1-a_0}{a_2}\,\epsilon + (\tfrac{a_1}{a_0}b - \tfrac{a_1+a_2}{a_0}\mathbf{u})/a_0\,\epsilon^2 + O(\epsilon^3)\\
x_4 &= -\frac{a_3 - a_2 - a_1 + a_0}{a_2 + a_1 + a_0}\,\frac{a_0}{\epsilon} + \ldots
\end{aligned}$$

Now recall that for the original autonomous equation we had a finite x_4 (it started like $\mathbf{u} + \ldots$). In order to preserve this property, i.e., to get singularity confinement at this *same* step we must require

$$a_{n+3} - a_{n+2} - a_{n+1} + a_n = 0,\ \forall n.$$

This equation has the solution

$$a_n = \alpha + \beta n + \gamma\,(-1)^n,$$

and with this choice for a_n the singularity is indeed confined, and

$$x_4 := \frac{\mathbf{u}(\alpha+\gamma) + 2b\beta}{\alpha + 3\beta - \gamma} + O(\epsilon),$$

in particular, if $\beta = \gamma = 0$ (=autonomous case), $x_4 = \mathbf{u} + \ldots$. For $\gamma = 0$ we recover the discrete first Painlevé equation (7). [Later confinement is also possible, but does not seem to be connected with integrability[11].]

3.4 Singularity confinement is not sufficient

Unfortunately singularity confinement is only part of the story and in particular it is not sufficient for regularity. A counterexample is provided by the following equation[12]

$$x_{n+1} + x_{n-1} = x_n + \frac{1}{x_n^2}. \tag{9}$$

Epsilon analysis indicates that it passes the singularity confinement test: Assume $x_{-1} = \mathbf{u}$, $x_0 = \epsilon$, then

$$\begin{aligned}
x_1 &= \epsilon^{-2} - \mathbf{u} + \epsilon, \\
x_2 &= \epsilon^{-2} - \mathbf{u} + \epsilon^4 + O(\epsilon^6), \\
x_3 &= -\epsilon + 2\epsilon^4 + O(\epsilon^6), \\
x_4 &= \mathbf{u} + 3\epsilon + O(\epsilon^3),
\end{aligned}$$

Thus singularity is confined with singularity pattern $\ldots, 0, \infty, \infty, 0, \ldots$. Furthermore, the initial information $\mathbf{u}$ is recovered in x_4. However, it turns out that this equation shows numerical chaos[12].

3.5 Singularity confinement: summary

- Singularity confinement is *necessary* for a well defined evolution.
- It is easy to verify.
- It can be used effectively to de-autonomize a given map.
- It is not sufficient for integrable evolution.

4 Complexity and integrability

Since singularity confinement in not sufficient we should also consider other tests. One closely related property is *complexity*, which we will now discuss.

4.1 Analysis in projective space

In order to see better what happens under iterations it is useful to convert the rational map in normal Euclidean space into a polynomial map in the projective space.

As an example consider dP-I (7) and write it as a first order system

$$\begin{cases} x_{n+1} &= -x_n - y_n + \frac{a_n}{x_n} + b, \\ y_{n+1} &= x_n, \end{cases}$$

then homogenize it with $x_n = u_n/f_n$, $y_n = v_n/f_n$, which yields

$$\begin{cases} \frac{u_{n+1}}{f_{n+1}} &= -\frac{u_n}{f_n} - \frac{v_n}{f_n} + a_n \frac{f_n}{u_n} + b, \\ \frac{v_{n+1}}{f_{n+1}} &= \frac{u_n}{f_n}. \end{cases}$$

Next clear the denominators by defining f_{n+1} suitably, this yields

$$\begin{cases} u_{n+1} &= -u_n(u_n+v_n)+f_n(a_nf_n+bu_n), \\ v_{n+1} &= u_n^2, \\ f_{n+1} &= f_nu_n. \end{cases} \tag{10}$$

However, this equation is identical to the original equation only if we identify those vectors (u,v,f) that correspond to the same x,y: $(u,v,f)\sim(\lambda u,\lambda v,\lambda f)$. This means that (10) must be interpreted as a map in the complex projective space $\mathbb{CP}^2$. (Note: $(0,0,0)\notin\mathbb{CP}^2$.)

Let us now see how singularity confinement looks in this projective way of writing dP-I (10). The previously discussed sequence

$$x_{-1}=u,\ x_0=0,\ x_1=\infty,\ x_2=-\infty,\ x_3=-\infty+\infty=?$$

now looks as

$$\begin{pmatrix}0\\u\\1\end{pmatrix}\to\begin{pmatrix}1\\0\\0\end{pmatrix}\to\begin{pmatrix}-1\\1\\0\end{pmatrix}\to\begin{pmatrix}0\\1\\0\end{pmatrix}\to\begin{pmatrix}0\\0\\0\end{pmatrix}$$

and the singularity manifests in taking us out of $\mathbb{CP}^2$.

In order to see what really happens we do, as usual, the epsilon analysis. In the autonomous case it yields

$$\begin{pmatrix}\epsilon\\u\\1\end{pmatrix}\to\begin{pmatrix}a+(b-u)\epsilon+\dots\\ \epsilon^2\\ \epsilon\end{pmatrix}\to\begin{pmatrix}-a^2+\epsilon a(2u-b)+\dots\\ a^2+2\epsilon a(b-u)+\dots\\ \epsilon a+\epsilon^2(b-u)+\dots\end{pmatrix}$$

$$\to\begin{pmatrix}\epsilon^2a^3+\dots\\ a^4+2\epsilon a^3(b-2u)+\dots\\ -\epsilon a^3+\epsilon^2a^2(3u-2b)+\dots\end{pmatrix}\to\begin{pmatrix}-ua^6\epsilon^3+\dots\\ \epsilon a^6\epsilon^3+\dots\\ -a^6\epsilon^3+\dots\end{pmatrix}=\begin{pmatrix}u+\dots\\ -\epsilon+\dots\\ 1+\dots\end{pmatrix}.$$

The important thing here is the *projective cancellation* of $-a^6\epsilon^3$ in the last step. It is necessary for *exiting from the singularity*, but at the same time it is also crucial for *reducing growth of complexity*. This duality is the essential ingredient: singularity confinement is a yes/no test, complexity analysis yields further information.

4.2 Complexity analysis

Let us assume a generic $\mathbb{CP}^2$ map

$$\begin{cases} u_{n+1} &= P(u,v,f), \\ v_{n+1} &= Q(u,v,f), \\ f_{n+1} &= R(u,v,f), \end{cases}$$

where P, Q, R are homogeneous polynomials of degree κ.

Given some starting values $(u_0, v_0, f_0) := (a, b, c)$ the subsequent iterates u_n, v_n, f_n will have total degree (complexity) $d_n \leq \kappa^n$ as polynomials of (a, b, c). The actual degree will depend on the projective cancellations that can take place at various steps of iteration. In the worst case example (8) the singularity is not confined, in fact there are no cancellations ever, and $d_n = 2^n$.

Let us define two basic types of growth:

- If $d_n \sim \rho^n$ for some $\rho \leq \kappa$ we say the map has *exponential growth of complexity.*
- If $d_n \sim n^\sigma$, for some $\sigma > 0$ we say the map has *polynomial growth of complexity.*
- Main conjecture: Polynomial growth of complexity is associated with integrability[13].

Singularity confinement implies reduced growth, but does that guarantee integrability? To illustrate the above let us consider two examples[12]:

The integrable map

$$x_{n+1} + x_{n-1} = \frac{a}{x_n} + \frac{b}{x_n^2},$$

reads in projective coordinates as

$$\begin{cases} u_{n+1} &= -v_n u_n^2 + a f_n^3 + b f_n^2 u_n, \\ v_{n+1} &= u_n^3, \\ f_{n+1} &= f_n u_n^2. \end{cases}$$

The nominal growth of complexity is $d_n = 3^n$, but the actual growth is

$$1, 3, 9, 19, 33, 51, 73, 99, 129, 163, 201, \ldots$$

which is exactly

$$d_n = 2n^2 + 1.$$

This is an integrable equation and shows polynomial growth of complexity.

Next consider the nonintegrable HV-map (9), in $\mathbb{CP}^2$ it reads

$$\begin{cases} u_{n+1} &= -v_n u_n^2 + u_n^3 + f_n^3, \\ v_{n+1} &= u_n^3, \\ f_{n+1} &= f_n u_n^2. \end{cases}$$

Nominal growth is again $d_n = 3^n$, and the actual growth is slightly less:

$$1,\ 3,\ 9,\ 27,\ 73,\ 195,\ 513,\ 1347,\ 3529,\ 9243,\ 24201, \ldots$$

This sequence of numbers is described by

$$d_n = -2 - \tfrac{1}{5}(-1)^n + \tfrac{8}{5}\left[\left(\tfrac{3+\sqrt{5}}{2}\right)^n + \left(\tfrac{3-\sqrt{5}}{2}\right)^n\right],$$

and therefore asymptotically the growth is exponential,

$$d_n \approx 1.6 \times 2.62^n.$$

This is a nonintegrable equation and has exponential growth of complexity.

In all examples studied so far the polynomial growth of complexity has been found in integrable equations and exponential growth in non-integrable ones.

5 Conclusions

In this contribution we have discussed the problem of identifying integrable difference equations without solving them. Integrable equations have many nice properties and therefore they are interesting as such. But more importantly, even if a given equation is nonintegrable there may be an integrable difference equation nearby, around which one can start a perturbative expansion.

Here we have discussed two methods of identifying integrable systems: In Sec. 3, we considered the "singularity confinement criterion", which is relatively simple and therefore the method that one should apply first. If the equation passes that test, one should usually do next a complexity analysis, discussed in Sec. 4.

These are methods that should be in the toolbox of anybody who studies difference equations.

References

[1] F. Calogero, "Why are Certain Nonlinear PDEs Both Widely Applicable and Integrable", in *What is Integrability*, V.E. Zakharov (ed.), Springer-Verlag, (1991) 1-62.

[2] M.J. Ablowitz and P.A. Clarkson: *Solitons, Nonlinear Evolution Equations and Inverse Scattering*, Cambridge University Press (1991).

[3] T. Miwa, M. Jimbo and E. Date, *Mathematics of Solitons*, Cambridge University Press (1999).

[4] A. Nagai and J. Satsuma, Phys. Lett. A209 (1995) 305.

[5] G.R.W. Quispel, J.A.G. Roberts and C.J. Thompson, Physica D34 (1989) 183.

[6] A. Ramani, B. Grammaticos and J. Hietarinta, Phys. Rev. Lett. 67 (1991) 1829.

[7] M.J. Ablowitz, R. Halburd and B. Herbst, Nonlinearity 13 (2000) 889.

[8] *The Painlevé Property, One Century Later*, R. Conte (ed.), Springer (1999).

[9] P. Painlevé, Acta Mathematica 25, (1902) 1.

[10] B. Grammaticos, A. Ramani, and V. Papageorgiou, Phys. Rev. Lett. 67 (1991) 1825.

[11] J. Hietarinta and C. Viallet, Chaos, Solitons and Fractals 11 (2000) 29.

[12] J. Hietarinta and C. Viallet, Phys. Rev. Lett. 81 (1999) 325.

[13] V.I. Arnold, Bol. Soc. Bras. Mat. 21 (1990) 1; A.P. Veselov, Comm. Math. Phys. 145 (1992) 181; G. Falqui, C.-M. Viallet, Comm. Math. Phys. 154 (1993) 111; M.P. Bellon and C.-M. Viallet, Comm. Math. Phys. 204 (1999), 425.

Robustness in Difference Equations

Jack K. Hale

Abstract. Difference equations are used as models for determining the dynamics of various types of processes. In many situations, the delays (or differences) can represent observation times or the time that it takes to transport informatin in the system. A basic question is: What properties of the system are preserved when subjected to small variations in the delays? The purpose of this paper is to point out some positive and negative results for linear systems, to give some applications to control problems and mention some unsolved problems for nonlinear systems.

1. Introduction. The emphasis in these notes is on the effects of variations of the delays (or differences) upon the dynamics of difference equations. The dimension of the space in which the dynamics of a difference equation must be considered depends in a significant way on the delays.

To further appreciate this remark, consider a scalar equation

$$y(t) = f(y(t-r), y(t-1)), \tag{1.1}$$

where $0 < r < 1$, and $f : \mathbb{R}^2 \to \mathbb{R}$ is a continuous function. The delays are $\{r, 1\}$. If we considered a difference equation with two arbitrary delays $r_2 > r_1 > 0$, then a rescaling of time yields an equation (1.1) with $0 < r = r_1/r_2 < 1$.

If the function $f(x_1, x_2) = g(x_2)$ does not depend upon the first variable, then the dynamics of (1.1) is determined by a map from $\mathbb{R}$ to $\mathbb{R}$: $y \mapsto g(x)$.

If $f(x_1, x_2)$ explicitly depends upon both x_1 and x_2, then the value of r determines the amount of information that is needed to obtain a solution of equation (1.1). For example, if $r = 1/2$, then $y(t)$ at time t is determined by a point in $\mathbb{R}^2$ corresponding to the values $(y(t-1/2), y(t-1))$. The natural state space for the system is $\mathbb{R}^2$ since the evolution of the system at time t is determined from the knowledge of $(y(t-\frac{1}{2}), y(t-1))$. If we let $x_1(t) = y(t)$, $x_2(t) = y(t-1/2)$, then the equation is equivalent to

$$\begin{aligned} x_1(t) &= f(x_1(t-1/2), x_2(t-1/2)) \\ x_2(t) &= x_1(t-1/2)). \end{aligned} \tag{1.2}$$

If we let $F(x_1, x_2) = (f(x_1, x_2), x_1)$, then the dynamics of (1.2) is is equivalent to the dynamics of the planar map $x \equiv (x_1, x_2) \in \mathbb{R}^2 \mapsto F(x) \in \mathbb{R}^2$.

If we now replace the delays $\{1/2, 1\}$ by delays $\{0.49, 1\}$, then, in order to define the solution of (1.1) for all $t \geq 0$, we need to specify 100 points corresponding to the values of $y(\theta), \theta = -1, -1 + .01, \ldots, -1/2, -.49, \ldots -.01$. As before, it is possible to introduce some additional variables in order to see that the dynamics of the equation for the delays $\{0.49, 1\}$ is equivalent to a map on $\mathbb{R}^{100}$ which is uniquely determined by f. The dimension of the state space has changed from 2 to 100.

If $0 < r < 1$ is irrational, then, to define a solution of (1.1) for $t \geq 0$, we must specify a function on the interval $[-1, 0]$. As a consequence, the state space must be some infinite dimensional space X of functions mapping $[-1, 0]$ to $\mathbb{R}$. If we are to study the dependence of the dynamics upon r, then the comparison of the systems must be made in the space X. Therefore, even if the delay set is $\{\frac{1}{2}, 1\}$, we consider the equation as defined on X.

Of course, there are many possibilities for the choice of the space of initial functions. We restrict our attention to continuous functions.

Let $C = C([-1, 0], \mathbb{R})$. To obtain a solution of (1.1) through an element $\varphi \in C$ at $t = 0$ which is continuous for $t \geq 0$, we must impose the condition that $\varphi(0) = f(\varphi(-r), \varphi(-1))$. For this reason, for a given C^k-function $f : \mathbb{R}^2 \to \mathbb{R}$, $k \geq 1$, we let

$$C_{f,r} = \{\varphi \in C : \varphi(0) = f(\varphi(-r), \varphi(-1))\}.$$

If f is a linear function, then $C_{f,r}$ is a linear subspace of C. If f is a nonlinear function, then $C_{f,r}$ is a C^k-submanifold of C.

For any function $\varphi \in C_{f,r}$, we say that $x(t, \varphi)$ is a solution of (1.1) with initial value φ at $t = 0$ if there is an $\alpha > 0$ such that $x(t, \varphi)$ is a continuous function on the interval $[-1,\alpha)$, $x(\theta,\varphi)=\varphi(\theta)$ for $\theta\in[-1,0]$ and $x(t,\varphi)$ satisfies (1.1) for $t\in[0,\alpha)$.

It is clear that there is a unique solution $x(t,\varphi)$ of (1.1) with initial value $\varphi\in C_{f,r}$ at $t=0$ and that it exists on $[-1,\infty)$. Define

$$
\begin{aligned}
&T_{f,r}(t): C_{f,r}\to C_{f,r}, \quad t\geq 0\\
&(T_{f,r}(t)\varphi)(\theta)=x(t+\theta,\varphi), \quad \theta\in[-1,0].
\end{aligned}
\tag{1.3}
$$

One can show that $T_{f,r}(t)$ is a C^k-semigroup on $C_{f,r}$ (equivalently, a C^k dynamical system on $C_{f,r}$) and satisfies

(i) $T_{f,r}(t)\varphi$ is continuous in all variables and C^k in φ,
(ii) $T_{f,r}(0)=I$, the identity,
(iii) $T_{f,r}(t+s)=T_{f,r}(t)T_{f,r}(s)$, $t\geq s\geq 0$.

By considering (1.1) in the space $C_{f,r}$, we are now in a position to discuss variations in the delay r. The basic question is the following:

What properties of the dynamics of (1.1) for fixed delays $\{r_0,1\}$ are preserved under small variations in r_0?

Our objective here is to point out some of the difficulties involved in trying to make progress on this question, mentioning some specific precise results on the preservation of stability in linear systems, present a connection with linear neutral difference equations, give some applications to control problems and then discuss some of the diffiuculties that seem to be unresolved when we consider nonlinear difference equations and the behavior near equilibria.

2. A linear equation. Even though we are going to discuss a linear equation with only two delays $\{r,1\}$, we will see, in the next section, that many of the results are valid for general linear equations with an arbitrary number of delays.

Consider the linear equation

$$z(t)=az(t-r)+bz(t-1), \tag{2.1}$$

where a, b are real constants. This is a special case of (1.1) with $f(\varphi(-r), \varphi(-1)) = a\varphi(-r) + b\varphi(-1)$.

Since our objective is to emphasize the dependence of the solutions of (2.1) on the parameters $\{a, b, r\}$, we introduce a notation which brings out explicitly these parameters. Let $C_{a,b,r} = \{\varphi \in C : \varphi(0) = a\varphi(-r) + b\varphi(-1)\}$ and let $S_{a,b,r}(t)$, $t \geq 0$, be the linear semigroup on $C_{a,b,r}$ defined by (2.1).

The domain $\mathcal{D}(A_{a,b,r})$ of the infinitesimal generator $A_{a,b,r}$ of $S_{a,b,r}(t)$ is the set of $\varphi \in C_{a,b,r}$ such that

$$\lim_{t \to 0+} \frac{1}{t}[S_{a,b,r}(t)\varphi - \varphi]$$

exists and belongs to $C_{a,b,r}$. It is not difficult to show that

$$\mathcal{D}(A_{a,b,r}) = \{\varphi \in C_{a,b,r} : \varphi \in C^1([-1,0], \mathbb{R}), \dot{\varphi}(0) = a\dot{\varphi}(-r) + b\dot{\varphi}(-1)\}$$

and

$$A_{a,b,r}\varphi = \dot{\varphi}, \quad \varphi \in \mathcal{D}(A_{a,b,r}).$$

It also is easy to check that the spectrum $\sigma(A_{a,b,r})$ of $A_{a,b,r}$ is given by

$$\sigma(A_{a,b,r}) = \{\lambda : 1 = ae^{-r\lambda} + be^{-\lambda}\}. \tag{2.2}$$

A more difficult but important result due to Henry (1974) (see also Avellar and Hale (1982), Hale and Verduyn-Lunel (1993)) is

Theorem 2.1.

$$\sigma(S_{a,b,r}(1)) = \sigma_{\text{ess}}(S_{a,b,r}(1)) = \operatorname{Cl} e^{\sigma(A_{a,b,r})} \textit{ plus possibly zero.} \tag{2.3}$$

From (2.3), it is clear that the real parts of the spectrum of $A_{a,b,r}$ play an important role. The following result is in Avellar and Hale (1982).

Proposition 2.2. *If we define*

$$Z(a, b, r) = \operatorname{Cl} \operatorname{Re} \sigma(A_{a,b,r}), \tag{2.4}$$

then $Z(a,b,r)$ is always continuous in a, b and $Z(a,b,r)$ is continuous in r in the Hausdorff sense if r is irrational.

Let ρ_0 and ρ_2 be defined by the following expressions

$$
\begin{aligned}
1 &= |a|e^{-\rho_0 r} + |b|e^{-\rho_0} \\
|b|e^{-\rho_2} &= 1 + |a|e^{-\rho_2 r}.
\end{aligned}
\tag{2.5}
$$

We can now state the following result.

Proposition 2.3. *The set $\sigma(T_{a,b,r}(1))$ satisfies the following properties:*

(i) $\sigma(T_{0,b,r}(1)) = \{b\}$,

(ii) $\sigma(T_{a,0,r}(1)) = \{0, a\}$,

(iii) With ρ_o, ρ_2 defined as in (2.5), $\sigma(S_{a,b,r}(1)) \subset [e^{\rho_2}, e^{\rho_0}]$ and if, $a \neq 0$, $b \neq 0$ and r is irrational. then $\sigma(S_{a,b,r}(1)) = [e^{\rho_2}, e^{\rho_0}]$.

(iv)

$$\rho_0 < 0 \text{ if and only if } |a| + |b| < 1,$$

$$\rho_0 = 0 \text{ if and only if } |a| + |b| = 1,$$

$$\rho_0 > 0 \text{ if and only if } |a| + |b| > 1.$$

(v) $\rho_2 < 0$ *if* $|a| + |b| = 1$, $|a| \neq 0, |b| \neq 0$.

(vi) If $a \neq 0$, $b \neq 0$ and $r = p/q$, where $p < q$ are integers with no common factors, then

$$\sigma(S_{a,b,r}(1)) = \{\mu_1, \mu_2, \ldots, \mu_q\}, \tag{2.6}$$

where each μ_j is a solution of the equation

$$\mu^q - a\mu^{q-p} - b = 0. \tag{2.7}$$

Proof. As noted before, part (i) is trivial. If $b = 0$, we must remember that the equation is being defined on the space $C_{a,0,r}([-1,0]; \mathbb{R})$. The eigenvalues of $A_{a,0,r}$ are $\frac{1}{r} ln|a| + 2k\pi i/r$, $k = 0, \pm 1, \pm 2, \ldots$. Therefore,

$a \in \sigma(S_{a,0,r}(1))$. The solution through any function which vanishes on $[-r, 0]$ is equal to zero for all $t \geq 0$. Therefore, $\{0\}$ belongs to the spectrum of $\sigma(S_{a,b,r}(1))$. This proves (ii). Part (iii) is stated without proof with references given above. Part (iv) is clear from (2.5). For part (v), we note that, if $\rho_2 \geq 0$, then (2.5) implies that we must have $2 - |b| \leq |b|$ or $|b| \geq 1$, but this is impossible since we are assuming that $|a| > 0$. Part (vi) is obtained by letting $\lambda = q\sigma$, $\mu = e^{-\sigma}$.

Remark 2.4. *Dynamics of (2.1) for $b = 0$.* In case $b = 0$, it is possible and instructive to discuss the complete dynamics of (2.1). The space $C_{a,0,r}$ can be decomposed as

$$
\begin{aligned}
C_{a,0,r} &= P_{a,r} \oplus Q_{a,r} \\
P_{a,0,r} &= \{\varphi \in C_{a,0,r} : \varphi(\theta) = 0, \theta \in [-r, 0]\} \\
Q_{a,0,r} &= \{\varphi \in C_{a,0,r} : \varphi(\theta) = c, \theta \in [-1, -r], c \in \mathbb{R}\}.
\end{aligned}
$$

If $\varphi \in C_{a,0,r}$, then it can be written as

$$\varphi = \varphi_{P_{a,r}} + \varphi_{Q_{a,r}}$$

where

$$\varphi_{Q_{a,r}} = \begin{cases} \varphi(\theta) - \varphi(-r) & \text{if } \theta \in [-1, -r], \\ 0 & \text{if } \theta \in [-r, 0] \end{cases},$$

$$\varphi_{P_{a,r}} = \begin{cases} \varphi(-r) & \text{if } \theta \in [-1, -r]. \\ \varphi(\theta) & \text{if } \theta \in [-r, 0]. \end{cases}$$

The linear subspaces $P_{a,r}$, $Q_{a,r}$ are positively invariant under $S_{a,0,r}(t)$ and correspond, respectively, to the spectral points 0 and a for $S_{a,0,r}(1)$.

Note that, for any $\varphi \in Q_{a,r}$, $S_{a,0,r}(t)\varphi = 0$ for $t \geq 1 - r$ and, thus, $S_{a,0,r}(t)\varphi \in P_{a,r}$ for any $\varphi \in C_{a,0,r}$ and any $t \geq 1 - r$. The linear subspace $Q_{a,r}$ is the super exponentially stable manifold of the origin (solutions approach zero faster than any exponential).

If $|a| < 1$, then the origin is exponentially stable. If $|a| = 1$, then the origin is stable but not exponentially. If $|a| > 1$, then the origin is a hyperbolic saddle point with a super exponentially stable manifold $Q_{a,r}$ and unstable manifold $P_{a,r}$.

Remark 2.5. Exponential stability independent of delay. We say that the origin of (2.1) is *exponentially stable independent of the delays* $(r, 1)$ if and only if, for every $r > 0$, $|\sigma(S_{a,b,r}(1)) < 1$. From (iv), we see that we have exponential stability for irrational r if and only if $|a| + |b| < 1$. Thus, the zero solution of (2.1) is exponentially stable for every r if and only if it is exponentially stable at an irrational r. This type of stability depends only upon the coefficients.

2.6. An example. *Destruction of exponential stability.*

If $r = 1/2$ in (2.1), then part (vi) of Proposition 2.3 implies that $\sigma(S_{a,b,1/2}(1)) = \{\mu_1, \mu_2\}$ where μ_1, μ_2 are the solutions of the equation

$$\mu^2 - a\mu - b = 0. \tag{2.8}$$

If $a^2 + 4b < 0$, then μ_1, μ_2 are complex and $\sigma(S_{a,b,1/2}(1))$ lies on the circle with center zero and radius $|\mu_1|$. If $a^2 + 4b \geq 0$, then $\sigma(S_{a,b,1/2}(1))$ lies on the real axis and consists of the real numbers μ_1, μ_2.

If $a = b = -1/2$, then $a^2 + 4b < 0$ and $|\mu_1|^2 = 1/2$ which implies that $\sigma(S_{a,b,1/2}(1))$ lies on the circle with radius $2^{-1/2} < 1$. Therefore, solutions approach zero exponentially.

If r is irrational, then part (iii) of Proposition 2.3 implies that $\sigma(S_{a,b,1/2}(1)) = [e^{\rho_2}, 1]$ since ρ_0 is obviously zero. Also, $\rho_2 < 0$. The exponential stability is destroyed by making a very small change in the delay.

In simulations, one would never encounter irrational r. It also happens that exponential stability also is destroyed by some rational numbers arbitrarily close to 1/2. In fact, if $r = \frac{1}{2}(1 - \frac{1}{2n+3})$ and n is an odd integer, then there is a solution of the difference equation given by the function $z(t) = \sin{(n + \frac{3}{2})\pi t}$. We can get arbitrarily close to the delay 1/2 by taking n large.

3. Remarks on several delays

Appropriate extensions of the results in section 2 hold for the case of several delays provided that the delays are independent. Consider the equation

$$z(t) = \Sigma_{j=1}^{N} a_j z(t - r_j) \tag{3.1}$$

where each $a_j \neq 0$ is a real constant and $0 < r_1 < r_2 < \cdots < r_N$. Let $a = (a_1, a_2, \ldots, a_n)$, $r = (r_1, r_2, \ldots, r_N)$.

The eigenvalues of (3.1) are the solutions of the characteristic equation

$$1 = \Sigma_{j=1}^{N} a_j e^{-\lambda r_j}. \tag{3.2}$$

As before, we let

$$Z(a, r) = \text{Cl}\,\{\lambda : 1 = \Sigma_{j=1}^{N} a_j e^{-\lambda r_j}\}. \tag{3.3}$$

Let ρ_0 and ρ_N be the solutions of the equations

$$\begin{aligned} 1 &= \Sigma_{j=1}^{N} |a_j| e^{-\rho_0 r_j} \\ |a_N| e^{-\rho_N r_N} &= 1 + \Sigma_{j=1}^{N-1} |a_j| e^{-\rho_N r_j}. \end{aligned} \tag{3.4}$$

We have the following result.

Proposition 3.1. *If ρ_0, ρ_N are defined by (3.4), then*

$$Z(a, r) \subset [\rho_N, \rho_0].$$

Also, if the components of r are rationally independent, then ρ_0 and ρ_N belong to $Z(a, r)$.

Furthermore, if the components of r are rationally independent, then

$$\rho_0 < 0 \quad \Leftrightarrow \quad \Sigma_{j=1}^{N} |a_j| < 1. \tag{3.5}$$

From Proposition 3.1, we see that the system is exponentially stable for all values of delays if and only if (3.5) is satisfied (this is equivalent to being exponentially stable for all delays in a neighborhood of a given value of the delays).

If the delays are not independent, then the problem of stability becomes much more complicated. To describe a specific result, let $k_j = (k_{j1}, k_{j2}, \cdots, k_{jM})$, $1 \leq j \leq N$, where each k_{ji} is a nonnegative integer, $s = (s_1, s_2, \ldots, s_M)$ with each $s_j > 0$ and define

$$r_j = \Sigma_{i-1}^{M} k_{ji} s_i \equiv k_j \cdot s, \ 1 \leq j \leq N. \tag{3.6}$$

We order the constants r_j as before. Define

$$\begin{aligned} H(\rho, \theta, s) &= 1 + \Sigma_{j=1}^{N} a_j e^{-\rho k_j \cdot s} e^{i k_j \cdot \theta} \\ \theta &= (\theta_1, \theta_2, \ldots, \theta_M). \end{aligned} \tag{3.7}$$

The following result is due to Henry (1974).

Proposition 3.2. *If the components of s are rationally independent, then $\rho \in Z(a, r)$ if and only if there is a θ such that $H(\rho, \theta, s) = 0$.*

3.3. Example. *Effects of not varying the delays independently*

The details of the computations in this example are in Avellar and Hale (1982). Consider the equation

$$z(t) = -z(t - r_1) - z(t - r_2) - z(t - r_3) \tag{3.8}$$

where $r_1 < r_2 < r_3$.

If the delays are rationally independent then a few computations show that the smallest closed interval containing $Z(r_1, r_2, r_3)$ is approximately $[-0.56, 0.60]$.

If we choose $r_1 = 1, r_2 = 2, r_3 = \pi$, then a careful analysis of the function

$$H(\rho, \theta_1, \theta_2, 1, \pi) = 1 + e^{-\rho} e^{i\theta_1} + e^{-2\rho} e^{2i\theta_1} + e^{-\pi\rho} e^{\pi i \theta_2}$$

will yield the conclusion that the smallest closed interval containing $Z(1, 2, \pi)$ is approximately $[-0.56, 0.30]$ which is smaller than the ones for which the delays vary in an independent way.

3.4. Example. *Very bad instability*

Consider the equation

$$z(t) = -kz(t-r) + kz(t-2-r) - z(t-2) \tag{3.9}$$

where k is a positive constant and the delays are $\{r, 2, 2+r\}$. In this case, we have

$$H(\rho, \theta_1, \theta_2, r, 2) = 1 + ke^{-\rho r}e^{i\theta_1} - ke^{-\rho(r+2)}e^{2i\theta_2 + i\theta_1} + e^{-2\rho}e^{2i\theta_2}.$$

This function is zero if and only if

$$-ke^{-\rho r}e^{i\theta_1} = \frac{1 + e^{-2\rho}e^{2i\theta_2}}{1 - e^{-2\rho}e^{2i\theta_2}}.$$

If we choose θ_1, θ_2 so that $e^{2i\theta_2} = -1 = e^{i\theta_1}$, then, if r is irrational, ρ is in $Z(r, 2)$ if ρ satisfies the equation

$$ke^{-\rho r} = \frac{1 - e^{-2\rho}}{1 + e^{-2\rho}}.$$

For any $k > 0$, this equation has a unique solution $\rho(r, k) > 0$ and the corresponding difference equation is exponentially unstable. The function $\rho(r, k)$ is monotone increasing in r.

A more striking conclusion holds if $k \geq 1$. In fact, if $k \geq 1$, it is easy to see that $\rho(r, k) \to \infty$ as $r \to 0$. This implies that, for any $\tilde{\rho} > 0$, there is an $r(\tilde{\rho})$ which approaches zero as $\tilde{\rho} \to \infty$ and there is a solution which grows exponentially as $e^{\tilde{\rho} t}$ as $t \to \infty$.

Remark 3.5. Suppose that solutions of (3.2) lie on the imaginary axis. If an eigenvalue is multiple with nonsimple elementary divisors, then there are solutions of (3.1) which are bounded.

If all eigenvalues are simple and lie on the imaginary axis, it might be expected that all solutions of (3.1) are bounded. However, this may not be the case. Zverkin (1968) has given such an example where there are unbounded solutions.

It is interesting to consider compact perturbations of (3.1). If a is an L^2-function from $[-1,0]$ to $\mathbb{R}$, consider the equation

$$z(t) = \Sigma_{j=1}^{N} a_j z(t - r_j) + \int_{-1}^{0} a(\theta) z(t+\theta) d\theta. \tag{3.10}$$

The characteristic equation is

$$1 = \Sigma_{j=1}^{N} a_j e^{-\lambda r_j} + \int_{-1}^{0} a(\theta) e^{\lambda\theta} d\theta. \tag{3.11}$$

The essential spectrum of the semigroup for (3.10) coincides with the essential spectrum of (3.1).

Rabah, Sklyar and Rezounenko (2004) proved the following result.

Proposition 3.6. *If the eigenvalues of (3.1) lie on the imaginary axis, then the following assertions hold:*

(i) If the eigenvalues of (3.10) have negative real parts and are simple, then the origin is stable and every solution approaches zero as $t \to \infty$.

(ii) If there is an multiple eigenvalue of (3.1) with nonsimple elementary divisors, then there are unbounded solutions of (3.10).

(iii) If there is an eigenvalue of (3.1) of multiplicity 2 with simple elementary divisors, then there is an integral perturbation for which there is an unbounded solution of (3.10).

4. Linear Neutral Functional Differential Equations

In this section, we want to make a few remarks about the role of difference equations in the study of certain classes of neutral functional differential equations (NFDE). We only discuss scalar equations, but the remarks are valid for n-dimensional equations.

Suppose that $C = C([-1,0], \mathbb{R})$ and suppose that $L : C \to \mathbb{R}$, $D : C \to \mathbb{R}$,

$$D\varphi = \varphi(0) - \Sigma_{j=1}^{N} a_j \varphi(-r_j)$$

where a_j are constants and each $r_j > 0$. A linear NFDE is defined as

$$\frac{d}{dt} D x_t = L x_t, \tag{4.1}$$

where, if $x(t)$ is a continuous function on $[-1,\alpha)$, $\alpha > 0$, the symbol $x_t \in C$ denotes $x_t(\theta) = x(t+\theta)$ for $\theta \in [-1,0]$.

For a given $\varphi \in C$, it is not difficult to show that there is a unique continuous function $x(t,\varphi)$ defined on $[-1,\infty)$, $x(t,\varphi) = \varphi(t)$ for $t \in [-1,0]$, $\frac{d}{dt}Dx_t$ is continuously differentiable on $(0,\infty)$ with a continuous right hand derivative at $t=0$ and satisfies (4.1) for all $t \geq 0$. If we let $T(t)\varphi = x_t(\cdot,\varphi)$, then $T(t)$, $t \geq 0$, is a continuous linear dynamical system on C.

The infinitesimal generator A of $T(t)$ is given by

$$A\varphi = \dot{\varphi}$$
$$\mathcal{D}(A) = \{\varphi \in C^1 \cap C : D\dot{\varphi} = L\varphi\}\,.$$

The spectrum $\sigma(A)$ of A consists only of eigenvalues of finite multiplicity and the eigenvalues are the solutions of the characteristic equation

$$\det(\lambda D e^{\lambda\cdot} I - L e^{\lambda\cdot} I) = 0. \tag{4.2}$$

As we have noted before, we have the linear dynamical system $S(t)$, $t \geq 0$ on $C_0 = \{\varphi \in C : D\varphi = 0\}$ corresponding to the difference equation $Dz_t = 0$. For a proof of the following result, see, for example, Hale ane Verduyn-Lunel (1993).

Proposition 4.1. *With $T(t), S(t)$ defined as above,*

$$\sigma(T(1)) = \operatorname{Cl} e^{\sigma(A)} \textit{ plus possibly } \{0\}$$
$$\sigma_{ess}(T(1)) = \sigma_{ess}(S(1)) = \sigma(S(1))\,.$$

From this fact, we can make the following interesting remark. If $|\sigma(S(1))| < 1$, then the stability properties of (4.1) are determined by the eigenvalues of A. This also implies that bifurcation from the origin in nonlinear equations will be determined by eigenvalues of A. If the delays must be allowed to vary over an interval, then $\Sigma_{j=1}^{n}|a_j| < 1$ is necessary and sufficient to have this condition satisfied.

5. Applications to control

5.1. Feedback stabilization of difference equations is unrealistic

If the zero solution of the difference equation

$$x(t) - ax(t-r) = 0$$

is not exponentially stable, then $|a| \geq 1$. We can apply a feedback control $bx(t-r)$ with $|a+b| < 1$ so that the zero solution of the feedback system

$$x(t) - ax(t-r) - bx(t-r) = 0$$

is exponentially stable. It the control cannot be immplemented instantaneously, then there is small time delay in the feedback:

$$x(t) - ax(t-r) - bx(t-r-\epsilon) = 0.$$

If the zero solution is to be exponentially stable for all small ϵ (and, therefore, for all ϵ), then we must have $|a| + |b| < 1$ as noted in Propositon 2.3. This is impossible.

As a consequence, one cannot exponentially stabilize a non-exponentially stable solution in any realistic way. This is a general fact which we now formulate for n-dimensional systems.

Consider the following linear control system with time delays

$$x(t) - \Sigma_{j=1}^{M} A_j x(t-r_j) - \Sigma_{j=1}^{M} B_k u(t-r_j) = 0, \tag{5.1}$$

where $x \in \mathbb{R}^n$, each $r_j > 0$, each A_j is a real $n \times n$ matrix, each B_k is an $n \times m$ matrix and $u \in \mathbb{R}^m$. Let $r(\sigma_{ess}(T_0(1)))$ be the radius of the essential spectrum at time 1 of the semigroup $T_0(t)$ generated by

$$x(t) - \Sigma_{j=1}^{M} A_j x(t-r_k) = 0. \tag{5.2}$$

If $r(\sigma_{ess}(T_0(1))) \geq 1$, then the solutions of (5.1) are not exponentially stable. Suppose that the system is exponentially stabilizable by a feedback loop $u(t) = Fx(t)$; that is, the difference equation

$$x(t) - \Sigma_{j=1}^{M} (A_j - B_j F) x(t-r_j) = 0 \tag{5.3}$$

is exponentially stable; that is, radius of the essential spectrum at $t = 1$ of the semigroup $T_F(t)$ generated by (5.3) is < 1.

For realistic control, one must allow for small discrepancies in the implementation time of the control. Therefore, the equation (5.3) should be replaced by

$$x(t) - \Sigma_{j=1}^{M}(A_j x(t - r_j) - B_j F x(t - r_j - \epsilon_j)) = 0, \tag{5.4}$$

where each ϵ_j is a small quantity.

The following result due to Hale and Verduyn-Lunel (2001) asserts that feedback stabilization is not realistic for difference equations (see this paper for related results and refernces, especially, Logemann and Townley (1996)).

Proposition 5.1. *Suppose that the time delays in (5.1) are rationally independent and that $r(\sigma_{ess}(T_0(1))) \geq 1$ (resp. > 1). If the control system (5.2) is exponentially stabilizable by feedback control $u(t) = Fx(t)$, then there is a dense subset E of $[0, \infty)$ such that, for every $\epsilon_j \in E$, $1 \leq j \leq M$, the closed loop system (5.4) is such that the essential spectral radius at $t = 1$ of the corresponding semigroup is ≥ 1 (resp. > 1).*

The proof of this result involves characterization of the essential spectral radius of semigroups defined by difference equations when the delays are rationally independent. We have given a result for this in the previous section for a scalar equation. The matrix case is somewhat more difficult (see Hale and Verduyn-Lunel (2001)).

5.2. Numerical implementation of delayed control

Consider the linear n-dimensional system with delayed control

$$\dot{x}(t) = Ax(t) + Bu(t - 1), \tag{5.5}$$

where $x \in \mathbb{R}^n$, $u \in \mathbb{R}$, A is an $n \times n$ matrix, B is a $1 \times n$ matrix, the system $\dot{x} = Ax$ is not exponentially stable and the pair (A, B) is stabilizable.

An approach for the stabilization and control of (5.5), called *finite spectrum assignment* (see Manitius and Olbrot (1979), Artstein (1982)) may be interpreted as follows: a prediction of the state variable over one delay interval is generated first and then a feedback of the predicted state is applied, thereby compensating for the effect of the time delay. This results in a closed loop system with a finite number of eigenvalues which can be chosen to be any values.

More precisely, for any given function $u(t)$ defined for $t \geq -1$, we can obtain the solution $x(t)$ for $t \geq 0$ through the variation of constants formula

$$x(t) = e^{A(t-s)}x(s) + \int_s^t e^{A(t-\tau)}u(\tau - 1)d\tau. \tag{5.6}$$

Suppose that φ is an integrable function on $[-1, 0]$ and $x_0 \in \mathbb{R}^n$ is fixed. If we let $u(t) = \varphi(t)$ for $t \in [-1, 0]$, $x(0) = x_0$, then one can determine $x(t)$ on the interval $[0, 1]$ from (5.6). We now define $u(t)$ for $t \geq 0$ by the feedback relation $u(t) = K^T x(t + 1)$; that is, with $t \mapsto t + 1$ and $s \mapsto t$ in (5.6),

$$u(t) = K^T(e^A x(t) + \int_0^1 e^{A\theta}Bu(t - \theta)d\theta). \tag{5.7}$$

A few elementary calculations show that the solution of (5.5) with $u(t)$ given in (5.7) satisfies the ODE

$$\dot{x}(t) = (A + BK^T)x(t). \tag{5.8}$$

Since A, B is stabilizable, one can choose K so that all eigenvalues of (4.8) are in the left half plane.

Let us now be precise about the space of initial values for the system (5.5), (5.7). If $C = C([-1, 0], \mathbb{R})$, define

$$D\varphi = \varphi(0) - K^T \int_{-1}^0 e^{-A\theta}\varphi(\theta)d\theta, \quad \varphi \in C. \tag{5.9}$$

Also, define the spaces

$$\begin{aligned} \tilde{C} &= \{\varphi \in C : D\varphi = K^T e^A x_0, x_0 \in \mathbb{R}^n\}, \\ C_0 &= \{\varphi \in C : D\varphi = 0\} \end{aligned} \tag{5.10}$$

With this notation, we can rewrite (5.5), (5.7) as

$$
\begin{aligned}
\dot{x}(t) &= Ax(t) + Bu(t-1) \\
Du_t &= K^T e^A x(t) \qquad\qquad (5.11)\\
x(0) &= x_0, \quad u_0 = \varphi \in C, \quad D\varphi = K^T e^A x_0.
\end{aligned}
$$

It is not difficult to show that (5.11) has a unique solution $(x(t, x_0, \varphi), u(t, x_0, \varphi)) \in \mathbb{R}^n \times \tilde{C}$ for all $t \geq 0$. Furthermore, if we define $T(t)(x_0, \varphi) = (x(t, x_0, \varphi), u_t(\cdot, x_0, \varphi))$, then $T(t), t \geq 0$ is a linear dynamical system on $\mathbb{R}^n \times \tilde{C}$.

It is also not difficult to see that

$$
r(\sigma_{ess}(T(1))) = r(\sigma_{ess}(S(1))), \tag{5.12}
$$

where $S(t)$, $t \geq 0$ is the linear dynamical system on C_0 defined by the equation

$$
Du_t = 0, \quad t \geq 0, \quad u_0 \in C_0. \tag{5.13}
$$

As a consequence of the fact that the integral term in the definition of D is a compact operator on C_0, it follows that

$$
r(\sigma_{ess}(S(1))) = 0. \tag{5.14}
$$

A difficulty in applying the control law (5.7) consists in determining on line the function $u(t)$ satisfying (5.7). One such approach is to apply a numerical quadrature rule to the integral operator in (5.7). The new system will then involve the interaction of a differential equation with a difference equation. The operator D will become a difference operator and will have essential spectral radius $\neq 0$. If the new difference operator is exponentially stable, then the numerical scheme will be satisfactory for solving the original control problem. If it is not, then solutions of the equations will become unbounded and will not reflect the properties of the continuous system.

The above difficulties have been noted by several authors and the reader should consult Michiels, Mondié, Roose and Dambrine (2004) for references. The remaining remarks in this subsection are taken from this paper.

Let us approximate the operator D in (5.9) by

$$D_n\varphi = \varphi(0) - K^T\Sigma_{j=1}^n h_{jn}e^{Ar_{jn}}B\varphi(-r_{jn}), \tag{5.15}$$

where $h_{jn} > 0$, $r_{jn} \in [0,1]$ for all j, n. We assume that, for every $\varphi \in D$, $D_n\varphi \to D\varphi$ as $n \to \infty$. We now consider the solutions of the equation

$$\begin{aligned} \dot{x}(t) &= Ax(t) + Bu(t-1) \\ D_n u_t &= K^T e^A x(t) \\ x(0) &= x_0, \quad u_0 = \varphi \in C, \quad D_n\varphi = K^T e^A x_0. \end{aligned} \tag{5.16}$$

We obtain a dynamical system $T_n(t)$, $t \geq 0$, on $\mathbb{R}^n \times \tilde{C}_n$, $\tilde{C}_n = \{\varphi \in C : D_n\varphi = K^T e^A x_0\}$ and a corresponding dynamical system $S_n(t)$, $t \geq 0$, on $C_{0n} \equiv \{\varphi \in C : D_n\varphi = 0\}$. As before, we have

$$r(\sigma_{ess}(T_n(1))) = r(\sigma_{ess}(S_n(1))). \tag{5.17}$$

As remarked earlier, this radius in (5.12) is not zero as it was for the continuous case. If we want *realistic implemention of the control process using finite spectrum assignment*, then we must allow the r_{jn} to vary over some small interval (which depends upon n) and preserve stability. We have seen before that this will be true if and only if there is a $\delta < 1$ and n_0 such that

$$\Sigma_{j=1}^n |K^T h_{jn}| e^{Ar_{jn}} < \delta < 1, \quad n \geq n_0. \tag{5.18}$$

This imposes conditions on the applicability of the control process using finite spectrum assignment and suggests the following result which is stated in Michiels, Mondié, Roose and Dambrine (2004) and proved in Michiels, Mondié and Roose (2003).

Proposition 5.2. *There is a realistic implementation of the control process (5.11) if*

$$\Gamma \equiv \int_0^1 |K^T e^{A\zeta} B| d\zeta < 1. \tag{5.19}$$

If $\Gamma > 1$, then the control process (5.11) cannot be realistically implemented.

As an example, consider the system (x, u scalars)

$$\begin{aligned} \dot{x}(t) &= x(t) + u(t-1) \\ u(t) + k\int_{-1}^{0} e^{-\theta}u(t+\theta)d\theta &= -kex(t), \end{aligned} \tag{5.20}$$

where $k > 1$. This accomplishes a pole shift to $1 - k < 0$ and the system is exponentially stable. On the other hand,

$$\Gamma = k\int_0^1 e^{\zeta}d\zeta = k(e-1) > 1$$

and Proposition 5.2 implies that the control is not realistically implementable.

A discussion of how to obtain stable implementation by piecewise constant controls or the addition of low pass filters as well as references is given in Michiels, Mondié, Roose and Dambrine (2004).

5.3. Wave equation with boundary damping

Consider the wave equation

$$\partial_t^2 u - \partial_x^2 u = 0, \quad x \in (0,1), \tag{5.21}$$

with the boundary conditions

$$u(0,t) = 0, \quad \partial_x u(1,t) = -k\partial_t u(1, t-r), \tag{5.22}$$

where $k > 0$, $r \geq 0$ are constants.

The term $-k\partial_t u(1, t-r)$ is the control function implemented with a delay time $r \geq 0$.

For $r = 0$, (5.21), (5.22) generates a linear dynamical system $T_{0,k}(t)$, $t \geq 0$, on the space

$$X = \{(\varphi, \psi) \in H^1(0,1) \times L^2(0,1) : \varphi(0) = 0\}.$$

It is easy to verify that $(0,0)$ is exponentially stable; that is, there are postive constants α, β such that

$$\|T_{0,k}(t)\|_{\mathcal{L}(X,X)} \leq \beta e^{-\alpha t}, \quad t \geq 0. \tag{5.23}$$

From the practical point of view, it is not reasonable to expect that the control can be implemented exactly at time t. Therefore, it is necessary to investigate whether the exponential stability is preserved under small variations in the implementation time; that is, small values of r. For this situation, we consider (5.21), (5.22) on the space

$$Y = C([-r, 0], X).$$

The system defines a linear dynamical system $T_{r,k}(t)$, $t \geq 0$.

The eigenvalues λ of (5.21), (5.22) correspond to those values of λ for which $e^{\lambda t}v(x)$ is a solution (5.21), (5.22) and, therefore, satisfy

$$e^{\lambda} + e^{-\lambda} = -k(e^{\lambda} - e^{-\lambda})e^{-\lambda r}$$

which is equivalent to

$$1 + ke^{\lambda r} - ke^{-\lambda(2+r)} + e^{-2\lambda} = 0. \tag{5.24}$$

Equation (5.24) is the characteristic equation for the difference equation (3.9). As we have noted there, the zero solution of the equation can be very unstable. If we interpret these results for the wave equation (5.21), (5.22), we conclude that there are sequences $\rho_j \to \infty$, $r_j \to 0$ as $j \to \infty$ and a corresponding solution $u_j(t,x)$ of (5.21), (5.22) with $r = r_j$ which becomes unbounded as $t \to \infty$ at the rate $e^{\rho_j t}$. System (5.21), (5.22) is not robust with respect to small variations in the input time of the control. The reason that this can occur is because the control is not a compact perturbation of the system with control. This can have drastic effects on the essential spectrum of the corresponding semigroup.

A more realistic model for (5.21) would be to include internal damping; that is, a term $-c\partial_x^2\partial_t u$, $c > 0$, and apply the control function $-k\partial_t u(1, t-$

r) to the complete strain $\partial_x u(1,t) + c\partial^2_{xt}u(1,t)$. In this case, the control becomes a compact perturbation and the stability is insensitive to small variations r (see, for example, Morgül (1995), (1998), Hale and Verduyn-Lunel (2001)).

6. Bifurcation from equilibrium of nonlinear equations. Consider the parametrized family of difference equations

$$x(t) = f_\lambda(x(t-r), x(t-1)),$$

where $0 < r < 1$ and λ is a real parameter (one could consider λ in a complete metric space). Also, suppose that $f_0(0,0) = 0$; that is, the origin is an equilibrium of the equation $x(t) = f_0(x(t-r), x(t-1))$.

A fundamental problem is to discuss bifurcation at the origin when λ varies in a neighborhood of zero.

Suppose that the linearization of $f_0(0,0)$ about the origin is given by (2.1). If there is to be a bifurcation at $\lambda = 0$, then there is a $\mu \in \sigma(S_{a,b,r}(1))$ with $|\mu| = 1$. Since we know that $\sigma(S_{a,b,r}(1))$ coincides with the essential spectrum, we are faced with a difficult bifurcation problem that, to the author's knowledge, has not been discussed for general delays r.

If r is rational, for any $\varphi \in C_{f_0,r}$ and any $\theta \in [-1,0]$, we have noted that there is an integer m (which may be very large) for which the complete dynamics is determined by a finite dimensional map on $\mathbb{R}^m$. In this case, one treats the bifurcation problem by considering the bifurcation in the map on $\mathbb{R}^m$. This avoids the problem of bifurcation from the essential spectrum.

If r is irrational, the problem is infinite dimensional and cannot be reduced to a finite dimensional problem. One must discuss bifurcation from the essential spectrum. Unless one can find some mathematical way to overcome these difficulties, it is perhaps reasonable in many situations to reassess the model to see if there is another model for which more classical results can apply.

Another possibility is to abandon the use of only difference equations and allow the possibility of some distributed delays. Some conditions must be imposed on the part that consists only of discrete delays.

As an example, consider the equation

$$z(t) = az(t-r) + bz(t-1) + \int_{-1}^{0} c(\theta)z(t+\theta)d\theta, \tag{6.1}$$

where c is a continuous function on $[-1, 0]$. If $T_{a,b,c}(t)$, $t \geq 0$, is the corresponding dynamical system, then

$$r(\sigma_{\text{ess}}\, T_{a,b,c}(1))) = r(\sigma_{\text{ess}}\, T_{a,b,0}(1))); \tag{6.2}$$

that is, the same as the radius of the spectrum of the dynamical system given by (2.1).

If we assume that $|a| + |b| < 1$, then

$$r(\sigma_{\text{ess}}\, T_{a,b,c}(1))) < 1 \tag{6.3}$$

for all delays $r \in (0, 1)$. If we choose any $\epsilon > 0$ such that $\delta_\epsilon \equiv \epsilon + |a| + |b| < 1$, then the number of eigenvalues with real part $> c_\epsilon$ are finite in number and have finite multiplicity. They are continuous in r. As a consequence, any local bifurcation from the origin for the nonlinear equation

$$x(t) = ax(t-r) + bx(t-1) + \int_{-1}^{0} c(\theta)x(t+\theta)d\theta + g(x_t), \tag{6.4}$$

where g is small, are the result of a finite number of eigenvalues of (6.1) crossing the imaginary axis.

Under certain conditions, but for more general equations, Hale and Oliveira (1980) considered the equation

$$x(t) = g(x(t-r_1), \ldots, x(t-r_m), \int_{-r}^{0} A(\theta)x(t+\theta)d\theta),$$

where $x \in \mathbb{R}^n$, $r > 0$, $0 < r_j < r$, $j = 1, 2, \ldots, m$, and $A(\theta)$ is a continuous $n \times n$ matrix.

If we suppose that zero is a solution of the equation, then the linear variational equation about zero is

$$x(t) = \Sigma_{k=1}^{m} A_k x(t - r_k) + A_{m+1} \int_{-r}^{0} A(\theta)x(t + \theta)d\theta).$$

Hale and Oliveira (1980) studied bifurcation from the origin using $\{r_k, A_k, 1 \le k \le m, A_{m+1}\}$ as parameters. To include the discrete delays as bifurcation parameters, they made the assumption that the zero solution of the difference equation

$$y(t) = \Sigma_{k=1}^{m} A_k y(t - r_k)$$

is exponentially stable independently of the delay. In this case, bifurcations from the origin occur through bifurcations from eigenvalues of finite multiplicity. They proved that the Hopf bifurcation to periodic orbits occurs under the usual hypotheses that one assumes for ordinary differential equations. The proof given there consisted of first proving that any periodic solution must be smooth in time and then using estimates on Fourier series to obtain the existence. With the knowledge that we now have about center manifolds for difference equations, the proof could follow by more standard arguments using the flow on the center manifold.

7. The delayed logistic equation. We end this paper with the description of some known results about the delayed logistic equation for $r = 1/2$ (equivalent to a two dimensional map) pointing out that it is unknown if any of the results remain valid if r varies over a small interval containing 1/2.

The delayed difference equation is

$$y(t) = \alpha y(t - r)(1 - y(t - 1)), \tag{7.1}$$

where $\alpha > 0$ and $0 < r < 1$ are constants.

When the delays are $\{1/2, 1\}$, the dynamics of (7.1) can be determined from the map $F_\alpha : (x_1, x_2) \in \mathbb{R}^2 \mapsto (\alpha x_1(1 - x_2), x_1) \in \mathbb{R}^2$. The following result is well known (see, for example, Devaney (1986) or Hale and Kocak (1991, p. 455 and p.474)).

Proposition 7.1. *The map* $F_\alpha : (x_1, x_2) \in \mathbb{R}^2 \mapsto (\alpha x_1(1-x_2), x_1) \in \mathbb{R}^2$ *satisfies the following properties:*

(i) *For* $0 < \alpha < 1$*, the origin is an exponentially stable fixed point of* F_α*.*

(ii) *There is a bifurcation at* $\alpha = 1$ *to another fixed point* $Y_\alpha = (y_a, y_\alpha)$*,* $y_\alpha = 1 - \frac{1}{\alpha}$*, of* F_α *which is exponentially stable for* $1 < \alpha < 2$*,*

(iii) *There is a supercritical Hopf bifurcation for the map* F_α *at* $\alpha = 2$*; that is, there are a* $\delta > 0$ *and a closed curve* $\Gamma_\alpha \subset \mathbb{R}^2$*, existing for* $\alpha \in (2, 2+\delta)$*, which is invariant (that is,* $F_\alpha \Gamma_\alpha = \Gamma_\alpha$*) and* Γ_α *is exponential stable.*

In the spirit of our previous discussions, we ask if there is a neighborhood $U \subset \mathbb{R}$ of (1/2) such that properties (i)-(iii) hold for (7.1) for each $r \in U$.

For any $\alpha > 0$, the origin is an equilibrium of (7.1) with the linear variational equation about 0 being

$$
\begin{aligned}
&z(t) = \alpha z(t-r), \quad t \geq 0 \\
&z(t) = \varphi(t), \quad t \in [-1, 0], \quad \varphi \in C_{\alpha,r}.
\end{aligned} \tag{7.2}
$$

As we have noted in Section 2, the zero solution of (7.1) is exponentially stable for $0 < \alpha < 1$, is stable for $\alpha = 1$ and a hyperbolic saddle point for $\alpha > 1$. This implies that (i) remains valid for any r.

If $\alpha > 1$, the point $y_\alpha = 1 - \frac{1}{\alpha}$ is also an equilibrium of (7.1). The linear variational equation about y_α is

$$
z(t) = z(t-r) - (\alpha - 1)z(t-1). \tag{7.3}
$$

In this case, if r is irrational, it follows from Proposition 2.2 that the spectrum of the semigroup generated by (7.3) is the annulus in the complex plane with inner circle of radius $e^{\rho_2} < 1$ and outer radius $e^{\rho_0} > 1$. Therefore, the equilibrium y_α of (7.1) always is unstable. There are many solutions which approach zero exponentially and many which become unbounded at an exponential rate. However, the flow does not act like a saddle point. *What is happening to the flow?*

As we see from these remarks, we are not able to be very precise about what happens near $\alpha = 1$ and have no idea what to say about $\alpha = 2$ that occurs in Proposition 7.1.

We now make some trivial remarks about what happens if we replace the discrete delay r by a distributed delay around a point r; that is, consider the equation

$$y(t) = [\int_{-1}^{0} a(\theta)y(t+\theta)d\theta](1 - y(t-1)), \tag{7.4}$$

where a is a nonnegative continuous function with support in $(0, 1)$ and is not identically zero. Let $\bar{\alpha} \equiv \int_{-1}^{0} a(\theta)d\theta$ and consider (7.4) with this as the bifurcation parameter.

The linear variational equation about the origin is

$$z(t) = \int_{-1}^{0} a(\theta)z(t+\theta)d\theta. \tag{7.5}$$

The radius of the essential spectrum of the linear dynamical system defined by (7.5) is zero. From this, it is easy to see that the origin is exponentially stable if $\bar{\alpha} < 1$ and that 0 is a simple eigenvalue if $\bar{\alpha} = 1$. For $\bar{\alpha} > 1$, there is the equilibrium $y_\alpha = 1 - \frac{1}{\bar{\alpha}}$. Furthermore, the origin is unstable. The linear variational equation about $\bar{\alpha}$ is

$$z(t) = \frac{1}{\bar{\alpha}} \int_{-1}^{0} a(\theta)z(t+\theta)d\theta + (\bar{\alpha} - 1)z(t-1). \tag{7.6}$$

The radius of the essential spectrum of the linear dynamical system defined by (7.6) is $\bar{\alpha} - 1$, $\bar{\alpha} > 1$. From this, it is an easy analysis to show that y_α is exponentially stable for $\bar{\alpha} < 2$. It cannot be exponentially stable at $\bar{\alpha} = 2$ since the radius of the essential spectrum is one. We are back in the same situation as before with the problem of bifurcation from the essential spectrum.

Suppose that we put a distribulted delay around the point r and also around the point 1; that is, consider the equation

$$y(t) = \int_{-1}^{0} a(\theta)y(t+\theta)d\theta \int_{-1}^{0} b(\theta)(1 - y(t+\theta)d\theta, \tag{7.7}$$

where b is a continuous nonnegative function, not identically zero and $\mathrm{supp}(a) \cap \mathrm{supp}(b) = \emptyset$. Without loss of generality, we may assume that $\bar{\beta} = \int_{-1}^{0} b(\theta)d\theta = 1$. The radius of the essential spectrum of the linear dynamical system defined by (7.7) is zero. We are now in a position to discuss the bifurcations up through the Hopf bifurcation for continuous dynamical systems (not maps) in the standard way.

These remarks do not allow us to make any connection with the results in Proposition 7.1.

References

Avellar, C.E. and J.K. Hale (1980) On the zeros of exponential polynomials. *J. Math. Anal. Appl. 74*, 434-452.

Devaney, R.L. (1986) *An Introduction to Chaotic Dynamics.* Benjamin/ Cummings.

Hale, J.K. and H. Kocak (1991) *Dynamics and Bifrucations.* Springer-Verlag.

Hale, J.K. and J.C.F. de Oliveira (1980) Hopf bifurcation for functional equations. *J. Differential Eqns. 74*, 41-59.

Hale, J.K. and S.M. Verduyn-Lunel (1993) *Introduction to Functional Differential Equations.* Springer-Verlag.

Hale, J.K. and S.M. Verduyn-Lunel (2001) Effects of delays on stability and control. *Operator Theory: Advances and Applications 122*, 275-301.

Hale, J.K. and S.M. Verduyn-Lunel (2002) Strong stabilization of neutral functional differential equations. *IMA J. Math. Contr. Info. 19*, 5-23.

Henry, D. (1974) Linear autonomous neutral functional differential equations. *J. Differential Eqns. 15*, 106-128.

Henry, D. (1987) Topics in Analysis. *Pub. Mat. Univ. Auto. Barcelona 31*, 29-84.

Logemann, H. and S. Townley (1996) The effect of small delays in the feedback loop on the stability of neutral systems. *Systems and Control Letters 27*, 267-274.

Michiels, W., Mondié, S. and B. Roose (2003) Robust stabilization of time-delay systems with distributed control laws: necessary and sufficient conditions for a safe implementation. *Technical Report TW-363*. Dept. Comp. Sci. K.U. Leuven, Belgium.

Michiels, W., Mondié, S. Roose, D. and M. Dambrine (2004) The effect of approximating distributed delay control laws on stability. In *Advances in Time-Delay Systems* (Eds. S. Niculescu and K. Gu). Lect. Notes Comp. Sci. Eng., Vol. 38, Springer-Verlag.

Morgül, O. (1995) On stabilization and stability robustness against small delays of some damped wave equations. *IEEE Trans. Automat. Control 40*, 1626-1630.

Morgül, O. (1998) Stabilization and disturbance rejection for the wave equation. *IEEE Trans. Automat. Control 43*, 89-95.

Rabah, R., Sklyar, G.M. and A.V. Rezounenko (2004) Stability analysis of neutral type systems in Hilbert space. Preprint.

Zverkin, A.M. (1968) The connection between boundedness and stability of solutions of linear systems with an infinite number of degrees of freedom. (Russian) *Differencial'nye Uravnenija 4*, 366-377. English Transl. *Diff. Eqns. 4*, 196-197.

SOLVABILITY OF THE DISCRETE LQR-PROBLEM UNDER MINIMAL ASSUMPTIONS

ROMAN HILSCHER*
Department of Mathematical Analysis
Faculty of Science, Masaryk University
CZ-60200 Brno, Czech Republic
E-mail: hilscher@math.muni.cz
Phone: +420-549494226
Fax: +420-541210337

VERA ZEIDAN†
Department of Mathematics, Michigan State University
East Lansing, MI 48824-1027, USA
E-mail: zeidan@math.msu.edu
Phone: +1-517-353-0857
Fax: +1-517-432-1562

ABSTRACT. The purpose of this paper is to provide a solution to the classical discrete linear-quadratic regulator problem under minimal assumptions. In particular, we do not assume the positive or nonnegative definiteness of the coefficients. Instead, a natural condition is imposed which is necessary for minimizing the involved discrete quadratic functional. The optimal solution is constructed from a generalized discrete Riccati equation and has a feedback form.

2000 *Mathematics Subject Classification.* 39A12, 49K99.

Key words and phrases. Discrete linear regulator, Discrete quadratic functional, Riccati matrix equation.

* Corresponding author. Research supported by the Ministry of Education, Youth, and Sports of the Czech Republic under grant 1K04001, by the Grant Agency of the Academy of Sciences of the Czech Republic under grant KJB1019407, and by the Czech Grant Agency under grant 201/04/0580.

† Research supported by the National Science Foundation under grant DMS – 0306260.

1. The problem

1.1. Discrete LQR problem. In this note we consider a classical discrete linear-quadratic regulator problem

$$\text{minimize} \quad \mathcal{I}(x,u) := x_{N+1}^T \Gamma x_{N+1} + \sum_{k=0}^{N} \{x_{k+1}^T C_k x_{k+1} + u_k^T B_k u_k\} \tag{LQR}$$

subject to constraints

$$\Delta x_k = A_k x_{k+1} + B_k u_k, \quad k \in [0,N], \quad x_0 = x_0^*. \tag{1}$$

Here $n, N \in \mathbb{N}$ are given numbers, A_k, B_k, C_k, Γ are real $n \times n$ matrices such that B_k, C_k, Γ are symmetric, $\tilde{A}_k := (I - A_k)^{-1}$ exists, $x_0^* \in \mathbb{R}^n$ is a given initial state, and $\{x_k\}_{k=0}^{N+1}$, $\{u_k\}_{k=0}^{N}$ are the state and control sequences of n-vectors. In our notation, $\Delta x_k := x_{k+1} - x_k$ is the forward difference operator. Also, we will use $\operatorname{Ker} M$, $\operatorname{Im} M$, M^T, M^{T-1}, $M^\dagger$, $M \geq 0$, and $M > 0$ to denote the kernel, image, transpose, inverse of the transpose, Moore-Penrose generalized inverse, nonnegative definiteness, and positive definiteness of a given matrix M, respectively. The intervals $[0,N]$ and $[0,N+1]$ denote discrete sets with indicated endpoints.

Note that traditionally the quadratic functional $\mathcal{I}$ and the difference equation in (1) have x_k instead of x_{k+1}. However, the invertibility of the matrix $I - A_k$ implies that (1) reduces to

$$x_{k+1} = \tilde{A}_k x_k + \tilde{A}_k B_k u_k, \quad k \in [0,N],$$

which, when inserted in $\mathcal{I}$, yields the classical formulation with x_k. As in the recent literature (see e.g., [3]), we use in this paper the formulation with the shift in x_{k+1}.

In the last thirty years the discrete-time linear-quadratic regulator problem (LQR) as well as its continuous-time counter part, problem (P) below, have attracted the attention of numerous researchers in control theory and engineering (see e.g., [1,6–11,15–17]). However, discrete-time systems can be either inherent, such as digital computers, monetary systems, digital filters etc., or obtained from continuous-time systems as follows:

Consider the continuous-time linear-quadratic regulator problem

$$\text{minimize} \quad \mathcal{J}(x,u) := x^T(b)\,\Gamma\,x(b) + \int_a^b \{x^T(t)\,C(t)\,x(t) + u^T(t)\,B(t)\,u(t)\}\,\mathrm{d}t \tag{P}$$

subject to constraints

$$x'(t) = A(t)\,x(t) + B(t)\,u(t), \quad t \in [a,b], \quad x(a) = x_a^*. \tag{2}$$

Partition the (continuous-time) interval $[a, b]$ into $N+1$ subintervals

$$a = t_0 < t_1 < \cdots < t_N < t_{N+1} = b.$$

Now set $\mu_k := \Delta t_k$, $A_k := A(t_k)$, $B_k := B(t_k)$, and $C_k := C(t_k)$. Furthermore, with a pair $(x(t), u(t))$ satisfying (2) we set $x_k := x(t_k)$ and $u_k := u(t_k)$. Then the discretized problem of (P) is a variable stepsize (LQR) problem, i.e., the problem

$$\text{minimize} \quad \mathcal{I}(x,u) := x_{N+1}^T \Gamma \, x_{N+1} + \sum_{k=0}^{N} \left\{ x_{k+1}^T C_k x_{k+1} + u_k^T B_k u_k \right\} \mu_k \tag{μLQR}$$

subject to constraints

$$\frac{\Delta x_k}{\mu_k} = A_k x_{k+1} + B_k u_k, \quad k \in [0, N], \quad x_0 = x_0^*. \tag{3}$$

The sequence $\{\mu_k\}_{k=0}^N$ of positive numbers is called the *stepsize* or *graininess*. In this setting, we assume that $\tilde{A}_k := (I - \mu_k A_k)^{-1}$ exists. This holds whenever the stepsize μ_k is sufficiently small. Equation (3) can be solved for x_{k+1} and then we get,

$$x_{k+1} = \tilde{A}_k x_k + \mu_k \tilde{A}_k B_k u_k, \quad k \in [0, N].$$

Hence, the discretized problem (μLQR) of (P) is in the form of (LQR).

Therefore, it is of great interest to study the discrete-time problem (LQR). In particular, one would want to find *minimal* conditions on the data A_k, B_k, C_k, and Γ guaranteeing the existence of an optimal solution. Furthermore, in order not to re-solve the problem with each change of the initial state x_0^*, it is then important to find the optimal control u_k^* in a form of a *feedback* control (that is, in terms of the state x_k^*).

1.2. Assumptions. The standard assumptions on these coefficients of $\mathcal{I}$ are

$$C_k \geq 0, \quad B_k > 0, \quad \text{and} \quad \Gamma \geq 0, \tag{4}$$

see e.g. [14, pg. 490]. Under (4), it is shown in [14] that the optimal solution (x^*, u^*) of (LQR) is unique and the optimal control u^* is obtained from the *feedback law* $u_k^* = W_k x_k^*$ for all $k \in [0, N]$. Here, the (symmetric nonpositive definite) matrices W_k solve the Riccati matrix equation

$$W_k = \mathcal{D}_k^{-1} B_k \tilde{A}_k^T (W_{k+1} - C_k) \, \tilde{A}_k = \mathcal{D}_k^{-1} - B_k^{-1} \tag{5}$$

with the endpoint condition $W_{N+1} = -\Gamma$, where

$$\mathcal{D}_k := B_k - B_k \tilde{A}_k^T (W_{k+1} - C_k) \, \tilde{A}_k B_k > 0.$$

For the proof of $\mathcal{D}_k > 0$ see [12, Remark 3].

It is worth noting that under assumption (4), equation (5) is equivalent to

$$\Delta W_k = C_k - A_k^T W_k - (W_{k+1} - C_k)\,\tilde{A}_k(A_k + B_k W_k), \tag{6}$$

which is the Riccati matrix equation studied recently in the connection with the disconjugacy of the linear Hamiltonian difference system

$$\Delta x_k = A_k x_{k+1} + B_k u_k, \quad \Delta u_k = C_k x_{k+1} - A_k^T u_k, \tag{H}$$

and with the positivity of discrete quadratic functionals $\mathcal{I}$. See e.g. [3] for more details. Moreover, the matrix $\mathcal{D}_k$ plays an important role in these references, since it (among others) completes the quadratic functional $\mathcal{I}$ to a “square”. This completion to a ”square” is known as the discrete Picone identity. More precisely,

$$x_{k+1}^T C_k x_{k+1} + u_k^T B_k u_k = z_k^T \mathcal{D}_k z_k + \Delta(x_k^T W_k x_k),$$

where $z_k = u_k - W_k x_k$. Another appearance of $\mathcal{D}_k$ is in the focal points definition for conjoined bases (X, U) of (H), see [3, Lemma 2].

A weaker hypothesis on the matrices B_k is used in [5, 13], namely the conditions

$$C_k \geq 0 \quad \text{and} \quad B_k \geq 0 \tag{7}$$

are required. Moreover, in addition to (7), an extra assumption (called a “solvability condition”, which is in fact the invertibility of some matrix) is imposed in [5, 13] in order to ensure the *existence* of a minimal solution.

In this paper, we shall require neither (4) nor (7), but instead we assume *weaker* (and natural) conditions, which are shown to be *necessary* for minimality of the discrete quadratic functional $\mathcal{I}$ when the equation of motion is controllable (see Remark 5). These conditions are

$$\mathcal{D}_k \geq 0 \quad \text{and} \quad (I - \mathcal{D}_k \mathcal{D}_k^\dagger)\, B_k \tilde{A}_k^T (W_{k+1} - C_k)\,\tilde{A}_k = 0 \tag{8}$$

for all $k \in [0, N]$, where W_k solves a certain generalized Riccati matrix equation. The second condition in (8) will also be referred to as a “solvability condition”. In addition, we show that these two conditions yield that an optimal control u^* for (LQR) is *not* necessarily unique, but has to satisfy the *feedback law* $u_k^* = F_k x_k^* + (I - \mathcal{D}_k^\dagger \mathcal{D}_k)\,\gamma_k$, where γ_k are some vectors and F_k are uniquely determined matrices. See the next section (Corollary 1) for the precise statement and for the proof. A numerical example is then provided in order to illustrate the utility of our results in the sense that they apply when all the known results do not. Finally we state the obtained results for the variable stepsize problem (μLQR).

2. Solution to (LQR) problem

2.1. Dynamic programming approach. In this section we solve the linear-quadratic problem (LQR). We apply the method of dynamic programming as in [14] with some modifications from [4].

For $x \in \mathbb{R}^n$ and $k \in [0, N+1]$ we define the *value function* $V(x,k)$ by the following. We set $V(x, N+1) := x^T \Gamma\, x$ and for $k \in [0, N]$

$$V(x,k) := \min_{u \in \mathbb{R}^n} \Big\{ (\tilde{A}_k x + \tilde{A}_k B_k u)^T C_k (\tilde{A}_k x + \tilde{A}_k B_k u) + u^T B_k u \\ + V(\tilde{A}_k x + \tilde{A}_k B_k u, k+1) \Big\}, \tag{9}$$

provided the minimum exists. If we denote $x_{k+1}(u) := \tilde{A}_k x_k + \tilde{A}_k B_k u$, then the value function at (x_k, k) is

$$V(x_k, k) = \min_{u \in \mathbb{R}^n} \left\{ x_{k+1}^T(u)\, C_k x_{k+1}(u) + u^T B_k u + V\big(x_{k+1}(u), k+1\big) \right\}.$$

Remark 1. [4, Theorem 40.1, pg. 318] Equation (9) is known as the *Bellman equation*, but in our interpretation it is just the defining equation for $V(x,k)$. The Bellman principle of dynamic programming says:

> *If* (x^*, u^*) *is a pair satisfying* (1), *then it is optimal for* (LQR) *iff the minimum in* (9) *for* $V(x_k^*, k)$ *is attained at* u_k^* *for all* $k \in [0, N]$.

For our main result we will need the following recursive definitions. First we set $W_{N+1} := -\Gamma$. Next, suppose that W_{k+1} is defined and set

$$\mathcal{D}_k := B_k - B_k \tilde{A}_k^T (W_{k+1} - C_k)\, \tilde{A}_k B_k, \tag{10}$$

$$F_k := \mathcal{D}_k^{\dagger} B_k \tilde{A}_k^T (W_{k+1} - C_k)\, \tilde{A}_k, \tag{11}$$

$$W_k := (F_k^T B_k + I)\, \tilde{A}_k^T (W_{k+1} - C_k)\, \tilde{A}_k. \tag{GRE}$$

Note that all the matrices $\mathcal{D}_k$, F_k, and W_k are well defined (once $W_{N+1} := -\Gamma$ is given).

Remark 2. Equation (GRE) is called a *generalized Riccati matrix equation.* It generalizes the known Riccati equation (5) or (6) to the case when $\mathcal{D}_k$ is not invertible. Note that if the second condition of (8) holds, then (GRE) takes the form

$$W_k = F_k^T \mathcal{D}_k F_k + \tilde{A}_k^T (W_{k+1} - C_k)\, \tilde{A}_k.$$

Remark 3. [2, Corollary 2, pg.40] Let A be a matrix and x, b be vectors of appropriate dimensions. The system of linear equations $Ax = b$ has a solution iff $AA^\dagger b = b$. In this case, all solutions are given by the formula (for some vector γ)

$$x = A^\dagger b + (I - A^\dagger A)\,\gamma.$$

The proof of the following theorem is displayed after Corollary 2.

Theorem 1. *Let $\mathcal{D}_k$, F_k, and W_k be defined by* (10)–(GRE)*, respectively, with $W_{N+1} := -\Gamma$. Then the following statements are equivalent.*

(i) *For all $x \in \mathbb{R}^n$, the minimum in* (9) *is attained at*

$$\bar{u}_k(x) = F_k x + (I - \mathcal{D}_k^\dagger \mathcal{D}_k)\,\gamma_k, \tag{12}$$

for some vectors $\gamma_k \in \mathbb{R}^n$, and hence the value function takes the form $V(x,k) = -x^T W_k x$, for all $k \in [0, N]$.

(ii) *Condition* (8) *is satisfied for all $k \in [0, N]$.*

The following characterization of the optimal solutions to the (LQR) problem is an immediate consequence of Theorem 1.

Corollary 1. *Assume that* (8) *holds for all $k \in [0, N]$. Then the linear regulator problem* (LQR) *has an optimal solution (x^*, u^*) if and only if for all $k \in [0, N]$*

$$u_k^* = F_k x_k^* + (I - \mathcal{D}_k^\dagger \mathcal{D}_k)\,\gamma_k \tag{13}$$

for some vectors $\gamma_k \in \mathbb{R}^n$, and hence $V(x_k^*, k) = -(x_k^*)^T W_k x_k^*$.

Remark 4. (i) If $\mathcal{D}_k$ is invertible, then $F_k^T B_k = W_k B_k$. This follows immediately from (11), i.e., from $F_k^T \mathcal{D}_k = \tilde{A}_k^T (W_{k+1} - C_k)\,\tilde{A}_k B_k$, and by using (10).

(ii) If both $\mathcal{D}_k$ and B_k are invertible, then part (i) of this remark yields $F_k^T = W_k$ (so that F_k is symmetric). Hence, the control law (13) reduces to the well-known result $u_k^* = W_k x_k^*$, as we mentioned in the previous section.

If assumption (4) is not satisfied but $\mathcal{D}_k$ is still positive definite, then Corollary 1 yields the following.

Corollary 2. *Assume that $\mathcal{D}_k > 0$ for all $k \in [0, N]$. Then the linear regulator problem* (LQR) *has an optimal solution (x^*, u^*) if and only if $u_k^* = F_k x_k^*$ for all $k \in [0, N]$, where*

$$F_k = \mathcal{D}_k^{-1} B_k \tilde{A}_k^T (W_{k+1} - C_k)\,\tilde{A}_k,$$

and hence $V(x_k^*, k) = -(x_k^*)^T W_k x_k^*$.

Proof of Theorem 1. "(ii) $\Rightarrow$ (i)" Assume (8). The form $V(x,k) = -x^T W_k x$ holds for $k = N+1$ by the definition of the value function and by $W_{N+1} := -\Gamma$. Moreover, W_{N+1} is symmetric. Suppose that the above formula holds at an index $k+1$ and that W_{k+1} is symmetric. Then (9) yields

$$\begin{aligned} V(x,k) &= \min_{u\in\mathbb{R}^n} \left\{(\tilde{A}_k x + \tilde{A}_k B_k u)^T (C_k - W_{k+1})\,(\tilde{A}_k x + \tilde{A}_k B_k u) + u^T B_k u\right\} \\ &= \min_{u\in\mathbb{R}^n} \left\{u^T \mathcal{D}_k u - 2x^T \tilde{A}_k^T (W_{k+1} - C_k)\,\tilde{A}_k B_k u\right\} \\ &\qquad - x^T \tilde{A}_k^T (W_{k+1} - C_k)\,\tilde{A}_k x. \end{aligned} \tag{14}$$

Note that $\mathcal{D}_k$ is symmetric. If now $\mathcal{D}_k \geq 0$ is assumed as in (8), i.e., if the second derivative of the minimized expression with respect to u is nonnegative, then the minimum above is attained whenever, for some u, (the first derivative of the minimized expression with respect to u is zero)

$$\mathcal{D}_k u - B_k \tilde{A}_k^T (W_{k+1} - C_k)\,\tilde{A}_k x = 0,$$

i.e., whenever, for some u,

$$\mathcal{D}_k u = B_k \tilde{A}_k^T (W_{k+1} - C_k)\,\tilde{A}_k x. \tag{15}$$

By Remark 3, this equation has a solution iff

$$(I - \mathcal{D}_k \mathcal{D}_k^\dagger)\, B_k \tilde{A}_k^T (W_{k+1} - C_k)\,\tilde{A}_k x = 0.$$

Since (8) is assumed, equation (15) indeed has a solution and this solution has the form

$$\bar{u}_k(x) = \mathcal{D}_k^\dagger B_k \tilde{A}_k^T (W_{k+1} - C_k)\,\tilde{A}_k x + (I - \mathcal{D}_k^\dagger \mathcal{D}_k)\,\gamma_k = F_k x + (I - \mathcal{D}_k^\dagger \mathcal{D}_k)\,\gamma_k.$$

Note that by multiplying the above equation by $\mathcal{D}_k$ we get $\mathcal{D}_k \bar{u} = \mathcal{D}_k F_k x$. Calculate now the value of $V(x,k)$ by plugging $\bar{u}$ into (14). We use $\mathcal{D} = \mathcal{D}\mathcal{D}^\dagger\mathcal{D}$ and $\mathcal{D}^\dagger = \mathcal{D}^\dagger\mathcal{D}\mathcal{D}^\dagger$ to get

$$\begin{aligned} V(x,k) &= \bar{u}^T \mathcal{D}_k \bar{u} - 2x^T \tilde{A}_k^T (W_{k+1} - C_k)\,\tilde{A}_k B_k \bar{u} - x^T \tilde{A}_k^T (W_{k+1} - C_k)\,\tilde{A}_k x \\ &\overset{(15)}{=} \bar{u}^T \mathcal{D}_k \bar{u} - 2\bar{u}^T \mathcal{D}_k \bar{u} - x^T \tilde{A}_k^T (W_{k+1} - C_k)\,\tilde{A}_k x \\ &= -\bar{u}^T \mathcal{D}_k \mathcal{D}_k^\dagger \mathcal{D}_k \bar{u} - x^T \tilde{A}_k^T (W_{k+1} - C_k)\,\tilde{A}_k x \\ &\overset{(15)}{=} x^T \left\{-F_k^T \mathcal{D}_k F_k - \tilde{A}_k^T (W_{k+1} - C_k)\,\tilde{A}_k\right\} x \\ &= -x^T\,(F_k^T B_k + I)\,\tilde{A}_k^T (W_{k+1} - C_k)\,\tilde{A}_k\, x = -x^T\, W_k\, x. \end{aligned}$$

Note that the above computation also shows that W_k is symmetric. Hence, this implication is proven.

"(i) $\Rightarrow$ (ii)" Fix $k \in [0,N]$. Let $\bar{u}_k(x) = F_k x + (I - \mathcal{D}_k^\dagger \mathcal{D}_k)\gamma_k$ be a vector, for which the minimum of

$$M_k(u) := u^T \mathcal{D}_k u - 2x^T \tilde{A}_k^T (W_{k+1} - C_k)\,\tilde{A}_k B_k u$$

is attained and $V(x,k) = -x^T W_k x$. Since $M_k(u)$ is quadratic in u, it results that $\nabla^2 M_k(u) = \nabla^2 M_k(\bar{u}_k(x)) \geq 0$, that is, $\mathcal{D}_k \geq 0$. Furthermore, as the minimum of $M_k(u)$ is attained at $\bar{u}_k(x)$, we have that $\nabla M_k(\bar{u}_k(x)) = 0$, i.e., $\bar{u}_k(x)$ solves (15). By Remark 3, this is equivalent to

$$(I - \mathcal{D}_k \mathcal{D}_k^\dagger)\, B_k \tilde{A}_k^T (W_{k+1} - C_k)\, \tilde{A}_k = 0.$$

Therefore, condition (8) holds. ■

A consequence of the proof of Theorem 1 is the following.

Corollary 3. *A pair (x^*, u^*) is optimal for the problem* (LQR) *if and only if $\mathcal{D}_k \geq 0$ and*

$$(I - \mathcal{D}_k \mathcal{D}_k^\dagger)\, B_k \tilde{A}_k^T (W_{k+1} - C_k)\, \tilde{A}_k x_k^* = 0 \tag{16}$$

for all $k \in [0, N]$.

Remark 5. Assume that the system (1) is controllable, that is, the reachable set at k,

$$\mathcal{R}_k(x_0) := \left\{ x_k \in \mathbb{R}^n : \{x_j\}_{j=0}^k \text{ solves (1) for some } \{u_j\}_{j=0}^{k-1} \right\} = \mathbb{R}^n,$$

for all $k \in [0, N+1]$. Then condition (16) becomes the solvability condition in (8).

2.2. Numerical example. In the next example we wish to demonstrate that the assumptions of the (LQR) problem in this note, namely condition (8) with the matrices $\mathcal{D}_k$, F_k, and W_k given by (10)–(GRE), allow the inclusion of a class of problems that is larger than the one previously considered. More specifically, the matrices $\mathcal{D}_k$ do not need to be positive definite but only $\mathcal{D}_k \geq 0$ (hence $\mathcal{D}_k$ can be singular).

Example 1. Consider the (LQR) problem with

$$\mathcal{I}(x,u) := x_{N+1}^T \left(\begin{smallmatrix} -2 & 0 \\ 0 & 0 \end{smallmatrix}\right) x_{N+1} + \sum_{k=0}^{N} \left\{ x_{k+1}^T \left(\begin{smallmatrix} 4 & 0 \\ 0 & 0 \end{smallmatrix}\right) x_{k+1} + u_k^T \left(\begin{smallmatrix} -1 & 0 \\ 0 & 0 \end{smallmatrix}\right) u_k \right\}$$

subject to $\Delta x_k = \left(\begin{smallmatrix} -1 & 0 \\ 0 & 0 \end{smallmatrix}\right) u_k$, $k \in [0, N]$ for some initial state $x_0 \in \mathbb{R}^2$. This means that we take $n = 2$, $A_k \equiv 0$, $\tilde{A}_k \equiv I$, $B_k \equiv \left(\begin{smallmatrix} -1 & 0 \\ 0 & 0 \end{smallmatrix}\right)$, $C_k \equiv \left(\begin{smallmatrix} 4 & 0 \\ 0 & 0 \end{smallmatrix}\right)$, and $\Gamma = \left(\begin{smallmatrix} -2 & 0 \\ 0 & 0 \end{smallmatrix}\right)$. Note that $B_k \not\geq 0$ (in fact $B_k \leq 0$). With $W_{N+1} := -\Gamma = \left(\begin{smallmatrix} 2 & 0 \\ 0 & 0 \end{smallmatrix}\right)$ it follows that

$$\mathcal{D}_k \equiv \left(\begin{smallmatrix} 1 & 0 \\ 0 & 0 \end{smallmatrix}\right), \quad F_k = W_k \equiv \left(\begin{smallmatrix} 2 & 0 \\ 0 & 0 \end{smallmatrix}\right)$$

for all k. Consequently, $\mathcal{D}_k \geq 0$ for all $k \in [0, N]$. The solvability condition in (8) is in this case

$$(I - \mathcal{D}_k \mathcal{D}_k^\dagger)\, B_k (W_{k+1} - C_k) = \left(\begin{smallmatrix} 0 & 0 \\ 0 & 1 \end{smallmatrix}\right) \left(\begin{smallmatrix} -1 & 0 \\ 0 & 0 \end{smallmatrix}\right) \left(\begin{smallmatrix} -2 & 0 \\ 0 & 0 \end{smallmatrix}\right) = 0.$$

Thus, our assumption (8) holds even though $B_k \leq 0$. An application of Corollary 1 then yields that the optimal value is in this case

$$V(x_0^*, 0) = -(x_0^*)^T W_0 x_0^* = -(x_0^*)^T \left(\begin{smallmatrix} 2 & 0 \\ 0 & 0 \end{smallmatrix}\right) x_0^*$$

and the optimal control feedback law is

$$u_k^* = F_k x_k^* + (I - \mathcal{D}_k^\dagger \mathcal{D}_k)\, \gamma_k = \left(\begin{smallmatrix} 2 & 0 \\ 0 & 0 \end{smallmatrix}\right) x_0^* + \left(\begin{smallmatrix} 0 & 0 \\ 0 & 1 \end{smallmatrix}\right) \gamma_k$$

for some vectors $\gamma_k \in \mathbb{R}^2$. Thus, the optimal feedback control can be found through this paper even when all the known results in the literature fail to apply.

2.3. Variable stepsize LQR problem. The results of this section can be naturally applied to the problem (μLQR), which is presented in the introduction as the discretization of the contiuous-time linear regulator problem (P). Thus, replace in the results A_k, B_k, and C_k by $\mu_k A_k$, $\mu_k B_k$, and $\mu_k C_k$, respectively. Hence, define recursively $W_{N+1} := -\Gamma$, and if W_{k+1} is defined, then with $\tilde{A}_k := (I - \mu_k A_k)^{-1}$ set

$$\mathcal{D}_k := B_k - \mu_k B_k \tilde{A}_k^T (W_{k+1} - \mu_k C_k)\, \tilde{A}_k B_k, \tag{17}$$

$$F_k := \mathcal{D}_k^\dagger B_k \tilde{A}_k^T (W_{k+1} - \mu_k C_k)\, \tilde{A}_k, \tag{18}$$

$$W_k := (\mu_k F_k^T B_k + I)\, \tilde{A}_k^T (W_{k+1} - \mu_k C_k)\, \tilde{A}_k. \tag{μGRE}$$

Our assumption for the variable stepsize problem is then

$$\mathcal{D}_k \geq 0 \quad \text{and} \quad (I - \mathcal{D}_k \mathcal{D}_k^\dagger)\, B_k \tilde{A}_k^T (W_{k+1} - \mu_k C_k)\, \tilde{A}_k = 0. \tag{19}$$

By applying Corollaries 1, 2, 3 to the (μLQR) problem we can obtain the corresponding variable stepsize results.

Remark 6. The variable stepsize generalized Riccati equation (μGRE) can be written in the form

$$\frac{\Delta W_k}{\mu_k} - C_k + A_k^T W_k + W_k A_k + (I - \mu_k A_k^T)\, F_k^T \mathcal{D}_k F_k (I - \mu_k A_k) = \mu_k A_k^T W_k A_k.$$

Now, if $\mu_k \to 0$ (as the number of points t_k in the partition of the interval $[a, b]$ increases to infinity), then $\tilde{A}_k \to I$, $A_k \to A(t)$, $B_k \to B(t)$, $C_k \to C(t)$, and $\mathcal{D}_k \to B(t)$. In addition, if $\mathcal{D}_k^\dagger \to B^\dagger(t)$, then the discrete Riccati equation (μGRE) converges to the classical continuous-time Riccati differential equation

$$W' - C(t) + A^T(t)\, W + W A(t) + W B(t)\, W = 0.$$

Moreover, condition $\mathcal{D}_k \geq 0$ in our assumption (19) converges to $B(t) \geq 0$, while the second condition in (19) converges to the trivial equality $[I - B(t)\, B^\dagger(t)]\, B(t)\, W(t) = 0$. Condition $\mathcal{D}_k^\dagger \to B^\dagger(t)$ is satisfied e.g. when $B(t) > 0$ and $\mathcal{D}_k > 0$, i.e., $\mathcal{D}_k^{-1} \to B^{-1}(t)$.

References

[1] Y. Bar-Ness, Sufficient conditions for the solution of the discrete infinite-time, linear regulator, *Internat. J. Control* **24** (1976), no. 3, 335–343.

[2] A. Ben-Israel, T. N. E. Greville, *Generalized Inverses: Theory and Applications,* Robert E. Krieger Publishing Company, Huntington, NY, 1980.

[3] M. Bohner, Linear Hamiltonian difference systems: disconjugacy and Jacobi-type conditions, *J. Math. Anal. Appl.* **199** (1996), 804–826.

[4] V. G. Boltyanskii, *Optimal Control of Discrete Systems,* John Wiley & Sons, New York - Toronto, 1978.

[5] S. L. Campbell, Optimal control of discrete linear processes with quadratic cost, *Internat. J. Systems Sci.* **9** (1978), no. 8, 841–847.

[6] A. Czornik, On discrete-time linear quadratic control, *Systems Control Lett.* **36** (1999), no. 2, 101–107.

[7] M. H. A. Davis, M. Zervos, A new proof of the discrete-time LQG optimal control theorems, *IEEE Trans. Automat. Control* **40** (1995), no. 8, 1450–1453.

[8] P. Dorato, Theoretical developments in discrete-time control, *Automatica* **19** (1983), no. 4, 395–400.

[9] P. Dorato, A. H. Levis, Optimal linear regulators: the discrete time systems, *IEEE Trans. Automat. Control* **AC-16** (1971), 613–620.

[10] J. C. Engwerda, LQ-problem: the discrete-time time-varying case, In: "Robust Control of Linear Systems and Nonlinear Control" (Amsterdam, 1989), Progr. Systems Control Theory, Vol. 4, pp. 103–112. Birkhäuser Boston, Boston, MA, 1990.

[11] G. C. Goodwin, M. E. Salgado, Unified continuous and discrete LQG theory, In: "Mathematical System Theory", pp. 177–188. Springer, Berlin, 1991.

[12] R. Hilscher, V. Zeidan, Symplectic difference systems: variable stepsize discretization and discrete quadratic functionals, *Linear Algebra Appl.* **367** (2003), 67–104.

[13] V. Kučera, Discrete linear regulator revisited, *Kybernetika (Prague)* **17** (1981), no. 1, 62–70.

[14] H. Kwakernaak, R. Sivan, *Linear Optimal Control Systems,* Wiley-Interscience, New York, 1972.

[15] L.-Z. Liao, D. Li, Successive method for general multiple linear-quadratic control problem in discrete time, *IEEE Trans. Automat. Control* **45** (2000), no. 7, 1380–1385.

[16] J. O'Reilly, Optimal instantaneous output-feedback controllers for discrete-time linear systems with inaccessible state, *Internat. J. Systems Sci.* **9** (1978), no. 1, 9–16.

[17] N. S. Rousan, Necessary and sufficient conditions for global optimality for linear discrete time systems, *Automatica J. IFAC* **29** (1993), no. 2, 537–539.

Some Discrete Competition Models and the Principle of Competitive Exclusion

J. M. Cushing* and Sheree LeVarge†

Abstract

One of the fundamental tenets of ecology is the Competitive Exclusion Principle. According to this principle too much interspecific competition between two species results in the exclusion of one species. This Principle is supported by a wide variety of theoretical models, of which the Lotka/Volterra model based on differential equations is the most familiar. It is perhaps less well known that difference equations also played an important role in the historical development of the Competitive Exclusion Principle. The Leslie/Gower model was used in conjunction with influential competition experiments using species of *Tribolium* (flour beetles) carried out in the first half of the last century. This difference equation model exhibits the same dynamic scenarios as does the Lotka/Volterra model and also supports the Competitive Exclusion Principle. A recently developed competition for *Tribolium* species, however, exhibits a larger variety of dynamic scenarios and competitive outcomes, some of which seemingly stand in contradiction to the Principle. We discuss features of this model that differentiate it from the Leslie/Gower model. We give a simpler, lower dimensional "toy" model that illustrates some non-Lotka/Volterra dynamics.

AMS Nos. 39A11, 92D40

KEYWORDS: difference equations, competition models, Competitive Exclusion Principle, Leslie/Gower model, life cycle stages, competitive coexistence

1 INTRODUCTION

Mathematical models of competition play a central role in theoretical ecology. Historically, the famous Lotka/Volterra competition model (and many other similar models) focused the study of interspecific competition on the notions of competitive exclusion, limiting similarity and ecological niche (notions that in fact had been around at least since Charles Darwin). The overwhelming majority of competition models supports the Principle of Competitive Exclusion, which asserts that "too much" interspecific

*Supported by National Science Foundation grants DMS-0414212 and DMS 9973126.

†Supported by National Science Foundation IGERT grant DGE-9870659.

competition results in the elimination of a species [10], [34], [50]. According to this principle, coexistence is not possible when one species in some way dominates a competitive interaction by direct confrontation and interference or by more efficiently utilizing one or more limited resources (e. g., see [2], [3], [4], [29], [30], [32], [40], [41], [48], [49], [57], [59], [62], [63], [65]). Put another way, to survive a species must find ways to avoid interspecific competition [1], [51]. Despite early criticism of this principle (e.g. see, [5], [6], [8]), it underlies (not always so explicitly) most thinking about the structure of ecosystems.

The prototypical competition model is, of course, the Lotka/Volterra system of two differential equations that describes the dynamics of two competing species. That model shows – as do virtually all existing theoretical models of interference competition – that coexistence is not possible (i.e., one species asymptotically goes extinct) when the intensity of interspecific competition is high, as measured by the magnitude of competition coefficients in the equations. From the 1930's to 1960's several biologists addressed this competitive exclusion principle by means of laboratory experiments involving pairs of species. Perhaps the most notable of these competition experiments were those of G. F. Gause [33], using yeast and protozoa, and T. Park [52], [53], [54] using insects with a more complicated life cycle (flour beetles of the genus *Tribolium*). The experimental results, which are today considered classic and still appear in most ecology textbooks, are interpreted as supporting the principle of competitive exclusion. One of Park's later experimental studies, however, yielded a "difficult to interpret" result that caught his attention and that of his renown collaborator P. H. Leslie [55], [46]. This result was seemingly at odds with the competitive exclusion principle in that neither species was eliminated during the course of the experiment (over 32 generations). Furthermore, the experimental results imply a dynamic scenario not permitted by Lotka/Volterra theory (nor, for that matter, by any competition model known to us), namely, a scenario with three attractors, two of which imply competitive exclusion and one of which implies competitive coexistence. Park and his collaborators addressed this anomalous "coexistence case" (as Park referred to it) with both experimental and model studies, but in the end they offered no theoretical or biological explanation [28], [46].

Park and Leslie did not use the Lotka/Volterra model in their competition studies. Instead they used a system of difference equations that predict the population abundance from one census to the next. Their model is based on the discrete logistic equation [45], [56]

$$x_{t+1} = b\frac{1}{1+x_t}x_t, \quad b > 0 \tag{1}$$

This difference equation defines a monotone map and therefore implies all solutions monotonically equilibrate. (The difference equation (1) is related to the logistic differential equations as follows. If a solution of the logistic differential equation is evaluated at equally spaced time intervals, the resulting sequence of population densities will satisfy a difference equation of this form.) If $b < 1$ then $x(t) \to 0$; if $b > 1$ then $x(t) \to b - 1$. The Leslie/Gower competition model used by Park and

Leslie in their studies [47] is given by the difference equations

$$x_{t+1} = b_1 \frac{1}{1 + x_t + c_1 y_t} x_t \qquad (2)$$
$$y_{t+1} = b_2 \frac{1}{1 + c_2 x_t + y_t} y_t.$$

This model couples two discrete logistic populations by means of the interspecific competition terms $c_1 y_t$ and $c_2 x_t$, which serve to decrease the recruitment of each species (at the next time step) due to the presence of the other species.

The system of difference equations (2) defines a monotone (semi) flow and has exactly the same set of possible dynamic scenarios as the Lotka/Volterra model [20], [60], [61]. If there exists no positive equilibrium then all orbits in the positive cone tend to an exclusion equilibrium lying on one of the positive coordinate axes. If there is a positive equilibrium, then it is either globally asymptotically stable in the positive cone (competitive coexistence) or it is a saddle. In the latter so-called "saddle" case all orbits in the positive cone not lying on the stable manifold tend to an exclusion equilibrium. (Thus, the Leslie/Gower model is the appropriate discrete analog of the Lotka/Volterra competition model.) The coexistence case occurs only if the competition coefficients c_1 and c_2 are sufficiently small (this is the competitive exclusion principle). Moreover, if both coefficients c_1 and c_2 are sufficiently large, then the saddle (competitive exclusion) case occurs.

The anomalous result in Park's experiment arose in a study of the saddle case when one culture did not result in competitive exclusion. Edmunds et al. [28] put forth an hypothesis that explains this result. Their hypothesis is based on an interference competition model different from (2). Their model can exhibit complex, dynamic scenarios that are considerably different from the four scenarios implied by the classic Lotka/Volterra theory.

The competition model studied in [28] is built on a dynamic model for stage structured species (also see [27]). The state variables in the model are the numbers of individuals in each of three distinct life cycle stages: a growth (larval) stage, a quiescent (pupal) stage, and a reproductive (adult) stage. The "LPA model"

$$\begin{aligned} L_{t+1} &= bA_t \exp(-c_{EL}L_t - c_{EA}A_t) \\ P_{t+1} &= (1-\mu_L)L_t \\ A_{t+1} &= P_t \exp(-c_{PA}A_t) + (1-\mu_A)A_t \end{aligned} \qquad (3)$$

has exponential nonlinearities common to many population dynamics models. In the case of the *Tribolium* species used by Park in his experiments, these exponential nonliearities can be derived from an important mechanism that drives their dynamics, namely cannibalism [14], [19], [22]. The fact that movable stages of each species (larvae and adults) cannibalize both their own immovable stages (pupae and eggs) and those of the other species leads to an interference competitive interaction modeled by the equations

$$\begin{aligned}
L_{t+1} &= b_1 A_t \exp(-c_{EL} L_t - c_{EA} A_t) \exp(-c_{El} l_t - c_{Ea} a_t) \\
P_{t+1} &= (1 - \mu_L) L_t \\
A_{t+1} &= P_t \exp(-c_{PA} A_t) \exp(-c_{Pa} a_t) + (1 - \mu_A) A_t \\
l_{t+1} &= b_2 a_t \exp(-c_{el} l_t - c_{ea} a_t) \exp(-c_{eL} L_t - c_{eA} A_t) \\
p_{t+1} &= (1 - \mu_l) l_t \\
a_{t+1} &= p_t \exp(-c_{pa} a_t) \exp(-c_{pA} A_t) + (1 - \mu_a) a_t.
\end{aligned} \tag{4}$$

The LPA model (3) has a considerable track record of successes in describing the dynamics of *Tribolium* (and in particular the same species used by Park) [9], [11], [12], [13], [17], [18], [19], [22], [23], [24], [25], [26], [36], [37], [38], [39], [44]. Because of this record, we anticipate that one could successfully use the competition LPA model (4) to account for the results of Park's classic competition experiments and, perhaps, the anomalous result described above.

Edmunds [27] gives some fundamental analysis of (4). That analysis, together with computer explorations, shows that the competition model can exhibit the classic Lotka/Volterra dynamic scenarios for certain parameter ranges. However, the model can also exhibit many other dynamic scenarios unlike those of Lotka/Volterra. Edmunds et al. [28] use one such non-Lotka/Volterra scenario to offer a possible explanation of Park's anomalous result.

Using parameter values consistent with those estimated for *Tribolium casteneum* in many previous studies, Edmunds et al. found that coexistence was possible with *increases* in certain *inter*specific competition coefficients in the model (4), seemingly contrary to the competitive exclusion principle. (One intriguing finding of Park was that cannibalism rates for the anomalous "coexistence" culture had increased during the course of his experiment.) Specifically, as the interspecific competition coefficients c_{eL} and c_{El} increase, the equilibrium configuration in the model assumes that of the saddle case, as is typical in classic Lotka/Volterra theory. However, as these competition coefficients continue to increase, a critical point is reached where a sudden onset of stable coexistence 2-cycles occurs (by means of a saddle-node bifurcation). This dynamic scenario possesses three attractors: two exclusion equilibria and a coexistence 2-cycle. This configuration is robust against further increases in the competition coefficients, and the basins of attraction of the coexistence 2-cycles are significantly large. This bifurcation scenario has several interesting features that we would like to understand better (not only with regard to the *Tribolium* experiments, but with regard to the fact that (4) is a rather general model of competition between two species with a commonly occurring, three stage life cycle).

We would like to know what mechanisms, mathematical and biological, cause the following (non-Lotka/Volterra) phenomena to occur in a competition model:

(a) coexistence of two species is promoted by increasing the *inter*specific coefficients;

(b) multiple attractor scenarios in which competitive coexistence and exclusion attractors (not necessarily equilibria) appear together, with asymptotic outcomes dependent on initial conditions.

Property (a) stands out because it is seemingly at odds with a fundamental tenet underlying virtually all ecological studies involving competition and ecological niche. For example, a recent review of current theoretical and empirical thinking about competitive coexistence states that coexistence requires "that intraspecific competition is stronger than interspecific competition". In addition, coexistence necessitates "some form of niche difference or partitioning between species that increases the strength of intraspecific competition relative to that of interspecific competition" [1].

One approach to addressing phenomena (a) and (b) is to consider the differences between the Leslie/Gower competition model (discrete Lotka/Volterra model) and the competition LPA model and to investigate these differences individually (and in various combinations) in order to determine which promote (a) and/or (b). The three main differences between the two competition models are: the LPA model contains life cycle stages (time delays, to a mathematician); the LPA model has a "stronger" nonlinearity (an exponential or Ricker type nonlinearity); and in the LPA model there is explicit iteroparity (reproducing adults can live longer than one unit of time). We can ask: which of these properties when incorporated into the Leslie/Gower model will promote phenomena (a) and/or (b) and why?

We have only begun an investigation of these questions and we are far from having complete answers. However, in the next section we give an example that demonstrates how the addition of a life cycle stage (time delay) to the Leslie/Gower competition model can give rise to phenomena (a) and (b). In this example these phenomena are not as robust as they are in the competition LPA model. Nonetheless, the example demonstrates that the introduction of a single life cycle stage into one of the species in a competitive system is capable of producing both (a) and (b).

2 A STAGE STRUCTURED COMPETITION MODEL

The system of difference equations

$$\begin{aligned} J_{t+1} &= b_1 \frac{1}{1 + A_t + c_1 y_t} A_t \\ A_{t+1} &= (1 - \mu) J_t \\ y_{t+1} &= b_2 \frac{1}{1 + c_2 J_t + y_t} y_t \end{aligned} \tag{5}$$

is a modification of the Leslie/Gower competition model (2) in which species x has been given two life cycle stages, a juvenile (non-reproducing) stage J and an adult (reproducing) stage A. According to the first equation, juvenile recruits at time $t+1$ are produced by the adult stage with an inherent, per capita recruitment rate b_1 that is discounted by a fraction dependent on the adult population numbers and the number of species y. The second equation simply states that a fraction μ of juveniles die and do not become adults in one unit of time. The final equation is the discrete logistic equation for species y in which there are competitive effects from the juvenile class J of the competing species. Clearly other types of life cycle histories and interspecific couplings could be used to modify the unstructured Leslie/Gower model, but here we will restrict attention to model (5). (See Section 3.)

In (5) species y is governed by the discrete logistic equation in the absence of the J, A species. Thus, when the species J, A is absent, species y will equilibrate (going extinct if $b_2 < 1$ and surviving if $b_2 > 1$). In the absence of the y species, the dynamics of the J, A species are described by the difference equations

$$\begin{aligned} J_{t+1} &= b_1 \frac{1}{1+A_t} A_t \\ A_{t+1} &= (1-\mu) J_t. \end{aligned} \tag{6}$$

This "delayed logistic" model has two equilibria

$$(J_e, A_e) = (0,0) \quad \text{and} \quad \left(\frac{n-1}{1-\mu}, n-1 \right) \tag{7}$$

where we have defined

$$n \triangleq b_1(1-\mu).$$

Biologically n is the inherent net reproductive number, i.e., n is the expected number of juveniles produced by one adult during the course of its lifetime [16], [14], [66]. As functions of n these two equilibrium branches cross at $n = 1$ where a transcritical bifurcation occurs. Clearly positive initial conditions $J_0 > 0$, $A_0 > 0$ yield positive solutions $J_t > 0$, $A_t > 0$. Similarly, non-negative initial conditions $J_0 \geq 0$, $A_0 \geq 0$ yield non-negative solutions $J_t \geq 0$, $A_t \geq 0$. Furthermore, the union of the non-negative coordinate axes in the J, A plane is forward invariant.

For non-negative initial conditions we have

$$\begin{aligned} 0 &\leq J_{t+1} \leq b_1 A_t \\ 0 &\leq A_{t+1} = (1-\mu) J_t \end{aligned} \tag{8}$$

and a straightforward induction argument shows that $n < 1$ implies $\lim_{t\to\infty} (J_t, A_t) = (0,0)$. (If each adult does not at least replace itself over its lifetime, the population will go extinct.)

On the other hand, if $n > 1$ then the trivial equilibrium $(J, A) = (0,0)$ is unstable and the positive equilibrium in (7) is globally asymptotically stable (GAS) in the positive cone. To see this, we calculate the Jacobian

$$\begin{pmatrix} 0 & \frac{b_1}{(A+1)^2} \\ 1-\mu & 0 \end{pmatrix}$$

and evaluate it at the two equilibria. For the trivial equilibrium the Jacobian has eigenvalues $\pm\sqrt{n}$. For the positive equilibrium the Jacobian has eigenvalues $\pm 1/\sqrt{n}$. These eigenvalues imply an exchange of local asymptotic stability (LAS) between the two equilibrium branches as n increases through the bifurcation value $n = 1$. To see that the positive equilibrium globally attracts positive initial conditions when $n > 1$, we note that the composite map

$$J_{t+2} = n\frac{1}{1+(1-\mu)J_t}J_t \qquad (9)$$
$$A_{t+2} = n\frac{1}{1+A_t}A_t$$

consists of two uncoupled, discrete logistic equations. It follows that

$$\lim_{t\to\infty}(J_{2t}, A_{2t}) = \left(\frac{n-1}{1-\mu}, n-1\right)$$

and from (6) that

$$\lim_{t\to\infty}(J_{2t+1}, A_{2t+1}) = \lim_{t\to\infty}\left(b_1\frac{1}{1+A_{2t}}A_{2t}, (1-\mu)J_{2t}\right) = \left(\frac{n-1}{1-\mu}, n-1\right).$$

The positive equilibrium is not, however, a global attractor for non-negative initial conditions. (The coordinate axes are invariant.) Using the composite equations we find that an initial condition $J_0 = 0$, $A_0 > 0$ (or $J_0 > 0$, $A_0 = 0$) produces a solution that tends to a 2-cycle

$$\begin{pmatrix} 0 \\ n-1 \end{pmatrix} \to \begin{pmatrix} \frac{n-1}{1-\mu} \\ 0 \end{pmatrix} \to \begin{pmatrix} 0 \\ n-1 \end{pmatrix} \to \cdots \qquad (10)$$

In this cycle the juvenile and adult classes are temporally separated and, for this reason, we refer to the cycle as a *synchronous* 2-cycle [15], [21]. This synchronous 2-cycle is GAS within the invariant coordinate axes. It is, however, unstable within the two dimensional J, A plane. Nonetheless, this 2-cycle will play an important role in the dynamics of the competition model (5).

Next we turn our attention to the two species competition system (5). We begin with an investigation of equilibria. The system (5) has the following equilibrium points (J, A, y):

$$E_0 : (0,0,0), \qquad E_1 : (0,0,b_2-1), \qquad E_2 : \left(\frac{n-1}{1-\mu}, n-1, 0\right)$$

$$E_3 : \left(\frac{(b_2-1)\,c_1-(n-1)}{c_1c_2-(1-\mu)}, (1-\mu)\frac{(b_2-1)\,c_1-(n-1)}{c_1c_2-(1-\mu)}, \frac{(n-1)\,c_2-(b_2-1)\,(1-\mu)}{c_1c_2-(1-\mu)}\right).$$

The inequalities (8) and $0 \le y_{t+1} \le b_2 y_t$ hold for solutions of the competition model (5) with non-negative initial conditions. It follows that the J, A species goes extinct if $n < 1$ and the y species goes extinct if $b_2 < 1$. Therefore, we assume

$$n > 1, \quad b_2 > 1.$$

This assumption implies both equilibria E_1 and E_2 are non-negative. It also implies that the extinction equilibrium E_0 is unstable, since eigenvalues of the Jacobian evaluated at the equilibrium are b_2, $\pm\sqrt{n}$. Therefore, the isolated invariant point E_0 is a repellor and no orbit can approach it as $t \to +\infty$; that is to say, E_0 is equal to its own stable set (it is clear from equations (5) that no orbit can reach E_0 in finitely many steps). It follows that the competition model is uniformly persistence with respect to E_0 [43].

The Jacobian

$$M(J,A,y) = \begin{pmatrix} 0 & b_1 \frac{yc_1+1}{(A+yc_1+1)^2} & -\frac{c_1 b_1 A}{(A+yc_1+1)^2} \\ 1-\mu & 0 & 0 \\ -\frac{c_2 b_2 y}{(y+Jc_1+1)^2} & 0 & b_2 \frac{Jc_2+1}{(y+Jc_2+1)^2} \end{pmatrix}$$

evaluated at the competitive exclusion equilibria E_1 and E_2 has eigenvalues

$$\frac{1}{b_2}, \quad \pm\sqrt{\frac{n}{(b_2-1)\,c_1+1}}$$

and

$$\pm\sqrt{\frac{1}{n}}, \quad \frac{(1-\mu)\,b_2}{(n-1)\,c_2+1-\mu}$$

respectively. The stability properties summarized in Table 1 follow from these calculations. The local stability analysis using the Jacobian evaluated at E_3 was performed, using the Jury conditions and the help of a computer algebra program, by Professor Lih-Ing Roeger[1].

The local stability properties of the equilibria in Table 1 are exactly the same as those in classical Lotka/Volterra theory. Equilibrium coexistence is possible if and only if the interspecific coefficients c_1 and c_2 are sufficiently small. However, unlike the Lotka/Volterra equilibrium scenarios, the stability properties in Table 1 are local and are not necessarily global. To see this, we turn our attention to 2-cycle solutions of the competition model (5).

Equilibrium	Unstable	Locally asymptotically stable
E_1	$c_1 < c_1^*$	$c_1^* < c_1$
E_2	$c_2 < c_2^*$	$c_2^* < c_2$
E_3	$c_1^* < c_1,\ c_2^* < c_2$	$c_1 < c_1^*,\ c_2 < c_2^*$

TABLE 1. The local stability properties of the three nontrivial equilibria of competition model (5) are expressed in terms of the interspecific competition coefficients c_1 and c_2. The critical values of c_1 and c_2 in this table are $c_1^* \triangleq (n-1)\,/\,(b_2-1)$, $c_2^* \triangleq (1-\mu)\,(b_2-1)\,/\,(n-1)$.

[1]Department of Mathematics and Statistics, Texas Tech University, Lubbock, TX 79409

In this paper we will not present a complete study of the existence and stability of 2-cycle solutions of the competition model. Instead we will focus on a particular branch of synchronous 2-cycles. The inherent synchronous 2-cycle (10) gives the *exclusion* synchronous 2-cycle

$$\begin{pmatrix} 0 \\ n-1 \\ 0 \end{pmatrix} \to \begin{pmatrix} \frac{n-1}{1-\mu} \\ 0 \\ 0 \end{pmatrix} \to \begin{pmatrix} 0 \\ n-1 \\ 0 \end{pmatrix} \to \cdots \tag{11}$$

for the competition model (5) (in which species y is absent). The (local asymptotic) stability of this 2-cycle can be determined from the eigenvalues of the Jacobian of the composite map, which is the product

$$M(0,n-1,0)M\left(\frac{n-1}{1-\mu},0,0\right) = \begin{pmatrix} \frac{1}{n} & 0 & -\frac{n-1}{n}\frac{b_2c_1}{c_2(n-1)+1-\mu} \\ 0 & n & 0 \\ 0 & 0 & \frac{(1-\mu)b_2^2}{c_2(n-1)+1-\mu} \end{pmatrix} \tag{12}$$

whose eigenvalues appear along the diagonal. This 2-cycle is unstable since $n > 1$.

If we consider stability of the cycle (11) within the invariant, non-negative coordinate planes $J = 0$ or $A = 0$ of the J, A, y phase space, then the 2-cycle (11) is LAS provided the eigenvalue in the lower right corner is less than 1. We say equilibria or cycles that lie in these invariant coordinate planes are *synchronously LAS* (or *sLAS*) if they are LAS within the invariant $J = 0$ or $A = 0$ coordinate planes (i.e., with respect to solutions whose J, A components are synchronous). If they are unstable in these invariant planes, then we say they are *synchronously unstable* (or *s-unstable*).

Thus, the exclusion 2-cycle (11) is sLAS if competition is sufficiently intense in the sense that

$$c_2 > c^* \triangleq (1-\mu)\frac{b_2^2-1}{n-1},$$

and it is s-unstable if competition is weak: $c_2 < c^*$. (Note $c^* > c_2^*$ in Table 1.) This loss of synchronous stability occurs because an eigenvalue of the Jacobian (12) passes through $+1$, which suggests a transcritical bifurcation with another branch of 2-cycles. To find this bifurcating branch of synchronous 2-cycles, we examine the fixed points of the composite map

$$\begin{pmatrix} 0 \\ A \\ y \end{pmatrix} \to \begin{pmatrix} \frac{n}{1-\mu}\frac{1}{1+A+c_{12}y}A \\ 0 \\ b_2\frac{1}{1+y}y \end{pmatrix} \to \begin{pmatrix} 0 \\ n\frac{1}{1+A+c_1y}A \\ b_2^2\frac{1}{1+(1+b_2)y+c_2\frac{n}{1-\mu}\frac{1+y}{1+A+c_1y}A}y \end{pmatrix},$$

which yields the equations

$$\begin{gathered} A = n-1-c_1y \\ c_1c_2y^2 - ((1+b_2)(1-\mu)+c_2(n-c_1-1))\,y + \left(b_2^2-1\right)(1-\mu) - c_2(n-1) = 0 \end{gathered} \tag{13}$$

for A and y. A positive solution $y > 0$ of the uncoupled second equation in (13) yields a non-negative, coexistence synchronous 2-cycle provided $0 < y < (n-1)/c_1$ (so that $A > 0$).

We are interested in the case when there is strong competition between the species, i.e., when c_1 and c_2 are both large. We can facilitate a study of this case by introducing a single parameter that measures the intensity of interspecific competition as follows. We fix the ratio

$$r \triangleq \frac{c_1}{c_2}$$

between the competition coefficients and define

$$c \triangleq c_2.$$

In terms of r and c the 2-cycle equations (13) become

$$A = n - 1 - rcy$$

$$rc^2y^2 - \left((1+b_2)(1-\mu) + c(n - rc - 1)\right)y + \left(b_2^2 - 1\right)(1-\mu) - c(n-1) = 0. \quad (14)$$

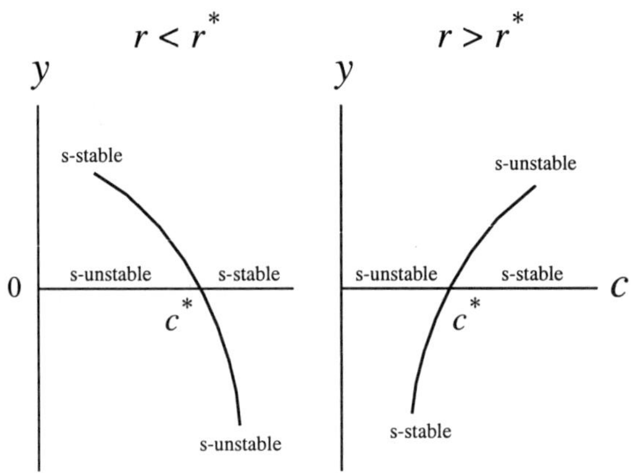

FIGURE 1. A branch $y = y(c)$ of solutions of equation (14) transcritically bifurcates with the trivial solution $y = 0$ at $c = c^*$. The positive solution branch yields coexistence synchronous 2-cycles of the competition model (5). In the "subcritical" bifurcation case $r < r^*$, the coexistence 2-cycles are synchronously LAS. They are unstable, however, with respect to non-synchronous solutions.

The second (quadratic) equation has a solution $y = y(c)$ that satisfies $y(c^*) = 0$ and, consequently, bifurcates from the exclusion synchronous 2-cycle (11) at $c = c^*$. The direction of bifurcation is determined by the sign of $y'(c^*)$, which can be calculated by an implicit differentiation of (14):

$$y'(c^*) < 0 \quad \text{if } r < r^*$$
$$y'(c^*) > 0 \quad \text{if } r > r^*$$
$$r^* \triangleq \frac{1}{1-\mu}\left(\frac{n-1}{b_2-1}\right)^2 \frac{b_2}{b_2+1}.$$

By the exchange of stability principle, the bifurcating coexistence synchronous 2-cycles are sLAS near the bifurcation point if $r < r^*$ and s-unstable if $r > r^*$. See Figure 1. (This can also be proved using a Liapunov-Schmidt analysis near the bifurcation point.) The coexistence 2-cycles are not LAS, however, with respect to nonsynchronous solutions near the bifurcation point (i.e., in the three dimensional J, A, y space).

When $r < r^*$ the branch of coexistence synchronous 2-cycles globally extends in one of two ways as shown in Figure 2. One possibility (Figure 2b) is that the bifurcating coexistence 2-cycle branch "turns around" and a saddle-node bifurcation of synchronous 2-cycles occurs. Numerical explorations indicate that s-LAS is lost along the branch when this occurs (i.e., the upper branch is s-unstable). The saddle-node bifurcation of 2-cycles occurs in a parameter region in which the equilibrium configuration is that of the competitive exclusion saddle case of Lotka/Volterra theory. The result is an interval of parameter values for which there are three attractors, two of which are exclusion equilibria and one of which is a coexistence 2-cycle. A numerical example is shown in Figure 3. This is the same scenario observed in the competition LPA model by Edmunds et al. [28].

It is interesting to note that it is possible for the synchronous coexistence 2-cycles to be LAS in the three dimensional J, A, y phase space. This occurs in the case shown in Figure 2b and 2c near each of the two saddle-node bifurcations. (Also see Figure 3b.)

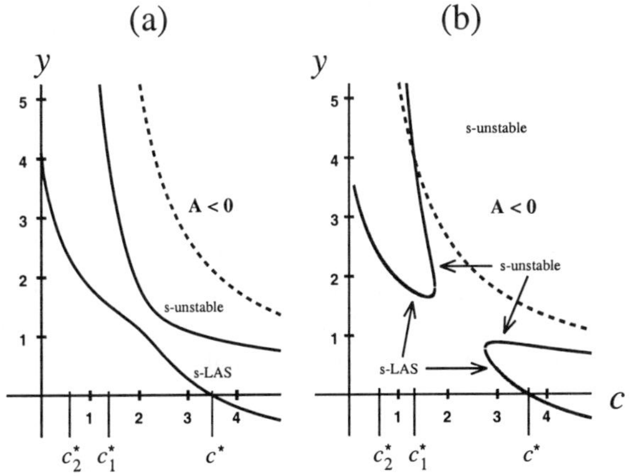

(c)

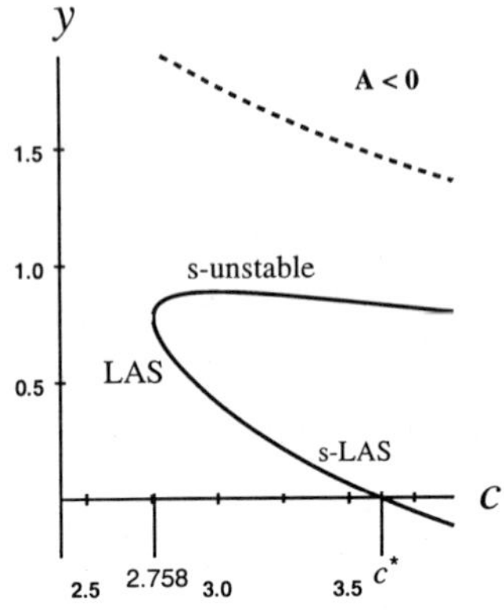

FIGURE 2. In (a) and (b) appear two examples of the subcritical bifurcation case ($r < r^*$) that illustrate the possible global geometry of the coexistence synchronous 2-cycle branch that bifurcates from $y = 0$ at $c = c^*$. Also shown is another branch of coexistence synchronous 2-cycles. Parameter values are $\mu = 0.2$, $b_2 = 5$, and $r = 1$. The broken line indicates the curve $A = n - 1 - rcy = 0$ below which $A > 0$. For $c > c_1^*$ the equilibrium configuration is the saddle competitive exclusion case of Lotka/Volterra theory (the exclusion equilibria E_1 and E_2 are LAS and the coexistence equilibrium E_3 is unstable).

In (a) $n = 6.5$ ($b_1 = 8.125$), $c_2^* = 0.5818$, $c_1^* = 1.375$, and $c^* = 3.491$. The bifurcating branch extends to the vertical axis where $c = 0$. Numerical simulations indicate that the 2-cycles from the other branch are s-unstable. In (b) $n = 6.3$ ($b_1 = 7.875$), $c_2^* = 0.6038$, $c_1^* = 1.325$, and $c^* = 3.623$. The bifurcating branch "turns around" to form a saddle-node bifurcation of 2-cycles at $c = 2.758$. Numerical simulations indicate that the 2-cycles from the upper branch are s-unstable. A similar, but reverse saddle-node bifurcation also occurs at $c = 1.741$.

(c) Numerical simulations show that the synchronous 2-cycles are fully LAS in J, A, y phase space near the both saddle-node bifurcation points in (b). The parameter intervals of LAS are small, however: $2.758 < c < 2.835$ for the saddle-node bifurcation at $c = 2.758$ (shown) and $1.693 < c < 1.741$ for the saddle-node bifurcation at $c = 1.741$ (not shown).

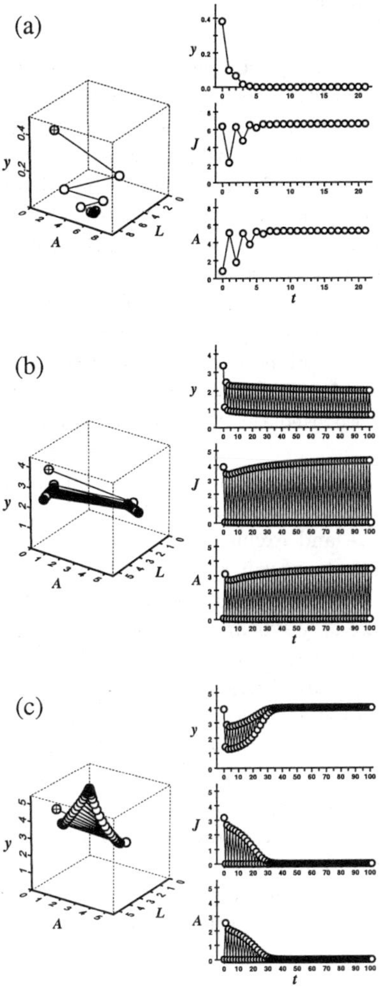

FIGURE 3. Three initial conditions produce solutions of the competition model (5) with three different attractors. (a) The solution with initial conditions $(J_0, A_0, y_0) = (6.362, 0.8290, 0.3819)$ tends to the exclusion equilibrium $E_2 = (6.625, 5.3, 0)$. (b) The solution with initial conditions $(J_0, A_0, y_0) = (3.881, 0.07095, 3.371)$ tends to a coexistence synchronous 2-cycle. **(c)** The solution with initial conditions $(J_0, A_0, y_0) = (3.173, 0.02234, 3.912)$ tends to the exclusion equilibrium $E_1 = (0, 0, 4)$. Parameter values are $\mu = 0.2$, $b_2 = 5$, $n = 6.3$. $(b_1 = 7.875)$, $r = 1$, $c = 2.8$ $(c_1 = c_2 = 2.8)$.

3 CONCLUSIONS

In theoretical models of interference competition between two biological species, large values of interspecific competition coefficients (relative to intraspecific competition coefficients) typically imply that one species will go extinct. This is the basis of the principle of competitive exclusion, which states that in order for two species to coexist they must find a way to decrease their competitive interactions (i.e., find their own "niche"). A large number of mechanisms utilized by species to avoid competition has been identified (for a list of 120 such mechanisms see [51]), most of which involve spatial, temporal, or resource separation). None of these mechanisms are applicable to the coexistence case observed in Park's experiment with *Tribolium* or to the explanation based on the competition LPA model given in [28]. In the latter explanation, coexistence was promoted by an increase in interspecific competition coefficients and the onset of non-equilibrium coexistence attractors (properties (a) and (b)). In this paper we used the "toy" model example (5) to illustrate these phenomena.

The models we examined do not, however, exhibit the phenomena (a) and (b) in as robust a way as does the competition LPA model. A multiple attractor scenario of mixed coexistence and exclusion attractors can arise from the model (5) on an interval of sufficiently large values of the interspecific competition coefficients. However, unlike for the LPA model, this parameter interval is of finite length, and the coexistence 2-cycles are only synchronous stability (except on a small subinterval where they are LAS). Moreover, simulations show that the basins of attraction of the coexistence cycles of (5) are restricted to an open region close to the invariant coordinate planes. The coexistence cycles in the LPA model, on the other hand, have basins of attraction that are significantly large in phase space.

Nonetheless, example (5) does at least illustrate that the introduction of a life cycle stage into (even only one species in) a model that predicts the principle of competitive exclusion (i.e., that asymptotically has only the Lotka/Volterra scenarios) can exhibit the properties (a) and (b).

Properties (a) and (b) can appear more robustly in models that include other nonlinear interactions among the life cycle stages J and A and the competing species y :

$$\begin{aligned}
J_{t+1} &= b_1 \frac{1}{1 + c_{11}J_t + c_{12}A_t + c_{13}y_t} A_t \\
A_{t+1} &= (1-\mu) \frac{1}{1 + c_{21}J_t + c_{22}A_t + c_{23}y_t} J_t \\
y_{t+1} &= b_2 \frac{1}{1 + c_{31}J_t + c_{32}A_t + c_{33}y_t} y_t .
\end{aligned} \tag{15}$$

We presented the special case (5) in Section 2 because of its analytic tractability.

For example, a modification of (5) that includes a juvenile density effect in adult reproduction, namely the equations

$$J_{t+1} = b_1 \frac{1}{1 + c_{11}J_t + A_t + rcy_t} A_t$$
$$A_{t+1} = (1 - \mu)J_t$$
$$y_{t+1} = b_2 \frac{1}{1 + cJ_t + y_t} y_t$$

with $c_{11} > 0$, exhibits the same bifurcation diagrams as appear in Figure 2. However, the interval for the interspecific competition coefficient c on which the coexistence synchronous 2-cycles are three dimensionally LAS is lengthened and the basins of attraction of the 2-cycles are considerably increased in size.

These examples demonstrate how life history characteristics can play an important role in the dynamics of interacting populations and, in particular, how they can promote competitive coexistence and hence ecological diversity (see [64], [7] and the references cited therein). An interesting open question is whether the phenomena (a) and (b) that arise in the competition LPA model, and in simpler competition models of the form (15), are in contradiction to the principle of competitive exclusion or whether they might, in some way, be reconciled with that principle.

References

[1] P. Amarasekare, Competitive coexistence in spatially structured environments: a synthesis, *Ecology Letters* 6 (2003), 1109-1122

[2] R. A. Armstrong and R. McGehee, Coexistence of two competitors on one resource, *Journal of Theoretical Biology* 56 (1976), 499-502

[3] R. A. Armstrong and R. McGehee, Coexistence of species competing for shared resources, *Theoretical Population Biology* 9 (1976), 317-328

[4] R. A. Armstrong and R. McGehee, Competitive exclusion, *The American Naturalist*, 115 (1980), no. 2, 151-170

[5] F. J. Ayala, Experimental invalidation of the principle of competitive exclusion, *Nature* 224 (1969), 1076-1079

[6] F. J. Ayala, Competition between species, *American Scientist* 60 (1972), 348-357

[7] M. B. Bonsall, V. A. A. Jansen, M. P. Hassell, Life history trade-offs assemble ecological guilds, *Science* 306 (October 2004), 111-114

[8] J. H. Brown, Two decades of homage to Santa Rosalia: Toward a general theory of diversity, *American Zoologist* 21 (1981), 877-888

[9] H. Caswell, *Matrix Population Models: Construction, Analysis and Interpretation*, Second edition, Sinauer Associates, Inc. Publishers, Sunderland, Massachusetts, 2001

[10] L. C. Cole, Competitive exclusion, *Science* 132 (1960), 348-349

[11] R. F. Costantino, J. M. Cushing, B. Dennis and R. A. Desharnais, Experimentally induced transitions in the dynamic behavior of insect populations, *Nature* 375 (1995), 227–230F

[12] R. F. Costantino, R. A. Desharnais, J. M. Cushing and B. Dennis, Chaotic dynamics in an insect population, *Science* 275 (1997), 389–391

[13] R. F. Costantino, J. M. Cushing, B. Dennis, R. F. Desharnais and S. M. Henson, Resonant population cycles in temporally fluctuating habitats, *Bulletin of Mathematical Biology* 60 (1998), 247–273

[14] J. M. Cushing, *An Introduction to Structured Population Dynamics*, CBMS-NSF Regional Conference Series in Applied Mathematics, Vol. 71, SIAM, Philadelphia, 1998

[15] J. M. Cushing, Cycle chains and the LPA model, *Journal of Difference Equations and Applications* 9, No. 7 (2003), 655-670

[16] J. M. Cushing and Zhou Yicang, The net reproductive value and stability in structured population models, *Natural Resource Modeling* 8, No. 4 (1994), 1-37

[17] J. M. Cushing, B. Dennis, R. A. Desharnais and R. F. Costantino, Moving toward an unstable equilibrium: saddle nodes in population systems, *Journal of Animal Ecology* 67 (1998), 298-306.

[18] J. M. Cushing, S. M. Henson, R. A. Desharnais, B. Dennis, R. F. Costantino and A. King. 2001. A chaotic attractor in ecology: theory and experimental data, *Chaos, Solitons and Fractals* 12 (2001), 219–234

[19] J. M. Cushing, J. M., R. F. Costantino, B. Dennis, R. A. Desharnais and S. M. Henson, *Chaos in Ecology: Experimental Nonlinear Dynamics*, Academic Press, New York, 2002

[20] J. M. Cushing, S. LeVarge, N. Chitnis, S. M. Henson, Some discrete competition models and the competitive exclusion principle, to appear in the *Journal of Difference Equations and Applications*

[21] N. Davydova, *Old and young: can they coexist?*, Ph.D. dissertation, University of Utrecht, The Netherlands, 2004

[22] B. Dennis, R. A. Desharnais, J. M. Cushing and R. F. Costantino, Nonlinear demographic dynamics: Mathematical models, statistical methods, and biological experiments, *Ecological Monographs* 65 (1995), 261–281

[23] B. Dennis, R. A. Desharnais, J. M. Cushing and R. F. Costantino, Transitions in population dynamics: Equilibria to periodic cycles to aperiodic cycles. *Journal of Animal Ecology* 66 (1997), 704–729

[24] B. Dennis, R. A. Desharnais, J. M. Cushing, S. M. Henson and R. F. Costantino, Estimating chaos and complex dynamics in an insect population, *Ecological Monographs* 71 (2001), 277–303

[25] R. A. Desharnais, R. F. Costantino, J. M. Cushing and B. Dennis, Estimating chaos in an insect population, Science 276 (1997), 1881–1882

[26] R. A.Desharnais, R. F. Costantino, J. M. Cushing, S. M. Henson and B. Dennis, Chaos and population of insect outbreaks, *Ecology Letters* 4 (2001), 229–235

[27] J. Edmunds, *A Study of a Stage-Structured Model of Two Competing Species*, Ph.D. dissertation, University of Arizona, Tucson, 2001

[28] J. Edmunds, J. M. Cushing, R. F. Costantino, S. M. Henson, B. Dennis and R. A. Desharnais, Park's *Tribolium* competition experiments: a non-equilibrium species coexistence hypothesis, *Journal of Animal Ecology* 72 (2003), 703-712

[29] J. E. Franke and A.-A. Yakubu, .Mutual exclusion versus coexistence for discrete competitive systems, *Journal of Mathematical Biology* 30 (1991), no. 2, 161-168

[30] J. E. Franke and A.-A. Yakubu, .Global attractors in competitive systems, *Nonlinear Analysis* 16 (1991), no. 2, 111-129

[31] J. E. Franke and A.-A. Yakubu, .Geometry of exclusion principles in discrete systems, *Journal of Mathematical Analysis and Applications* 168 (1992), no. 2, 385-400

[32] J. E. Franke and A.-A. Yakubu,.Exclusionary population dynamics in size-structured, discrete competitive systems, *Journal of Difference Equations and Applications* 5 (1999), no. 3, 235-249

[33] G. F. Gause, *The Struggle for Existence*, Hafner press, New York, 1934 (reprinted 1964)

[34] G. Hardin, The competitive exclusion principle, *Science* 131 (1960), 1292-1297

[35] U. G. Haussman, On the principle of competitive exclusion, *Theoretical Population Biology* 4 (1973), 31-41

[36] S. M. Henson, J. M. Cushing., R. F. Costantino, B. Dennis and R. A. Desharnais, Phase switching in population cycles, *Proceedings of the Royal Society, London B* 265 (1998) 2229–2234

[37] S. M. Henson,, R. F. Costantino, J. M. Cushing, B. Dennis and R. A. Desharnais, Multiple attractors, saddles, and population dynamics in periodic habitats. *Bulletin of Mathematical Biology* 61 (1999), 1121–1149

[38] S. M. Henson,, R. F. Costantino, J. M. Cushing, R. A. Desharnais and A. A. King, Lattice effects observed in chaotic dynamics of experimental populations, *Science* 294 (2001), 602–605

[39] S. M. Henson,, R. F. Costantino, R. A. Desharnais, J. M. Cushing and B. Dennis, Basins of attraction: population dynamics with two stable 4-cycles, *Oikos* 98 (2002), 17–24

[40] J. Hofbauer, V. Hutson and W. Jansen, Coexistence for systems governed by difference equations of Lotka-Volterra type, *Journal of Mathematical Biology* 25 (1987), 553-570

[41] S. B. Hsu, Limiting behavior for competing species, *SIAM Journal of Applied Mathematics* 34 (1978), 760-763

[42] S. B. Hsu, S. P. Hubbell and P. Waltman, A mathematical theory for single-nutrient competition in continuous cultures of micro-organisms, *SIAM Journal of Applied Mathematics* 32 (1976), 366-383

[43] Josef Hofbauer and Joseph W.-H. So, Uniform persistence and repellors for maps, *Proc. AMS* 107, No. 4 (1989), 1137-1142

[44] A. A. King, R. F. Costantino, J. M. Cushing, S. M. Henson, R. A. Desharnais and B. Dennis, Anatomy of a chaotic attractor: subtle model-predicted patterns revealed in population data, *Proceedings of the National Academy of Sciences* 101, No. 1 (2003), 408-413

[45] P. H. Leslie, Some further notes on the use of matrices in population mathematics, *Biometrika* 35 (1948), 213-245

[46] P. H. Leslie, T. Park and D. M. Mertz, The effect of varying the initial numbers on the outcome of competition between two Tribolium species, *Journal of Animal Ecology* 37 (1968), 9–23

[47] P. H. Leslie and J. C. Gower, The properties of a stochastic model for two competing species, *Biometrika* 45 (1958), 316-330

[48] S. A. Levin,. Community equilibria and stability, and an extension of the competitive exclusion principle, *The American Naturalist* 104 (1970), 413-423

[49] R. MacArthur and R. Levins, Competition, habitat selection, and character displacement in a patchy environment, *Proceedings of the National Academy of Sciences* 51 (1964), 1207-1210

[50] R. M. May, *Stability and Complexity in Model Ecosystems*, 3rd edition, Princeton Landmarks in Biology, Princeton University Press, Princeton, New Jersey, 2001

[51] M. W. Palmer, Variation in species richness: towards a unification of hypotheses, *Folia Geobot. Phytotax., Praha*, 29 (1994), 511-530

[52] T. Park, Experimental studies of interspecies competition. I, Competition between populations of the flour beetles *Tribolium confusum* Duval and *Tribolium castaneum* Herbst, *Ecological Monographs* 18 (1948), 265-308

[53] T. Park, Experimental studies of interspecies competition. II, Temperature, humidity and competition in two species of *Tribolium*, *Physiological Zoology* 27 (1954), 177-238

[54] T. Park, Experimental studies of interspecies competition. III, Relation of initial species proportion to the competitive outcome in populations of *Tribolium*, *Physiological Zoology* 30 (1957), 22-40

[55] T. Park, P. H. Leslie and D. B. Mertz, Genetic strains and competition in populations of *Tribolium*, *Physiological Zoology* 37 (1964), 97–162

[56] E. C. Pielou, *Mathematical Ecology*, John Wiley & Sons, New York, 1977

[57] A. Rescigno and I. W. Richardson, On the competitive exclusion principle, *Bulletin of Mathematical Biophysics*, Supplement 27 (1965), 85-89

[58] D. Simberloff, The status of competition theory in ecology, *Ann. Zool. Fennici* 19 (1982), 241-253

[59] S. Smale, On the differential equations of species in competition, *Journal of Mathematics Biology* 3 (1976), 5-7

[60] H. L. Smith, Planar competitive and cooperative difference equations, *Journal of Difference Equations and Applications* 3 (1998), 335-357

[61] H. L. Smith and H. R. Thieme, Stable coexistence and bi-stability for competitive systems on ordered Banach spaces, Journal of Differential Equations 176 (2001), 195-222.

[62] H. L. Smith and P. Waltman, *The Theory of the Chemostat: Dynamics of Microbial Competition*, Cambridge Studies in Mathematical Biology 13, Cambridge University Press, Cambridge, 1995

[63] P. Waltman, *Competition Models in Population Biology*, CBMS-NSF Regional Conference Series in Applied Mathematics 45, SIAM, Philadelphia, 1983

[64] E. E. Werner and J. F. Gilliam, The ontogenetic niche and species interactions in size-structured populations, *Ann. Rev. Ecol. Syst.* 15 (1984), 393-425

[65] G. S. Wolkowicz and Z. Lu, Global dynamics of a mathematical model of competition in the chemostat: general response functions and differential death rates, *SIAM Journal on Applied Mathematics* 52 (1992), 222-233

[66] Zhou Yicang and J. M. Cushing, Stability conditions for equilibria of nonlinear matrix population models, *Journal of Difference Equations and Applications* 4 (1998), 95-126

Stability under constantly acting perturbations for difference equations and averaging

Vladimir Burd
Yaroslavl State University, Russia
vburd@member.ams.org

1 Introduction

In 1967 Banfi established in [3] that uniform asymptotic stability of a solution of an averaged ordinary differential equation implies closeness of solutions of an exact and an averaged equations on an infinite interval provided the solutions have close initial conditions. Similar results have been obtained for integro-differential equations by Filatov in [8], for ordinary differential equations with slow and fast time by Sethna in [13], and for functional differential equations of retarded type by Burd in [5].

In this paper we consider the problem of closeness of solutions of an exact and an averaged difference equations on an infinite interval. Appropriate assertions are derived from one special theorem on the stability under constantly acting perturbations. We note that the stability under constantly acting perturbations is sometimes called the total stability (see, Agarwal [1], Hahn [9]).

2 Theorem on the stability under constantly acting perturbations

Basic notation. We will use the following notation: $|x|$ is a norm of vector $x \in \mathcal{R}^m, \mathcal{N}$ is the set of nonnegative integers, $B_x(K) = \{x : x \in \mathcal{R}^m, |x| \leq K\}$, $G = \mathcal{N} \times B_x(K)$. Let $f(n,x)$ be a function that is defined on G with values in $\mathcal{R}^m$ and is bounded in the norm. Let $N \in \mathcal{N}$. Let us assume

$$S_{x,N}(f) = \sup_{n\in\mathcal{N}} \left| \sum_{k=n}^{n+N-1} f(k,x) \right|, \quad x \in B_x(K).$$

Lemma. *Let the function $f(n,x)$ be continuous in x uniformly with respect to $n \in \mathcal{N}$. Assume that values of the function $x(n), (n \in \mathcal{N})$ belong to $B_x(K)$.*

Then for any $\eta > 0$ there exists a number $\varepsilon > 0$ such that

$$\sup_{n\in\mathcal{N}} \left| \sum_{k=n}^{n+N-1} f(k, x(k)) \right| < \eta,$$

if

$$S_{x,N}(f) < \varepsilon.$$

Proof. Based on the conditions of the Lemma there exists $\delta > 0$ such that $|f(n,x_1) - f(n,x_2)| < \eta$, if $|x_1 - x_2| < \delta$. We denote by $x^0(n)$ a function with values in $B_x(K)$ that has a finite set of values and satisfies $|x(n) - x^0(n)| < \delta$, $n \in \mathcal{N}$. Such function $x^0(n)$ exists as $B_x(K)$ is compact. The function $x^0(n)$ has no more than l different values, where number l depends only on δ. The statement of the Lemma follows from the inequality

$$\left| \sum_{k=n}^{n+N-1} f(k, x(k)) \right| \leq \left| \sum_{k=n}^{n+N-1} [f(k, x(k)) - f(k, x^0(k))] \right| + \left| \sum_{k=n}^{n+N-1} f(k, x^0(k)) \right|.$$

3 The main theorem

We consider the following difference equation in $\mathcal{R}^m$

$$\Delta x(n) = X(n,x) + R(n,x), \quad n = n_0,\, n_0+1, \ldots, \tag{1}$$

where $\Delta x(n) = x(n+1) - x(n)$ and functions $X(n,x)$ and $R(n,x)$ are defined on G.

Alongside with the equation (1) we consider the unperturbed difference equation

$$\Delta y(n) = X(n,y). \tag{2}$$

We suppose that equation (2) has a solution $\psi(n, n_0, \xi_0)$ ($\psi(n_0, n_0, \xi_0) = \xi_0$) which is defined for all $n \geq n_0$ and $\psi(n, n_0, \xi_0)$ together with its ρ-neighborhood ($\rho > 0$) remains in the interior of the set G.

Theorem 1. *Let function $X(n,x)$ be bounded on G and satisfies the Lipschitz condition*

$$|X(n,x_1)-X(n,x_2)|\le L|x_1-x_2|,\quad x_1,x_2\in B_x(K). \tag{3}$$

Let function $R(n,x)$ be continuous in x uniformly with respect to $n\in\mathcal{N}$ and be bounded on G. Let a solution $\psi(n,n_0,\xi_0)$ of equation (2) be uniformly asymptotically stable.

Then for any $\varepsilon>0\,(0<\varepsilon<\rho)$ there exists $N\in\mathcal{N}$ and numbers $\eta_1(\varepsilon)$, $\eta_2(\varepsilon)$ such that for all solutions $x(n,n_0,x_0)\in B_x(K)\,(x(n_0,n_0,x_0)=x_0)$ of equation (1) with initial values satisfying inequality

$$|x_0-\xi_0|<\eta_1(\varepsilon)$$

and for all functions $R(n,x)$ satisfying inequality

$$S_{x,N}(R)<\eta_2(\varepsilon)$$

the inequality

$$|x(n,n_0,x_0)-\psi(n,n_0,\xi_0)|<\varepsilon,\quad n\ge n_0 \tag{4}$$

holds.

Proof. Let $y(n,n_0,x_0)$ be a solution of equation (2) with the same initial condition as the solution $x(n,n_0,x_0)$ of equation (1). These solutions satisfy the following equations respectively

$$y(n,n_0,x_0)=x_0+\sum_{k=n_0}^{n-1}X(k,y(k,n_0,x_0)),$$

$$x(n,n_0,x_0)=x_0+\sum_{k=n_0}^{n-1}[X(k,x(k,n_0,x_0))+R(k,x(k,n_0,x_0))].$$

It follows the inequality

$$|x(n,n_0,x_0)-y(n,n_0,x_0)|\le\sum_{k=n_0}^{n-1}|X(k,x(k,n_0,x_0))-X(k,y(k,n_0,x_0))|+$$

$$\left|\sum_{k=n_0}^{n-1}R(k,x(k,n_0,x_0))\right|.$$

Using condition (3) of Theorem we obtain

$$|x(n,n_0,x_0)-y(n,n_0,x_0)| \le L\sum_{k=n_0}^{n-1}|x(k,n_0,x_0)-y(k,n_0,x_0)|+f(n),$$

where

$$f(n)=\left|\sum_{k=n_0}^{n-1}R(k,x(k,n_0,x_0))\right|=\left|\sum_{k=n_0}^{n_0+N-1}R(k,x(k,n_0,x_0))\right|,$$

and $N=n-n_0$. A well known inequality (see, for example, [2]) implies

$$|x(n,n_0,x_0)-y(n,n_0,x_0)| \le f(n)+L\sum_{k=n_0}^{n-1}f(k)(1+L)^{n-1-k}.$$

Therefore for $n_0 \le n \le n_0+N$ the upper bound for

$$|x(n,n_0,x_0)-y(n,n_0,x_0)|$$

depends on the values $f(n)$ $(n=n_0,\dots,n_0+N-1)$. From uniform asymptotic stability of solution $\psi(n,n_0,\xi_0)$ of equation (2) follows that there exist numbers $\delta<\varepsilon$ and $T\in\mathcal{N}$ such that inequality $|x_0-\xi_0|<\delta$ implies

$$\begin{array}{l} |y(n,n_0,x_0)-\psi(n,n_0,\xi_0)|<\frac{\varepsilon}{2} \quad n\ge n_0, \\ |y(n_0+T,n_0,x_0)-\psi(n_0+T,n_0,\xi_0)|<\frac{\delta}{2}. \end{array} \tag{5}$$

We now set $N=T$. The Lemma implies that we can find a number $\eta_2(\varepsilon)$ such that

$$|x(n,n_0,x_0)-y(n,n_0,x_0)|<\frac{\delta}{2}, \quad n_0\le n\le n_0+T \tag{6}$$

holds. Then

$$|x(n,n_0,x_0)-\psi(n,n_0,\xi_0)|<\frac{\varepsilon}{2}+\frac{\delta}{2}<\varepsilon, \quad n_0\le n\le n_0+T.$$

Furthermore, (5) and (6) imply

$$|x(n_0+T,n_0,x_0)-\psi(n_0+T,n_0,\xi_0)|<\delta.$$

Hence for the interval $[n_0, n_0+T]$ the solution $x(n,n_0,x_0)$ remains in ε-neighborhood of the solution $\psi(n,n_0,\xi_0)$ and at the moment $n=n_0+T$ belongs to the δ-neighborhood of $\psi(n,n_0,\xi_0)$.

We now consider $n = n_0 + T$ as an initial moment. Using the same arguments as above we obtain

$$|x(n, n_0, x_0) - \psi(n, n_0, \xi_0)| \le \varepsilon, \quad n_0 + T \le n \le n_0 + 2T$$

and

$$|x(n_0 + 2T, n_0, x_0) - \psi(n_0 + 2T, n_0, \xi_0| \le \delta.$$

Repetitive application of the same argument completes the proof of the Theorem.

We note that last part of proof of the Theorem 1 uses the reasoning similar to the Lemma 6.3 from [4].

Remark 1. The statement of Theorem 1 differs from the statements of known theorems on the stability under constantly acting perturbations [1, 9 - 12] in using a more general assumption on the perturbation $R(n, x)$. Theorem 1 implies Halanay's theorem, if number $\eta_2(\varepsilon) = N\delta_2(\varepsilon)$, where $\delta_2(\varepsilon)$ is a number from definition 5.13.1 [1]. If we assume

$$S_{x,N}(f) = \sup_{n\in\mathcal{N}} \sum_{k=n}^{n+N-1} |f(k,x)|, \quad x \in B_x(K),$$

then we obtain a difference analog of the Theorem 24.1 from [11].

Remark 2. We start with the following definition.

Definition. The solution $\psi(n, n_0, \xi_0)$ is called uniformly asymptotically stable with respect to a part of the variables $\psi_1, \ldots, \psi_k$, $k < m$, if its asymptotically stable in the sense of Lyapunov with respect to a part of the variables $\psi_1, \ldots, \psi_k$, $k < m$ and if for any number $\gamma > 0$ there exists a number $T(\gamma) \in \mathcal{N}$, such that for the solution $y(n, n_0, x_0)$ of the equation (2) is satisfied inequality

$$|y_i(n, n_0, x_0) - \psi_i(n, n_0, \xi_0)| < \gamma, \quad n \ge n_0 + T(\gamma), \quad i = 1, \ldots, k,$$

for any initial moment n_0 and any initial values x_0 from the domain of the attraction of solution $\psi(n, n_0, \xi_0)$ with respect to a part of the variables (i.e. from domain where is satisfied the limit equality

$$\lim_{n\to\infty} |y_i(n, n_0, x_0) - \psi_i(n, n_0, \xi_0)| = 0, \quad i = 1, \ldots, k.)$$

If $k = m$, the definition above coincides with the definition of uniform asymptotic stability (see, for example, [6]). A detailed discussion of the stability theory with respect to a part of the variables is given, for example, in [14].

If the solution $\psi(n, n_0, \xi_0)$ of equation (2) uniformly asymptotically stable only with respect to a part of the variables $\psi_1, \ldots, \psi_k$, $k < m$, then inequality (4) in the statement of the Theorem 1 can be replaced with the inequality

$$|x_i(n, n_0, x_0) - \psi_i(n, n_0, \xi_0)| < \varepsilon, \quad i = 1, \ldots, k.$$

4 Averaging on an infinite interval

Theorem 1 is applicable to the problem of averaging on an infinite interval for difference equations.

We consider the following difference equation in $\mathcal{R}^m$

$$\Delta x(n) = \varepsilon X(n, x), \tag{7}$$

where $\varepsilon > 0$ is a small parameter, $X(n, x)$ is defined for $(n, x) \in G$.

Theorem 2. *Let*
1) function $X(n, x)$ be continuous in x uniformly with respect to $n \in \mathcal{N}$;
2) $|X(n, x)| \leq M_1 < \infty, \quad (n, x) \in G$;
3) the limit

$$\lim_{N \to \infty} \frac{1}{N} \sum_{k=n}^{n+N-1} X(k, x) = \bar{X}(x)$$

exists uniformly with respect to n for any $(n, x) \in G$ and $\bar{X}(x)$ be bounded in the norm

$$|\bar{X}(x)| \leq M_2 < \infty, \quad x \in B_x(K);$$

4) function $\bar{X}(x)$ satisfies the Lipschitz condition

$$\left|\bar{X}(x_1) - \bar{X}(x_2)\right| \leq L|x_1 - x_2|, \quad x_1, x_2 \in B_x(K),$$

5) the averaged equation

$$\Delta x(n) = \varepsilon \bar{X}(x)$$

has a uniformly asymptotically stable solution $\psi(n, n_0, \xi_0)$ (uniformly asymptotically stable with respect to a part of the variables $x_1, \dots, x_k$, $k < m$), which with its ρ-neighborhood ($\rho > 0$) belong to G.

Then for any $\alpha\,(0 < \alpha < \rho)$ there exists $\varepsilon_1(\alpha)\,(0 < \varepsilon_1 < \varepsilon_0)$ and $\beta(\alpha)$ such that for all $0 < \varepsilon < \varepsilon_1$ the solution $\varphi(n, n_0, x_0) \in B_x(K)$ of equation (7) with an initial condition satisfying inequality

$$|x_0 - \xi_0| < \beta(\alpha) \quad (|x_{0i} - \xi_{0i}| < \beta(\alpha),\, i = 1, \dots, k < m)$$

we have

$$|\psi(n, n_0, \xi_0) - \varphi(n, n_0, x_0)| < \alpha, \quad n \geq n_0$$
$$(|\psi_i(n, n_0, \xi_0) - \varphi_i(n, n_0, x_0)| < \alpha, \quad i = 1, \dots, k < m \quad n \geq n_0).$$

The equation (7) can be written in the form

$$\Delta x(n) = \varepsilon \bar{X}(x) + \varepsilon R(n, x),$$

where $R(n, x) = X(n, x) - \bar{X}(x)$. We show that Theorem 2 follows from Theorem 1. Given α we choose $N(\alpha) = [\frac{1}{\varepsilon}]$ where $[x]$ is the integer part of x. Then

$$\left| \varepsilon \sum_{k=n}^{n+[\frac{1}{\varepsilon}]-1} R(k, x) \right| \leq \frac{1}{[\frac{1}{\varepsilon} - 1]} \left| \sum_{k=n}^{n+[\frac{1}{\varepsilon - 1}]} R(k, x) \right|.$$

Therefore condition 3) of the Theorem 2 implies that for a sufficiently small ε the function $R(n, x)$ satisfies of conditions of Theorem 1.

5 Averaging on an infinite interval of systems with the right-hand side that vanishes over time

We now consider the following difference equation in $\mathcal{R}^m$

$$\Delta x(n) = \frac{1}{n} X(n, x), \quad n = n_0, n_0 + 1, \dots, \tag{8}$$

where $X(n, x)$ is defined for $(n, x) \in G$.

Theorem 3. *Let*
1) function $X(n,x)$ *be continuous in* x *uniformly with respect to* $n \in \mathcal{N}$*;*
2) $|X(n,x)| \le M_1 < \infty, \quad (n,x) \in G$*;*
3) the limit

$$\lim_{N\to\infty} \frac{1}{N} \sum_{k=n}^{n+N-1} X(k,x) = \bar{X}(x)$$

exist uniformly with respect to n *for any* $(n,x) \in G$ *and* $\bar{X}(x)$ *be bounded in the norm*

$$|\bar{X}(x)| \le M_2 < \infty, \quad x \in B_x(K);$$

4) function $\bar{X}(x)$ *satisfies the Lipschitz condition*

$$\left|\bar{X}(x_1) - \bar{X}(x_2)\right| \le L|x_1 - x_2|, \quad x_1, x_2 \in B_x(K),$$

5) the averaged equation

$$\Delta x(n) = \frac{1}{n}\bar{X}(x)$$

has a uniformly asymptotically stable solution $\psi(n, n_0, \xi_0)$ *(uniformly asymptotically stable with respect to a part of variables* $x_1, \ldots, x_k$*,* $k < m$*), which with its* ρ*-neighborhood (*$\rho > 0$*) is contained in* G*.*

Then for any $\alpha\,(0 < \alpha < \rho)$ *there exists* $\varepsilon_1(\alpha)\,(0 < \varepsilon_1 < \varepsilon_0)$ *and* $\beta(\alpha)$ *such that for all* $0 < \varepsilon < \varepsilon_1$ *the solution* $\varphi(n, n_0, x_0) \in B_x(K)$ *of equation (8), with initial condition satisfying inequality*

$$|x_0 - \xi_0| < \beta(\alpha) \quad (|x_{0i} - \xi_{0i}| < \beta(\alpha),\, i = 1, \ldots, k)$$

we have

$$|\psi(n, n_0, \xi_0) - \varphi(n, n_0, x_0)| < \alpha, \quad n \ge n_0$$

$$(|\psi_i(n, n_0, \xi_0) - \varphi_i(n, n_0, x_0)| < \alpha, \quad i = 1, \ldots, k < m \quad n \ge n_0).$$

The proof of the Theorem 3 is quite similar to the proof of the Theorem 2.

6 Dynamics of selection of genetic population in a varying environment

As an example we consider dynamics of a selection of a Mendelian population with a genetic pool made of only two alleles, that we'll call A and a . We assume that the fitness of the genotypes AA, Aa, aa are $1-\varepsilon\alpha(n)$, 1, $1-\varepsilon\beta(n)$ respectively. Here n is number of the generation, $\varepsilon > 0$ is a small parameter, $\alpha(n)$, $\beta(n)$ are periodic functions with period $l \in \mathcal{N}$ and positive mean values. Let p_n, q_n be the frequencies of alleles A, a in generation n respectively. The evolution equation has the form (see, for example, [7])

$$\Delta p_n = \varepsilon p_n(1-p_n)\frac{\beta(n)-(\alpha(n)+\beta(n))p_n}{1-\varepsilon[(\alpha(n)+\beta(n))p_n^2+2\beta(n)p_n-\beta(n)]}. \qquad (9)$$

The averaged equation

$$\Delta\bar{p}_n = \varepsilon\bar{p}_n(1-\bar{p}_n)(\beta_0-(\alpha_0+\beta_0)\bar{p}_n)$$

has a unique asymptotically stable equilibrium

$$\bar{p} = \frac{\beta_0}{\alpha_0+\beta_0},$$

where α_0, β_0 are mean value of periodic functions $\alpha(n)$, $\beta(n)$ accordingly.

Then for a sufficiently small ε equation (9) has an asymptotically stable periodic solution with period $l \in \mathcal{N}$ (see [10]). Theorem 2 implies that for any $\delta > 0$ there exists $\eta(\delta)$ such that the solution $p_n(0,x_0)$ of equation (9), with initial condition satisfying inequality

$$|x_0-\xi_0| < \eta(\delta),$$

where $\xi_0 > 0$, $\xi_0 \neq \frac{\beta_0}{\alpha_0+\beta_0}$ the inequality

$$|p_n(0,x_0)-\bar{p}_n(0,\xi_0| < \delta, \quad n \geq 0$$

holds.

References

1. R. Agarwal, Difference equations and inequalities: theory, methods, and applications, Marcel Dekker, New York, 2000.

2. R. Agarwal, M. Bohner, A. Peterson, Inequalities on time scales: A survey, Mathematical inequalities and Applications, 4(4), (2001), 535–557.

3. C. Banfi, Sull'approssimazione di processi non stazionari in mecanica non lineare, Bolletino della Unione Matematica Italiana, 22 (1967), 442–450.

4. E.A. Barbashin, Vvedenie v teoriyu ustoichivosti, Nauka, Moscow, 1967 (in Russian).

5. V.Sh. Burd, Stability under constantly acting disturbances and principle of averaging on an infinite interval for systems with time lag, Functional Differential equations, 4 (1997), n. 3-4, 257–264.

6. S. Elaydi, An introduction to difference equations. Springer-Verlag, New York, 1996.

7. W.J. Ewens, Mathematical population genetics, Springer-Verlag, Berlin; New York, 1979.

8. A.N. Filatov, Methods of averaging in differential and integro-differential equations, Fan, Tashkent, 1971.

9. W. Hahn, Theory and applications on Liapunov's direct method, Prentice-Hall, Englwood Cliffs, N.J., 1963.

10. A. Halanay, D. Wexler, Teoria calitativa a sistemelor cu impulsuri, Editura Academiei Republicii Socialiste Romania, Bucuresti, 1968.

11. N.N. Krasovskii, Stability of motion, Stanford University Press, Stanford, California, 1963.

12. I.G. Malkin, Theory of stability of motion, Atomic Energy Commission, Translation No. 3352, Washington, D.C., 1958.

13. P.R. Sethna, Systems with fast and slow time, The 5th international conference on nonlinear oscillations, Ukrainian Academy of Sciences, Kiev, 1 (1971), 505–521.

14. V.I. Vorotnikov, Partial stability and control, Birkhäuser, Boston, 1997.

Symbolic dynamics in the study of bursting electrical activity

Jorge Duarte, Luís Silva, and J. Sousa Ramos

Departamento de Eng. Química,Secção de Matemática, Instituto Superior de Engenharia de Lisboa,Rua Conselheiro Emídio Navarro 1, 1949-014 Lisboa, Portugal
E-mail address: jduarte@deq.isel.pt

Departamento de Matemática,Universidade Évora, Rua Romão Ramalho, 59, 7000-671 Évora,Portugal
E-mail address: lfs@uevora.pt

Departamento de Matemática, InstitutoSuperior Técnico, Av. Rovisco Pais, 1, 1049-001 Lisboa, Portugal
E-mail address: sramos@math.ist.utl.pt
URL: http://www.math.ist.utl.pt/˜sramos

Abstract. Many cells exhibit a complex behavior, characterized by brief bursts of oscillatory activity interspersed with quiescent periods during which the membrane potential changes only slowly. This behavior is called bursting. The interpretation of bursting in terms of nonlinear dynamics is one of the recent success stories of mathematical physiology and provides an excellent example of how mathematics can be used to understand complex biological dynamical systems. In the present paper we study a map, that replicates the dynamics of bursting cells, presented in [**16**]. Using symbolic dynamics we characterize the topological entropy of the chaotic bursts and we analyse the variation of this important numerical invariant with the parameters of the system. This procedure allows us to distinguish different chaotic scenarios.

1. Motivation and preliminaries

Bursting behavior is ubiquitous in physical and biological systems, specially in neural systems where it plays an important role in information processing (see [**12**], [**6**], [**7**] and [**8**]).

Physiological questions concerning the dynamics of bursting cells lead to challenging mathematical problems. The complex bursting activity of individual biological neurons is the result of high-dimensional dynamics of nonlinear processes responsible for variations in the ionic currents across the membrane. The numerical studies of such neural activity are usually based on either realistic ionic-based models or phenomenological models. The ionic-based models proposed for a single

---

2000 *Mathematics Subject Classification.* Primary 37N25, 37B10; Secondary 37E05, 37B40

Key words and phrases. Bursting behavior, difference equations, symbolic dynamics, topological entropy, chaos

Partially supported by Instituto Superior de Engenhariade Lisboa

Partially supported by FCT/POCTI/FEDER

neuron are designed to replicate the physiological processes in the membrane. These models are usually given by many nonlinear equations. The strong nonlinearity and high dimensionality of the phase space is a significant obstacle in understanding the collective behavior of such dynamical systems [**5**]. Important mechanisms are hidden behind the complexity of the equations. The phenomenological models are designed to capture the most important features of neural behavior with minimal complexity of the model [**4**]. Recently a special type of phenomenological models based on a low-dimensional map was proposed. Indeed, the study of dynamical principles and mechanisms behind the bursting behavior is hardly possible without numerical studies of discrete-time models. There have been proposed only few explicit maps capable of generating bursting activity (for instance, see [**16**], [**17**], [**1**] and [**9**]).

In this work we address a contribution for the detailed study of the two-dimensional bursting map introduced in [**16**]. More precisely, using techniques of symbolic dynamics theory, we define the topological entropy of a subsystem which represents the fast dynamical variable of the map. This quantitative measure of the amount of chaos in a dynamical system gives us a finer distinction between different states of the system.

In order to facilitate the analysis and make this paper self-contained, we outline briefly some important aspects of the discrete-time model replicating chaotic bursting (for more details see [**16**]).

A group of irregularly bursting cells with different individual properties can be modeled using maps of the form

$$\begin{cases} x(i,n+1) = \frac{\alpha_i}{1+x(i,n)^2} + y(i,n) + \frac{\epsilon}{N}\sum_{j=1}^{N} x(j,n) \\ y(i,n+1) = y(i,n) - \sigma_i x(i,n) - \beta_i \end{cases} \tag{1.1}$$

where $x(i,n)$ and $y(i,n)$ are, respectively, the fast and slow dynamical variables of the ith oscillator, and ϵ is the strength of global coupling. The slow evolution of $y(i,n+1)$ is a result of the small values of the positive parameters β and σ, which are on the order of 0.001.

We will restrict our attention to the individual behavior of the cells, which is described by the map (1.1) with $\epsilon = 0$, that is,

$$\begin{cases} x_{n+1} = \frac{\alpha}{1+x_n{}^2} + y_n \\ y_{n+1} = y_n - \sigma x_n - \beta. \end{cases} \tag{1.2}$$

The model consists of fast and slow subsystems coupled to each other, where the x-variable replicates the dynamics of the membrane potential. The value of the parameter α is selected on the interval $[1.5, 8.0]$. We note that the considered mechanism of bursting is essentially the same as in the well-known Hindmarsh-Rose model of biological neuron, where the role of parameter α is played by a hyperpolarization current I [**4**].

Depending on the value of parameter α, each cell demonstrates two qualitatively different regimes of behavior, namely *continuous oscillations* and *bursts*. Typical regimes of temporal behavior of the two-dimensional map are shown in Figs. 1-4.

As depicted in these figures, the mean duration of the bursts is very sensitive to the value of α. The slow evolution of y_n for the next m steps is given by

$$y_{n+m} = y_n - m(\beta + \sigma \overline{x}_{n,m}), \tag{1.3}$$

FIGURE 1. Wave forms of temporal behavior of individual cells, computed for $\alpha = 4.8$, with $\sigma = \beta = 0.001$.

FIGURE 2. Wave forms of temporal behavior of individual cells, computed for $\alpha = 4.5$, with $\sigma = \beta = 0.001$.

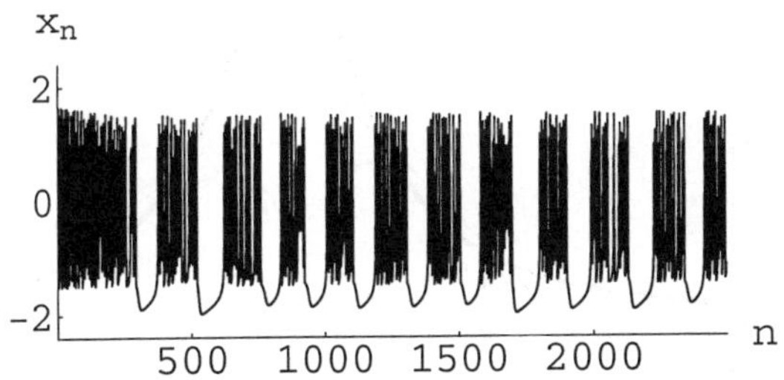

FIGURE 3. Wave forms of temporal behavior of individual cells, computed for $\alpha = 4.3$, with $\sigma = \beta = 0.001$.

where $\overline{x}_{n,m} = \left(\sum\limits_{j=n+1}^{n+m} x_j \right) / m$ is the mean value of x_n, computed for m consecutive iterations. According to (1.3), the value of y_n slowly increases during the next m steps if $\sigma \overline{x}_{n,m} < -\beta$, and decreases if $\sigma \overline{x}_{n,m} > -\beta$.

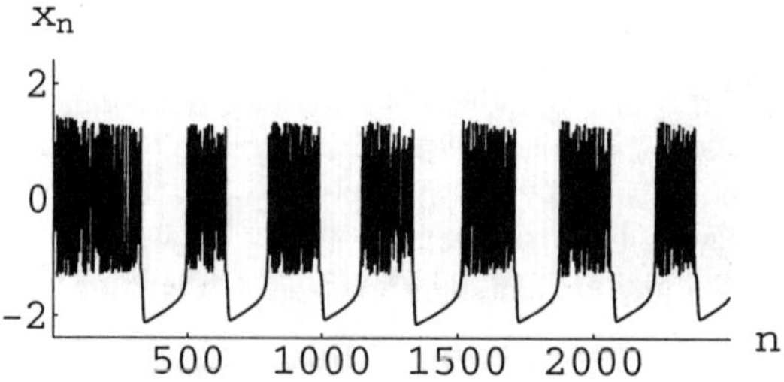

FIGURE 4. Wave forms of temporal behavior of individual cells, computed for $\alpha = 4.1$, with $\sigma = \beta = 0.001$.

As pointed out in [**16**], since y_n changes slowly, the time evolution of x_n can be considered independently of map $y(i, n+1) = y(i, n) - \sigma x(i, n) - \beta$ assuming that y_n is a control parameter $\gamma = y_n$. Therefore, important insights about the fast dynamics of the cell can be obtained from the analysis of the two-parameters family of unimodal maps

$$x_{n+1} = F_{\alpha,\gamma}(x_n) \equiv \frac{\alpha}{1 + {x_n}^2} + \gamma. \tag{1.4}$$

This family of maps is presented in Fig. 5.

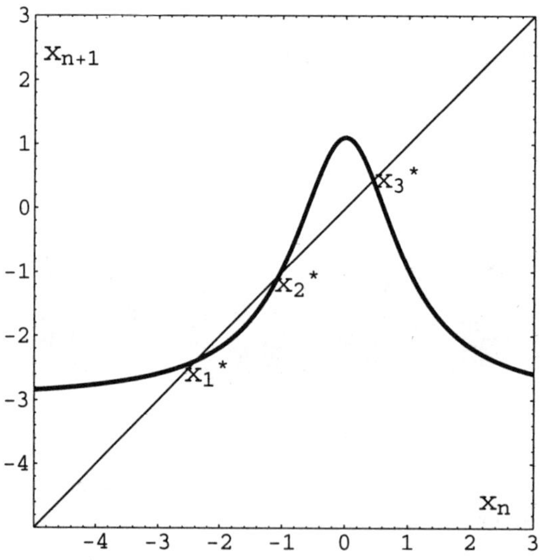

FIGURE 5. The shape of the function $F_{\alpha,\gamma}(x_n)$ plotted for $\alpha = 4.1$ and $\gamma = -3.0$. For these values of the parameters x_1^*, x_2^* and x_3^* are fixed points.

2. Symbolic dynamics. Topological entropy and chaos

In this paragraph, we describe techniques of symbolic dynamics, in particular some results concerning to Markov partitions associated to unimodal maps (family of continuous maps on the interval with two monotonic subintervals and one turning point). As we can see, the shape of the function $F_{\alpha,\gamma}(x_n)$ resembles the quadratic map. Therefore, the symbolic dynamics theory for unimodal maps is particularly important in our study (for more details see **[11]**, **[3]**, **[14]** and **[13]**). We characterize the topological entropy of $F_{\alpha,\gamma}(x_n)$, and we show situations of the variation of this numerical invariant with the parameters α and γ.

A unimodal map f on the interval $I = [c_0, c_2]$ is piecewise monotone and I is subdivided into two subintervals:

$$L = [c_0, c_1[, \qquad R =]c_1, c_2],$$

in such way that the restriction of f to interval L is strictly increasing and the restriction of f to interval R is strictly decreasing. Each such maximal intervals on which the function f is monotone is called a *lap* of f, and the number $\ell = \ell(f)$ of distinct laps is called the *lap number* of f.

Denoting by c the turning point (relative extremum) of f, we obtain the orbit

$$O(c) = \{x_i : x_i = f^i(c),\ i \in \mathbb{N}\}.$$

With the aim of studying the topological properties, we associate to orbit $O(c)$ a sequence of symbols $S = S_1S_2...S_j...$ where $S_j \in \Sigma = \{L, C, R\}$ and

$$\begin{cases} S_j = L & \text{if} \quad f^j(c) < c \\ S_j = C & \text{if} \quad f^j(c) = c \\ S_j = R & \text{if} \quad f^j(c) > c. \end{cases}$$

The point c plays an important role. The dynamics of the interval is characterized by the symbolic sequence associated to the orbit of point c, that is, the turning point itinerary. When $O(c)$ is a k-periodic orbit, we obtain a sequence of symbols that can be characterized by a block of length k, the kneading sequence $S^{(k)} = S_1S_2...S_{k-1}C$. We introduce, in the set of symbols, a order relation (R-parity of a sequence, meaning odd or even number of occurrence of a symbol R in the sequence):

$$\begin{aligned} L &< C < R \text{ in position } j \text{ if the number of } R,\ N_R(CS_1...S_{j-1}) \text{ is even,} \\ R &< C < L \text{ in position } j \text{ if the number of } R,\ N_R(CS_1...S_{j-1}) \text{ is odd.} \end{aligned}$$

The order of the symbols is extended to the symbolic sequences. Thus, for two of such sequences P and Q in $\Sigma^{\mathbb{N}}$, let i be such that $P_i \neq Q_i$ and $P_j = Q_j$ for $j < i$. If the R-parity of the block $P_1...P_{i-1} = Q_1...Q_{i-1}$ is even we say that $P < Q$ if $P_i < Q_i$ in the order $L < C < R$. If the R-parity of the same block is odd, we say that $P < Q$ if $P_i < Q_i$ in the order $R < C < L$. If no such index i exists, then $P = Q$.

The ordered sequence of elements x_i determines a partition $P^{(k)}$ of the interval $I = [f^2(c), f(c)] = [x_2, x_1]$ into a finite number of subintervals labeled by I_1, I_2,..., I_{k-1}. To this partition we associate a $(k-1) \times (k-1)$ Markov transition matrix $M = [a_{ij}]$ with

$$a_{ij} = \begin{cases} 1 & \text{if} \quad I_j \subset f(I_i) \\ 0 & \text{if} \quad I_j \subsetneq f(I_i) \end{cases}.$$

Take, for example, the period-5 kneading sequence $S = RLRRC$. Its successive points of the orbit are obtained by shifting the periodic sequence by one letter at a time, i.e.,

$$\begin{aligned} x_0 &\rightarrow CRLRR \\ x_1 &\rightarrow RLRRC \\ x_2 &\rightarrow LRRCR \\ x_3 &\rightarrow RRCRL \\ x_4 &\rightarrow RCRLR \\ x_5 &\rightarrow CRLRR. \end{aligned}$$

These points or the corresponding symbolic sequences are ordered in the following way: $x_2 < x_0 < x_3 < x_4 < x_1$. The dynamical invariant range is now divided into four subintervals, as shown in Fig. 6. By inspecting the partition $P^{(5)}$ of the

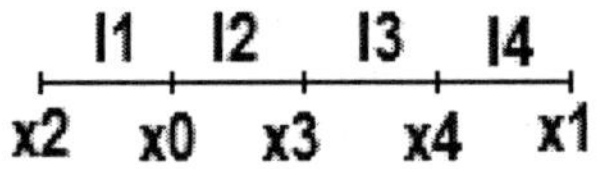

FIGURE 6. Partition of the interval corresponding to the kneading sequence $S = RLRRC$.

interval $I = [x_2, x_1]$, we write down the transition matrix

$$M = \begin{bmatrix} 0 & 0 & 1 & 1 \\ 0 & 0 & 0 & 1 \\ 0 & 1 & 1 & 0 \\ 0 & 0 & 0 & 1 \end{bmatrix}.$$

Now we consider the topological entropy. This numerical invariant allows us to quantify the complexity of the phenomenon. A possible definition of chaos in the context of one-dimensional dynamical systems states that a dynamical system is called chaotic if its topological entropy is positive. Thus, the topological entropy can be computed to express whether a map has chaotic behavior.

Let $F_{\alpha,\gamma}$ be the two-parameters family, representing the fast subsystem of the bursting map. The topological entropy of $F_{\alpha,\gamma}$, denoted by $h(F_{\alpha,\gamma})$, can be given by

$$h(F_{\alpha,\gamma}) = \log_2 \lambda_{\max}(\mathcal{M}(F_{\alpha,\gamma})) = \log_2 \, s(F_{\alpha,\gamma}),$$

where $\lambda_{\max}(\mathcal{M}(F_{\alpha,\gamma}))$ is the spectral radius of the transition matrix $\mathcal{M}(F_{\alpha,\gamma})$ and $s(F_{\alpha,\gamma})$, the growth rate

$$s(F_{\alpha,\gamma}) = \lim_{k\rightarrow\infty} \sqrt[k]{\ell(F^k_{\alpha,\gamma})}$$

of the lap number of $F^k_{\alpha,\gamma}$ (kth-iterate of $F_{\alpha,\gamma}$), is monotone. We have $s(F_{\alpha,\gamma}) = \lambda_{\max}(\mathcal{M}(F_{\alpha,\gamma}))$, see [**13**], [**10**] and [**15**].

To illustrate the previous considerations, we discuss the following example (for more details see [**2**]).

EXAMPLE 1. *Let us consider the orbit of the turning point defined by the period-7 kneading sequence $S = RLRRLRC$. Putting the points of the orbit in order we obtain:*

$$x_2 < x_5 < x_7 < x_3 < x_6 < x_4 < x_1.$$

The corresponding transition matrix is

$$\mathcal{M}(F_{\alpha,\gamma}) = \begin{bmatrix} 0 & 0 & 0 & 1 & 0 & 0 \\ 0 & 0 & 0 & 0 & 1 & 1 \\ 0 & 0 & 0 & 0 & 0 & 1 \\ 0 & 0 & 1 & 1 & 1 & 0 \\ 0 & 1 & 0 & 0 & 0 & 0 \\ 1 & 0 & 0 & 0 & 0 & 0 \end{bmatrix}$$

which has the characteristic polynomial

$$p(\lambda) = \det(\mathcal{M}(F_{\alpha,\gamma}) - \lambda I) = 1 - \lambda - \lambda^2 + \lambda^3 - \lambda^4 - \lambda^5 + \lambda^6.$$

The growth number $s(F_{\alpha,\gamma})$ (the spectral radius of matrix $\mathcal{M}(F_{\alpha,\gamma})$) is 1.55603.... *Therefore, the value of the topological entropy can be given by*

$$h_{top}(F_{\alpha,\gamma}) = \log_2 s(F_{\alpha,\gamma}) = 0.63787....$$

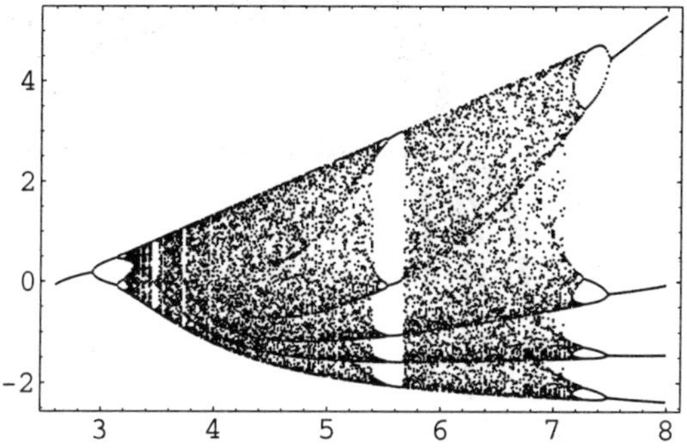

FIGURE 7. Bifurcation diagram for x_n as a function of α, with $\gamma = -2.65$ and $\alpha \in [2.6, 8.0]$.

To see the long term behavior for different values of the parameters at once, we plot, in Fig. 7 and Fig. 8, typical bifurcation diagrams. This bifurcations diagrams suggest the existence of an inversion in the usual chaos order (for instance, notice the inverted period-doubling bifurcations).

After these considerations it is easier to understand Fig. 9 and Fig. 10 that represent several situations of the variation of the topological entropy with each of the parameters α and γ.

In all situations, the topological entropy $h(F_{\alpha,\gamma})$ has an absolute maximum value. This behavior is determined by the symbolic sequences ordering associated to the successive orbits of the turning point $c = 0$.

Now we consider the region of the parameter space

$$\Omega = \{(\alpha, \gamma) \in \mathbb{R}^2 : 1.5 < \alpha < 8.0, \ -4.4 < \gamma < 0.0\}.$$

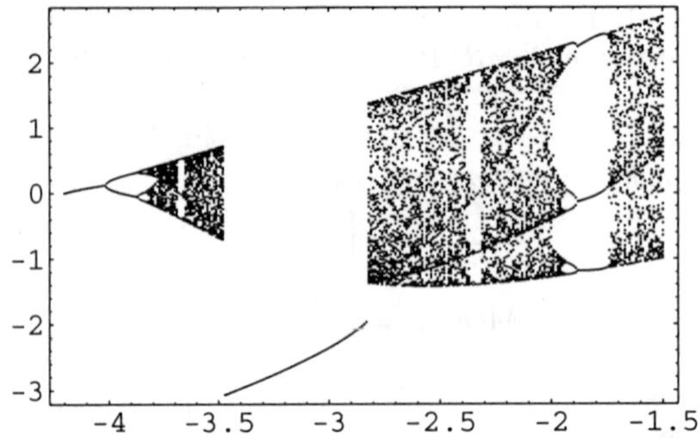

FIGURE 8. Bifurcation diagram for x_n as a function of γ, with $\alpha = 4.2$ and $\gamma \in [-4.2, -1.5]$.

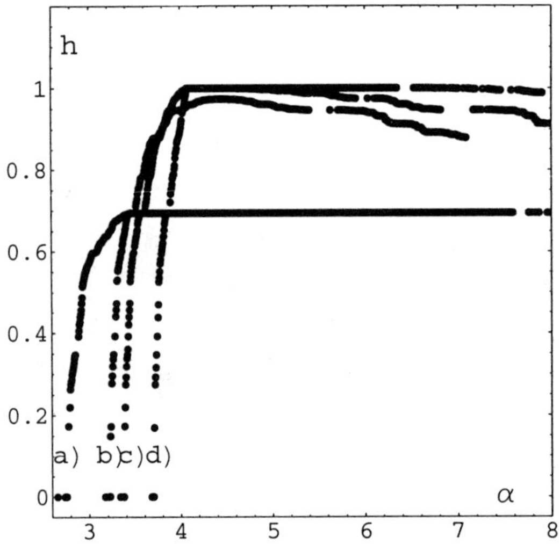

FIGURE 9. Variation of the topological entropy for $\alpha \in [2.6, 8.0]$ and four different values of γ: a) $\gamma = -1.85$, b) $\gamma = -2.65$, c) $\gamma = -2.85$, d)$\gamma = -3.25$.

The isentropic curves (the levels of topological entropy) given by

$$\mathcal{C}_{RL^\infty} = \{(\alpha, \gamma) \in \Omega : F_{\alpha,\gamma}(0) = x_1, \;\; F_{\alpha,\gamma}(x_1) = x_2, \;\; F_{\alpha,\gamma}(x_2) = x_2\}$$

and

$$\mathcal{C}_k = \{(\alpha, \gamma) \in \Omega : F^k_{\alpha,\gamma}(0) = 0\},$$

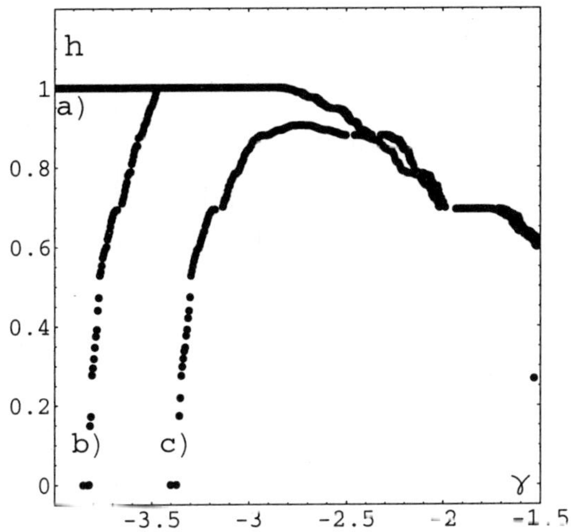

FIGURE 10. Variation of the topological entropy for $\gamma \in [-4.0, -1.5]$ and three different values of α: a) $\alpha = 4.6$, b) $\alpha = 4.2$, c) $\alpha = 3.8$.

for small n ($n \leq 5$), are shown in Fig. 11. The topological entropy remains constant over each curve. We remember that when we have the symbolic itinerary RL^{∞} of the turning point the dynamics of the iterates is a full shift of two symbols and the topological entropy is one. Regarding the previous considerations we derive the following result.

THEOREM 1. *Let $F_{\alpha,\gamma}(x)$ be the two-parameters family of unimodal maps defined by*

$$x_{n+1} = F_{\alpha,\gamma}(x_n) \equiv \frac{\alpha}{1 + x_n{}^2} + \gamma$$

and

$$\Omega = \{(\alpha, \gamma) \in \mathbb{R}^2 : 1.5 < \alpha < 8.0, \quad -4.4 < \gamma < 0.0\}$$

a region of the (α, γ)-parameter space. Then
a) In the subregion $C_{RL^{\infty}} \cup A$ (see Fig. 11), the topological entropy, $h(F_{\alpha,\gamma})$, is one.
b) When $\gamma = \gamma^*$ is fixed, the topological entropy $h(F_{\alpha,\gamma^*})$ has an absolute maximum value as a function of α, with $1.5 < \alpha < 8.0$. In the same way, when $\alpha = \alpha^*$ is fixed, the topological entropy $h(F_{\alpha^*,\gamma})$ has an absolute maximum value as a function of γ, with $-4.4 < \gamma < 0.0$ (see again Fig. 11).

PROOF. a) The topological entropy, $h(F_{\alpha,\gamma})$, is one when $(\alpha, \gamma) \in C_{RL^{\infty}}$ because the symbolic sequence RL^{∞} corresponds to the full shift. In subregion A the second iterate of $c = 0$ is greater than the fixed point x_1^* and less than the fixed point x_2^* (see Fig. 5). Therefore, the turning point has also the symbolic orbit

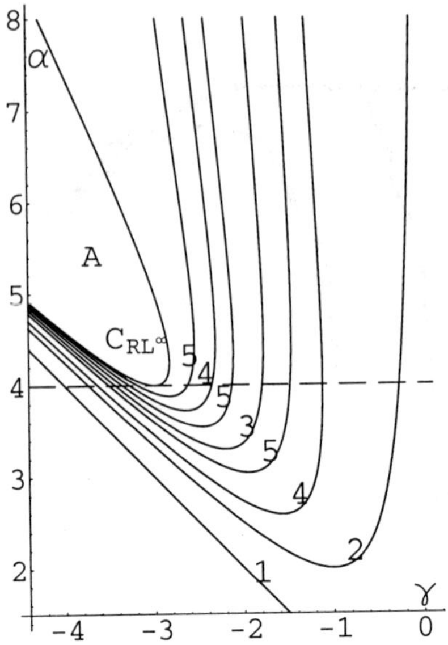

FIGURE 11. Curves in the parameter space corresponding to periodic orbits of the turning point $c = 0$ (periods $n \leq 5$). The labels are the respective periods.

RL^∞.
b) This result is a consequence of the geometry and of the ordering of the isentropic curves $\mathcal{C}_k$ and $\mathcal{C}_{RL^\infty}$ in the (α, γ)-parameter space. ∎

The evolution of the absolute maximum value of the topological entropy, $h_{\max}(F_{\alpha,\gamma})$ is depicted in Fig. 12 and Fig. 13.

3. Final considerations

In this work we have provided new insights into the study of a map which represents the fast dynamical variable of a model that replicates bursting behavior.

A rigorous approach of $F_{\alpha,\gamma}$ maps became possible using the techniques of symbolic dynamics. We studied the topological entropy and we introduced the parameter space ordering of the dynamics.

Indeed, the family of maps $F_{\alpha,\gamma}$ exhibit positive topological entropy, which means that the fast dynamics of the cell has a chaotic nature. Thus we can measure the complexity of chaotic physiological phenomena.

References

[1] Cazelles B., Courbage M. and Rabinovich M., Anti-phase regularization of coupled chaotic maps modelling bursting neurons, *Europhys. Lett.* **56** (2001), 504-509.

[2] Duarte, J. and Sousa Ramos, J., Topological entropy as a measure of chaos in forced excitable systems, *Int. J. Pure Appl. Math.* **4** (2003), no. 2, 165–180.

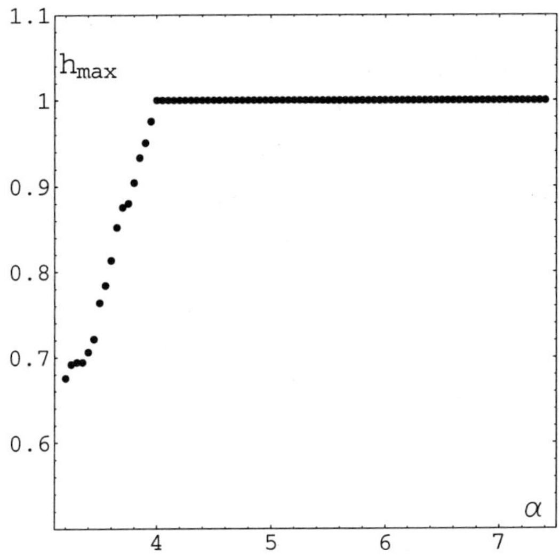

FIGURE 12. Graph of $h_{\max}(\alpha)$, with $\gamma \in [-4.4, 0.0]$. More precisely, for each value of $\alpha \in [3.2, 8.0]$, the maximum value of the topological entropy is computed when $\gamma \in [-4.4, 0.0]$.

[3] Hao, Bai-Lin and Zheng, Wei-Mou, Applied symbolic dynamics and chaos. *Directions in Chaos, 7.* World Scientific Publishing Co., Inc., River Edge, NJ, 1998.

[4] Hindmarsh J. L. and Rose R. M., A model of neuronal bursting using three coupled 1st order differential equations, *Proc. R. Soc. London* **B221** (1984), 87-102.

[5] Hodgkin A. L. and Huxley A. F., A quantitative description of membrane current and its application to conduction and excitation in nerve, *J. Physiol.* **117** (1952), 500-544.

[6] Izhikevich E. M., Neural excitability, spiking and bursting, *International Journal of Bifurcation and Chaos* **10** (2000), 1171-1266.

[7] Izhikevich E. M., Resonance and selective communication via bursts in neurons having subthreshold oscillations, *BioSystems* **67** (2002), 95-102.

[8] Izhikevich E. M., Desai N., Walcott E. and Hoppensteadt F. C., Bursts as a unit of neural information: selective communication via resonance, *Trends in Neurosci.* **26** (2003), 161-167.

[9] Laing C. R. and Longtin A., A two variable model of somaticdendritic interactions in a bursting neuron, *Bull. Math. Biol.* **64** (2002), 829-860.

[10] Lampreia, J. P. and Sousa Ramos, J., Computing the topological entropy of bimodal maps, *European Conference on Iteration Theory* (Caldes de Malavella, 1987), 431–437, World Sci. Publishing, Teaneck, NJ, 1989.

[11] Lampreia, J. P. and Sousa Ramos, J., Symbolic dynamics of bimodal maps, *Portugal. Math.* **54** (1997), no. 1, 1–18.

[12] Lisman J. E., Bursts as a unit of neuronal information: making unreliable synapses reliable, *Trends in Neurosci.* **20** (1997), 28-43.

[13] Milnor, J. and Thurston, W., On iterated maps of the interval I and II, *Lect. Notes in Math.*, No. 1342, Springer-Verlag (1988), 465-563.

[14] Mira, C., Chaotic dynamics, *World Scientific*, Singapore (1987).

[15] Misiurewicz, M. and Szlenk, W., Entropy of piecewise monotone mappings, *Studia Math.* **67** (1980), 45-63.

[16] Rulkov N. F., Regularization of synchronized chaotic bursts, *Phys. Rev. Lett.* **86** (2001), 183-186.

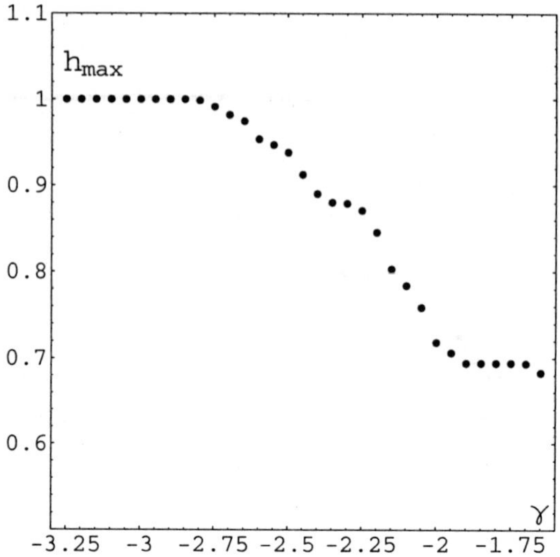

FIGURE 13. Graph of $h_{\max}(\gamma)$, with $\alpha \in [1.5, 8.0]$. More specifically, for each value of $\gamma \in [-3.25, -1.65]$, the maximum value of the topological entropy is computed when $\alpha \in [1.5, 8.0]$.

[17] Rulkov N. F., Modeling of spiking-bursting neuronal behavior using two-dimensional map, *Phys. Rev. E.* **65** (2002), 041922.